Robert Sobot

Wireless Communication Electronics by Example

Second Edition

 Springer

Robert Sobot
École Nationale Supérieure de l'Electronique et
de ses Applications
Cergy-Pontoise, France

ISBN 978-3-030-59500-5   ISBN 978-3-030-59498-5   (eBook)
https://doi.org/10.1007/978-3-030-59498-5

This Springer imprint is published by the registered company Springer Nature Switzerland AG
The registered company address is: Gewerbestrasse 11, 6330 Cham, Switzerland

To Allen

Preface

Preface to the Second Edition

This tutorial book is the updated, significantly modified, and reorganized version of the first edition. Some material was removed, and some added, all in the attempt to produce a more focused book with the explanations better suited for the students who are entering this topic.

In the first chapter of this edition, a short review of basic engineering mathematical definitions and techniques is given. Hopefully, these exercises will help the readers to refresh some of the basic topics in complex algebra and calculus that are relevant for all problems in engineering and science, not only in electronics.

The subsequent chapters parallel the second edition of my textbook "Wireless Communication Electronics: Introduction to RF Circuits and Design Techniques", while problems are systemically solved in tutorial manner. Although as a convenience some of the key formulas are given at the beginning of each chapter, it is assumed that the reader has already studied the same chapters in the textbook and learned the background theory. It should be noted that the problem examples are gradually developed where the previous solutions are reused and further advanced in each subsequent case. Therefore, it is recommended that the problems and chapters are not skipped during the study.

The timeframe limitations restricted the total number of problems that I could include in this already increased volume, which I hope to expand in the future editions of this book.

Cergy-Pontoise, France
July 24, 2020

Robert Sobot

Preface to the First Edition:

This tutorial book comes as a supplement to my textbook "Wireless Communication Electronics. Introduction to RF Circuits and Design Techniques", which resulted from my lecture notes in "Communication Electronics I" undergraduate course that I was offering over the last 7 years to students at the Western University in London Ontario, Canada. My main inspiration to write this tutorial book again came from my students who would always ask "How do I practice for this course?" while being frustrated for having to browse through large number of books in order to find a few applicable examples for practice. In addition, most of the modern undergraduate textbooks follow the same approach of giving a number of solved examples, accompanied by a number of problems with no solutions. Feedback that I received from my students is that without being able to verify both the final results and the methodology it is very discouraging to practice the unsolved problems from most textbooks. Consequently, many students become discouraged and never take that first most critical step.

In this tutorial book, I choose to give not only the complete solutions to the given problems but also detailed background information about the relevance of the problem and the underlying principles. By doing so, my hope is that my students will be able to make their first steps with my help and then to develop their own confidence and understanding of modern electronics. In order to solve engineering problems, one must first understand the underlying principles, and one must know the basic set of methods of how to solve the problems.

The intended audience of this book are primarily senior engineering undergraduate students who are just entering the field of wireless electronics after only first courses in electronics. At the same time, my hope is that graduate engineering students will find this book useful reference for some of the details that have been either only touched upon in the previous stages of their education or explained from a different point of view. Finally, the practicing junior RF engineers may find this book handy source for quick answers that are routinely omitted in most textbooks.

Paris, France   Robert Sobot
Winter, Spring, and Summer of 2013

Acknowledgements

I would like to acknowledge all those wonderful books that I used as the source of my knowledge and say thank you to their authors for providing me with the insights that otherwise I would not have been able to acquire. Under their influence, I was able to expand my own horizons, which is what acquiring of the knowledge is all about. Hence, I do want to acknowledge their contributions that are clearly visible throughout this book, which are now being passed on to my readers.

In professional life, one learns both from written sources and from experience. The experience comes from the interaction with people whom we meet and projects that we work on. I am grateful to my former colleagues who I was fortunate to have as my technical mentors on really inspirational projects, first at the Institute of Microelectronic Technologies and Single Crystals, University of Belgrade, former Yugoslavia, and then at PMC–Sierra Burnaby, BC, Canada, where I gained most of my experiences of the real engineering world.

I would like to acknowledge the contributions of my colleagues at the Western University in Canada, specifically of Professor John MacDougall, who initialized and restructured the course into the form of "design and build", and of Professors Alan Webster, Zine Eddine Abid, and Serguci Primak who taught the course at various times.

I would like to thank all of my former and current students who attended my classes in Canada, China, and France for relentlessly asking "Why?" and "How did you get this?". I hope that the material compiled in this book contains answers to at least some of those questions and that it will encourage them to keep asking questions with unconstrained curiosity about all the phenomena that surround us.

Sincere gratitude goes to my publisher and editors for their support and making this book possible.

Most of all, I want to thank my son for patiently growing up along with this book, for hanging around my desk, asking questions and having long discussions, and for making me laugh.

Contents

Acronyms

A/D	Analogue to digital
AC	Alternate current
ADC	Analogue to digital converter
AF	Audio frequency
AFC	Automatic frequency control
AGC	Automatic gain control
AM	Amplitude modulation
BiCMOS	Bipolar–CMOS
BJT	Bipolar junction transistor
BW	Bandwidth
CMOS	Complementary metal–oxide semiconductor
CRTC	Canadian Radio-Television and Telecommunications Commission
CW	Continuous wave
D/A	Digital to analogue
DAC	Digital to analogue converter
dB	Decibel
dBm	Decibel with respect to 1 mW
DC	Direct current
ELF	Extremely low frequency
EM	Electromagnetic
eV	Electron volts
FCC	Federal Communications Commission
FET	Field-effect transistor
FFT	Fast Fourier transform
FM	Frequency modulation
GHz	Gigahertz
HF	High frequency
Hz	Hertz
$\Im(z)$	Imaginary part of a complex number z
IC	Integrated circuit
I/O	Input–output
IF	Intermediate frequency
IR	Infrared
JFET	Junction field-effect transistor
KCL	Kirchhoff's current law
KVL	Kirchhoff's voltage law
LC	Inductive–capacitive

LF	Low frequency
LNA	Low noise amplifier
LO	Local oscillator
MOS	Metal–oxide semiconductor
MOSFET	Metal–oxide semiconductor field-effect transistor
NF	Noise figure
PCB	Printed circuit board
PLL	Phase locked loop
PM	phase modulation
pp	Peak-to-peak
ppm	Parts per million
pwl	Piecewise linear
Q	Quality factor
$\Re(z)$	Real part of a complex number z
RADAR	RAdio Detection and Ranging
RF	Radio frequency
RMS	Root mean square
S/N	Signal-to-noise
SAW	Surface acoustic wave
SHF	Super high frequency
SINAD	Signal-to-noise plus distortion
SNR	Signal-to-noise ratio
SPICE	Simulation Program with Integrated Circuit Emphasis
TC	Temperature coefficient
THD	Total harmonic distortion
UHF	Ultra high frequency
UV	Ultraviolet
V/F	Voltage to frequency
V/I	Voltage/current
VCO	Voltage-controlled oscillator
VHF	Very high frequency
VLF	Very low frequency
VSWR	Voltage standing wave ratio

Engineering Mathematics

1

Being fluent in the very basic techniques and definitions in engineering mathematics is prerequisite if one is to work on circuit design and analysis, specifically, the basic operations with complex numbers, vectors, and trigonometry identities and their equivalent geometrical interpretations. Mastery of these mathematical techniques is essential for rapid analysis of circuits and systems that are routinely used by working engineers.

1.1 Important to Know

RF circuit analysis employs both time domain and frequency domain analyses, therefore operations with sinusoidal functions (i.e. signals). The use of logarithmic function enables us to show both small and large numbers in the same graph and with same resolution. In addition, the complex numbers embed both exponential and sinusoidal functions (as shown by Euler's formula (1.12)), i.e. they contain in the same equation both the signal amplitude and its associated angle ("phase"). Physical interpretation of the phase is the one of "time (propagation) delay" measured relative to the duration of one sinusoidal period.

We recall that sin and cos functions have the same form where the only difference is the delay of $T/4$ units relative to each other. What is more, the multiplication of two sinusoidal forms (e.g. $\sin a \times \sin b$, see (1.19)–(1.21)) is the fundamental operation of RF communications, when the abstract arguments a and b are replaced with physical argument of frequency.

1. Logarithmic function definitions and identities:

$$\log_b(1) = 0 \quad \text{and} \quad \log_b(b) = 1 \tag{1.1}$$

$$a^b = x \quad \therefore \quad b = \log_a(x) \tag{1.2}$$

$$a^{\log_a(b)} = b \tag{1.3}$$

$$\log x^y = y \log x \tag{1.4}$$

$$\log_b(xy) = \log_b(x) + \log_b(y) \tag{1.5}$$

© Springer Nature Switzerland AG 2021
R. Sobot, *Wireless Communication Electronics by Example*,
https://doi.org/10.1007/978-3-030-59498-5_1

$$\log_b \left(\frac{x}{y}\right) = \log_b(x\,y^{-1}) = \log_b(x) - \log_b(y) \tag{1.6}$$

$$\log_b(a \pm c) = \log_b \left[a\left(1 \pm \frac{c}{a}\right)\right] = \log_b a + \log_b \left(1 \pm \frac{c}{a}\right) \tag{1.7}$$

$$\log_b a = \frac{\log_{10}(a)}{\log_{10}(b)} \tag{1.8}$$

2. Complex number definitions and identities:

$$j^2 = -1 \tag{1.9}$$

$$z = a + jb \quad \therefore \quad |z| = \sqrt{a^2 + b^2} \quad \text{and} \quad \phi = \arctan\frac{b}{a} \tag{1.10}$$

$$z^* = a - jb \quad \therefore \quad |z|^2 = z\,z^* \quad \text{and} \quad \angle(z) = -\angle(z^*) \tag{1.11}$$

$$A\,e^{j\phi} = A\,(\cos\phi + j\sin\phi), \tag{1.12}$$

where angle ϕ (a.k.a. "phase") of a complex number must be calculated accounting for its correct quadrant.

3. The geometrical equivalence among Pythagoras' theorem, vector addition, and complex numbers, Fig. 1.1:

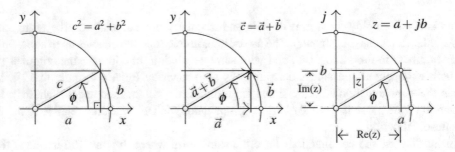

Fig. 1.1 Pythagoras' theorem, vector addition, and complex numbers

$$a = |\mathbf{a}| = \Re(z) = c\cos\phi = |\mathbf{a} + \mathbf{b}|\cos\phi = |z|\cos\phi \tag{1.13}$$

$$b = |\mathbf{b}| = \Im(z) = c\sin\phi = |\mathbf{a} + \mathbf{b}|\sin\phi = |z|\sin\phi \tag{1.14}$$

$$c^2 = |\mathbf{a} + \mathbf{b}|^2 = |z|^2 = a^2 + b^2 = |\mathbf{a}|^2 + |\mathbf{b}|^2 = [\Re(z)]^2 + [\Im(z)]^2 \tag{1.15}$$

$$= c^2\cos^2\phi + c^2\sin^2\phi \tag{1.16}$$

$\therefore$ in the special case of $c = 1$ (i.e. the "unity circle"), it follows

$$1 = \cos^2\phi + \sin^2\phi, \tag{1.17}$$

where in any case, the associated angle (a.k.a. "phase") is by definition

$$\tan \phi = \frac{b}{a} = \frac{|\mathbf{b}|}{|\mathbf{a}|} = \frac{\Im(z)}{\Re(z)} \tag{1.18}$$

4. Trigonometric identities:

$$\sin a \sin b = \frac{1}{2} \left[\cos(a-b) - \cos(a+b) \right] \tag{1.19}$$

$$\sin a \cos b = \frac{1}{2} \left[\sin(a+b) + \sin(a-b) \right] \tag{1.20}$$

$$\cos a \cos b = \frac{1}{2} \left[\cos(a+b) + \cos(a-b) \right] \tag{1.21}$$

(Note: in each identity, on the right side, there is always the "sum" and "difference" of the two arguments a and b in the products on the left side.)

1.2   Exercises

1.1 *   Logarithmic Functions

1. Solve the following equations (without a calculator):

   (a) $x = \log_{10} 10$
   (b) $x = \log_{10} 100$
   (c) $x = \log_{10} 1000$

   (d) $x = \log_2 16$
   (e) $x = \log_{25} \frac{1}{625}$
   (f) $x = \log_8 \frac{1}{4}$

   (g) $x = \log_8 \left(\log_4 \left(\log_2 (16) \right) \right)$

2. Solve the following equations (without a calculator):

   (a) $x = 25^{\log_5 3}$
   (b) $x = \log_5 \sqrt{5}$
   (c) $x = \log_{2/3} \frac{243}{32}$

   (d) $x = \sqrt{\log_{1/2}^2(4)}$
   (e) $x = \sqrt{\log_{1/2}(4)}$

1.2 *   Complex Numbers

1. Calculate $|z|$ and $\angle(z)$.

   (a) $z = j$
   (b) $z = j^2$
   (c) $z = j^3$
   (d) $z = j^4$
   (e) $z = j^5$
   (f) $z = j^{2020}$

2. Calculate $|z|$ and $\angle(z)$ in each case and show the following complex numbers z as vectors in the complex plane.

   (a) $z = 1$
   (b) $z = \frac{\sqrt{3}}{2} + j\frac{1}{2}$
   (c) $z = \frac{\sqrt{2}}{2} + j\frac{\sqrt{2}}{2}$

   (d) $z = \frac{1}{2} + j\frac{\sqrt{3}}{2}$
   (e) $z = j$
   (f) $z = -\frac{\sqrt{2}}{2} + j\frac{\sqrt{2}}{2}$

   (g) $z = -\frac{\sqrt{3}}{2} + j\frac{1}{2}$
   (h) $z = -1$
   (i) $z = -\frac{\sqrt{2}}{2} - j\frac{\sqrt{2}}{2}$

(j) $\quad z = -j$

(k) $\quad z = \dfrac{\sqrt{2}}{2} - j\dfrac{\sqrt{2}}{2}$

(l) $\quad z = \dfrac{\sqrt{3}}{2} - j\dfrac{1}{2}$

3. Show the following complex numbers as vectors in the complex plane. Then, calculate $|z|$ and $\angle(z)$ in each case.

(a) $\quad z = e^{j\,0}$

(b) $\quad z = e^{j\,\frac{\pi}{6}}$

(c) $\quad z = e^{j\,\frac{\pi}{4}}$

(d) $\quad z = e^{j\,\frac{\pi}{3}}$

(e) $\quad z = e^{j\,\frac{\pi}{2}}$

(f) $\quad z = e^{j\,\frac{3\pi}{4}}$

(g) $\quad z = e^{j\,\frac{5\pi}{6}}$

(h) $\quad z = e^{j\,\pi}$

(i) $\quad z = e^{j\,\frac{5\pi}{4}}$

(j) $\quad z = e^{j\,\frac{3\pi}{2}}$

(k) $\quad z = e^{j\,\frac{7\pi}{4}}$

(l) $\quad z = e^{j\,\frac{11\pi}{6}}$

4. Calculate $\angle(z)$ and $|z|$ of the following complex numbers.

(a) $\quad z = 2\,e^{j\,\frac{\pi}{4}}$

(b) $\quad z = \dfrac{1}{2}\,e^{j\,\frac{\pi}{3}}$

(c) $\quad z = -2\,e^{-j\,\frac{5\pi}{6}}$

(d) $\quad z = -\dfrac{1}{2}\,e^{j\,\frac{7\pi}{4}}$

(e) $\quad z = \left(\sqrt{2}\,e^{j\,\frac{\pi}{4}}\right)^2$

(f) $\quad z = \left(3\,e^{j\,\frac{\pi}{3}}\right)^3$

(g) $\quad z = \left(4\,e^{-j\,\frac{5\pi}{6}}\right)^5$

(h) $\quad z = \left(0.2\,e^{j\,\frac{7\pi}{4}}\right)^4$

(i) $\quad z = \left(-3\,e^{-j\,\frac{\pi}{6}}\right)^3$

1.3 * Trigonometry Identities

1. Calculate the following trigonometry functions (without a calculator):

(a) $\cos\left(\dfrac{\pi}{3}\right)$

(b) $\tan\left(\dfrac{\pi}{4}\right)$

(c) $\cos(150°)$

(d) $\sin(120°)$

(e) $\tan(-45°)$

(f) $\tan(-60°)$

(g) $\tan(300°)$

(h) $\tan(225°)$

(i) $\tan(480°)$

(j) $\tan(840°)$

(k) $\cos(1320°)$

(l) $\sin\left(\dfrac{\pi}{6}\right)\sin^2\left(\dfrac{\pi}{3}\right)$

2. Calculate the following expressions (without a calculator):

(a) $-4\,\sin\left(\dfrac{7\pi}{12}\right)\sin\left(\dfrac{13\pi}{12}\right)$

(b) $4\,\sin\left(\dfrac{5\pi}{8}\right)\cos\left(\dfrac{\pi}{8}\right)$

(c) $4\cos\left(\dfrac{5\pi}{8}\right)\cos\left(\dfrac{3\pi}{8}\right)$

(d) $\tan 20°\,\tan 40°\,\tan 80°$

1.4 ** Basic Integrals

1. Calculate integrals of the following functions:

(a) $\displaystyle\int \sin(2x)\,dx$

(b) $\displaystyle\int \sin^2(x)\,dx$

(c) $\displaystyle\int_0^1 f(x)^2\,dx$ given that $f(x)$ is a piecewise linear function as defined by its graph, see Fig. 1.2.

Fig. 1.2 Example 1.4-1 (c)

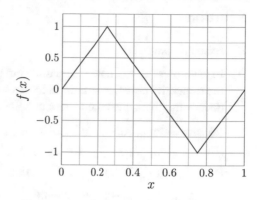

1.5 **   Basic Forms of Complex Functions $z(x)$

1. For each of $z(x)$ functions, derive $H(x) = 20 \log |z(x)|$ and $P(x) = \angle z(x)$ functions in $\log(x)$ scale. Then, calculate the following limits:

$$\lim_{x \to +\infty} H(x) \quad \lim_{x \to 0} H(x) \quad \lim_{x \to x_0} H(x) \quad \text{and}$$

$$\lim_{x \to +\infty} P(x) \quad \lim_{x \to 0} P(x) \quad \lim_{x \to x_0} P(x)$$

where x_0 is the number in each of respective x/x_0 fractions. Finally, calculate $H(x = 0.1x_0)$, $H(x = 10x_0)$, $P(x = 0.1x_0)$, and $P(x = 10x_0)$.

(a) $z(x) = j\dfrac{x}{2}$ (b) $z(x) = j\dfrac{x}{10}$ (c) $z(x) = 1 + j\dfrac{x}{2}$

(d) $z(x) = 1 + j\dfrac{x}{10}$ (e) $z(x) = 1 - j\dfrac{x}{2}$ (f) $z(x) = 1 - j\dfrac{x}{10}$

(g) $z(x) = \dfrac{1}{1 + j\dfrac{x}{2}}$ (h) $z(x) = \dfrac{1}{1 + j\dfrac{x}{10}}$ (i) $z(x) = \dfrac{1}{1 - j\dfrac{x}{10}}$

(j) $z(x) = \dfrac{1}{1 - j\dfrac{x}{2}}$

1.6 ***   Piecewise Linear Approximation of $z(x)$

1. For each $z(x)$ function in Example 1.5, using $\log(x)$ scale, show the piecewise linear approximation graphs of $H(x) = 20 \log |z(x)|$ and $P(x) = \angle z(x)$.

1.7 ****   Complex Functions: Case Study

1. Show piecewise linear approximation graphs of $H(x)$ and $P(x)$ functions for the following complex function:

$$z(x) = 20\,000 \, \frac{2 + jx}{220\,jx + 4\,000 - x^2} \tag{1.22}$$

Solutions

Exercise 1.1, page 5

1. In examples (a)–(c), note the relationship between the numbers of zeros in 10, 100, 1 000, . . ., and the respective equation solutions. In this special case of logarithmic functions with the base 10 and decade numbers, the noted shortcut helps us to perform mental calculations of these logarithms.

   (a) $x = \log_{10} 10$ $\therefore$ $10^x = 10$ $\therefore$ $x = 1$

   (b) $x = \log_{10} 100$ $\therefore$ $10^x = 100$ $\therefore$ $x = 2$

   (c) $x = \log_{10} 1\,000$ $\therefore$ $10^x = 1\,000$ $\therefore$ $x = 3$

   In addition, writing "log" without specific index number assumes the base 10, while writing "ln" assumes the base e. For example, $x = \log 1\,000\,000$ is found by counting the zeros, and therefore $x = 6$.

   Examples (d)–(g) illustrate some simple techniques for calculating various logarithms based on the knowledge of relationships between numbers, i.e. squares, roots, the multiplication/division table, and the basic log identities

   (d) $x = \log_2 16$ $\therefore$ $2^x = 16$ $\therefore$ $2^x = 2^4$ $\therefore$ $x = 4$

   (e) $x = \log_{25} \dfrac{1}{625} = \log_{25} \dfrac{1}{25^2} = \log_{25} 25^{-2} = -2 \underbrace{\log_{25} 25}_{1} = -2$

   (f) $x = \log_8 \dfrac{1}{4} = \log_8 \dfrac{1}{\sqrt[3]{64}} = \log_8 \dfrac{1}{\sqrt[3]{8^2}} = \log_8 8^{-2/3} = -\dfrac{2}{3} \underbrace{\log_8 8}_{1} = -\dfrac{2}{3}$

   (g)   Given $x = \log_8 \left(\log_4 \left(\log_2 (16) \right) \right)$, we start from the right side, as

   $x = \log_8 \left(\log_4 \left(\log_2 (2^4) \right) \right) = \log_8 \left(\log_4 \left(4 \underbrace{\log_2 (2)}_{1} \right) \right)$

   $= \log_8 \left(\underbrace{\log_4 4}_{1} \right) \Rightarrow x = \log_8(1) \Rightarrow 8^x = 8^{\log_8(1)} \Rightarrow 8^x = 1$

   $\therefore$

   $x = 0$

2. Exercises with problems combining log and $\sqrt[n]{a}$ functions, and absolute values of numbers.

   (a) $x = 25^{\log_5 3} = \left(5^2 \right)^{\log_5 3} = 5^{(2\,\log_5 3)} = 5^{\log_5 9} = 9$

   (b) $x = \log_5 \sqrt{5} = \log_5 5^{1/2} = \dfrac{1}{2} \underbrace{\log_5 5}_{1} = \dfrac{1}{2}$

   (c) $x = \log_{2/3} \dfrac{243}{32} = \log_{2/3} \dfrac{3^5}{2^5} = \log_{2/3} \left(\dfrac{2}{3} \right)^{-5} = -5 \underbrace{\log_{2/3} \dfrac{2}{3}}_{1} = -5$

   (d) $x = \sqrt{\log_{1/2}^2 (4)} = \left| \log_{1/2}(4) \right| = \left| \log_{1/2} \left(\dfrac{1}{2} \right)^{-2} \right| = \left| -2 \underbrace{\log_{1/2} \dfrac{1}{2}}_{1} \right| = 2$

(e) $\quad x = \sqrt{\log_{1/2}(4)} = \sqrt{\log_{1/2}\left(\dfrac{1}{2}\right)^{-2}} = \sqrt{-2\log_{1/2}\dfrac{1}{2}} = \sqrt{-2} = j\sqrt{2}$

Exercise 1.2, page 5

1. We note that powers of a complex number in effect force z to rotate around the circle; therefore, there is periodicity in the results. Calculate $|z|$ and $\angle(z)$.

(a) $\quad z = j$; therefore, $\quad |z| = 1$, and $\quad \angle z = \pi/2$

(b) $\quad z = j^2 = -1$; therefore, $\quad |z| = 1$, and $\quad \angle z = \pi$

(c) $\quad z = j^3 = j^2 \times j = -j$; therefore, $\quad |z| = 1$, and $\quad \angle z = 3\pi/2$

(d) $\quad z = j^4 = j^2 \times j^2 = 1$; therefore, $\quad |z| = 1$, and $\quad \angle z = 2\pi = 0$

(e) $\quad z = j^5 = j^4 \times j = j$; therefore, $\quad |z| = 1$, and $\quad \angle z = \pi/2$

(f) $\quad z = j^{2020} = j^{4\times505} = \left(j^4\right)^{505} = 1^{505} = 1$;

therefore, $\quad |z| = 1$, and $\quad \angle z = 2\pi = 0$

2. Here, complex number modules are found simply by Pythagoras' theorem for relation among the three sides of a right triangle, $|z| = \sqrt{\Re(z)^2 + \Im(z)^2}$.

(a) Given $z = 1$, we write
$$\left.\begin{array}{l} \Re(z) = 1 > 0 \\[4pt] \Im(z) = 0 \end{array}\right\} \text{(I)},$$
then (Fig. 1.3), the module is

$$|z| = \sqrt{1^2 + 0^2} = 1$$

and the phase is

$$\angle z = \arctan\frac{\Im(z)}{\Re(z)} = \arctan\frac{0}{1}$$
$$= \arctan(0) = 0°$$

(b) Given $z = \sqrt{3}/2 + j1/2$, we write
$$\left.\begin{array}{l} \Re(z) = \sqrt{3}/2 > 0 > 0 \\[4pt] \Im(z) = 1/2 > 0 \end{array}\right\} \text{(I)},$$
then (Fig. 1.4), the module is

$$|z| = \sqrt{\left(\sqrt{3}/2\right)^2 + (1/2)^2} = 1$$

$$\angle z = \arctan\frac{\Im(z)}{\Re(z)} = \arctan\frac{1/2}{\sqrt{3}/2}$$
$$= 30° = \frac{\pi}{6}$$

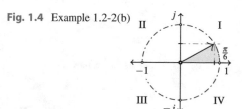

Fig. 1.3 Example 1.2-2(a)

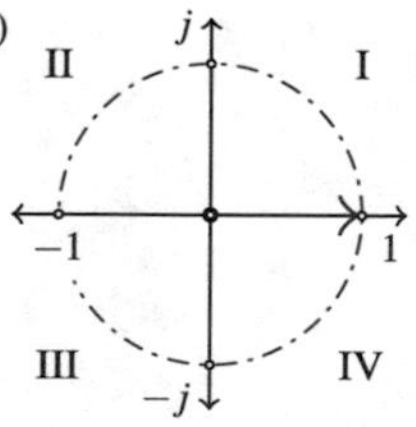

Fig. 1.4 Example 1.2-2(b)

(c) Given $z = \sqrt{2}/2 + j\sqrt{2}/2$, we write
$$\left.\begin{aligned}\Re(z) &= \sqrt{2}/2 > 0\\ \Im(z) &= \sqrt{2}/2 > 0\end{aligned}\right\} \text{(I)},$$
then (Fig. 1.5), the module is

$$|z| = \sqrt{\left(\sqrt{2}/2\right)^2 + \left(\sqrt{2}/2\right)^2} = 1$$

$$\begin{aligned}\angle z &= \arctan\frac{\Im(z)}{\Re(z)} = \arctan\frac{\frac{\sqrt{2}}{2}}{\frac{\sqrt{2}}{2}}\\ &= \arctan(1) = 45^\circ = \frac{\pi}{4}\end{aligned}$$

Fig. 1.5 Example 1.2-2(c)

(d) Given $z = 1/2 + j\sqrt{3}/2$, we write
$$\left.\begin{aligned}\Re(z) &= 1/2 > 0\\ \Im(z) &= \sqrt{3}/2 > 0\end{aligned}\right\} \text{(I)},$$
then (Fig. 1.6), the module is

$$|z| = \sqrt{(1/2)^2 + \left(\sqrt{3}/2\right)^2} = 1$$

$$\begin{aligned}\angle z &= \arctan\frac{\Im(z)}{\Re(z)} = \arctan\frac{\frac{\sqrt{3}}{2}}{\frac{1}{2}}\\ &= \arctan(\sqrt{3}) = 60^\circ = \frac{\pi}{3}\end{aligned}$$

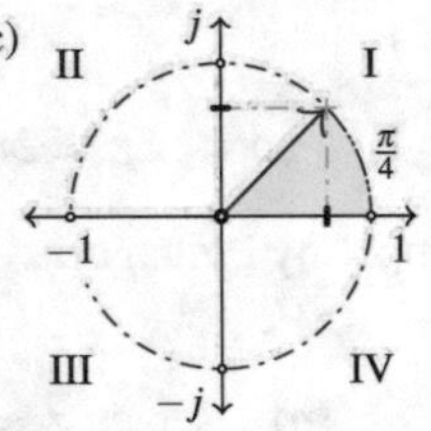

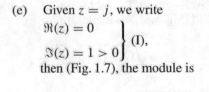

Fig. 1.6 Example 1.2-2(d)

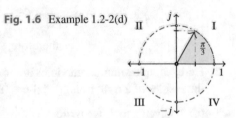

(e) Given $z = j$, we write
$$\left.\begin{aligned}\Re(z) &= 0\\ \Im(z) &= 1 > 0\end{aligned}\right\} \text{(I)},$$
then (Fig. 1.7), the module is

$$|z| = \sqrt{(0)^2 + (1)^2} = 1$$

$$\begin{aligned}\angle z &= \arctan\frac{\Im(z)}{\Re(z)} = \arctan\frac{1}{0}\\ &= \arctan(\infty) = 90^\circ = \frac{\pi}{2}\end{aligned}$$

Fig. 1.7 Example 1.2-2(e)

(f) Given $z = -\sqrt{2}/2 + j\sqrt{2}/2$, we write
$$\left.\begin{aligned}\Re(z) &= -\sqrt{2}/2 < 0\\ \Im(z) &= \sqrt{2}/2 > 0\end{aligned}\right\} \text{(II)},$$
then (Fig. 1.8), the module is

$$|x| = \sqrt{\left(-\sqrt{2}/2\right)^2 + \left(\sqrt{2}/2\right)^2} = 1$$

$$\begin{aligned}\angle z &= \arctan\frac{\Im(z)}{\Re(z)} = \arctan\frac{\frac{\sqrt{2}}{2}}{-\frac{\sqrt{2}}{2}}\\ &= \arctan\left(\frac{1}{-1}\right) = 135^\circ = \frac{3\pi}{4}\end{aligned}$$

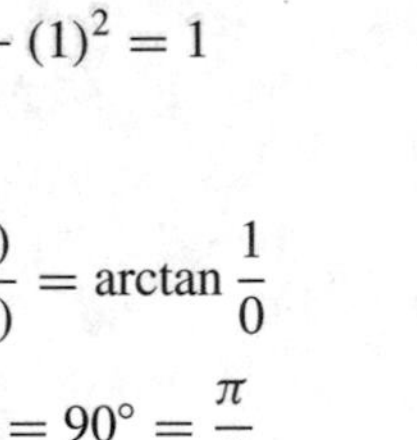

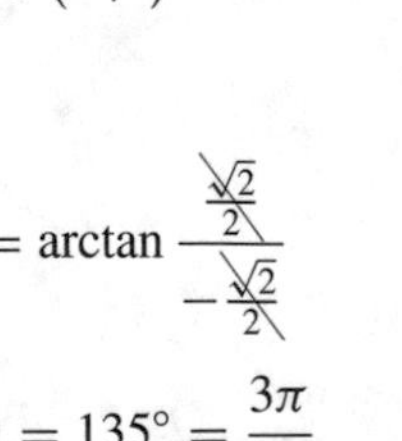

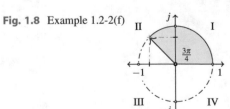

Fig. 1.8 Example 1.2-2(f)

(g)   Given $z = -\sqrt{3}/2 + j^{1}/2$, we write
$$\left.\begin{array}{l}\Re(z) = -\sqrt{3}/2 < 0 \\ \Im(z) = {}^{1}/2 > 0\end{array}\right\} \text{(II)},$$
then (Fig. 1.9), the module is

$$|x| = \sqrt{({}^{1}/2)^2 + \left(-\sqrt{3}/2\right)^2} = 1$$

$$\angle z = \arctan \frac{\Im(z)}{\Re(z)} = \arctan \frac{\frac{1}{2}}{-\frac{\sqrt{3}}{2}}$$

$$= \arctan ({}^{1}/-\sqrt{3}) = 150° = \frac{5\pi}{6}$$

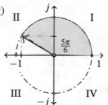

Fig. 1.9 Example 1.2-2(g)

(h)   Given $z = -1$, we write
$$\left.\begin{array}{l}\Re(z) = -1 < 0 \\ \Im(z) = 0\end{array}\right\} \text{(II)},$$
then (Fig. 1.10), the module is

$$|x| = \sqrt{(-1)^2 + (0)^2} = 1$$

$$\angle z = \arctan \frac{\Im(z)}{\Re(z)} = \arctan \frac{0}{-1}$$

$$= 180° = \pi$$

Fig. 1.10
Example 1.2-2(h)

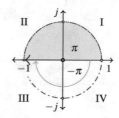

(i)   Given $z = -\sqrt{2}/2 - j\sqrt{2}/2$, we write
$$\left.\begin{array}{l}\Re(z) = -\sqrt{2}/2 < 0 \\ \Im(z) = -\sqrt{2}/2 < 0\end{array}\right\} \text{(III)},$$
then (Fig. 1.11), the module is

$$|x| = \sqrt{\left(-\sqrt{2}/2\right)^2 + \left(-\sqrt{2}/2\right)^2} = 1$$

$$\angle z = \arctan \frac{\Im(z)}{\Re(z)} = \arctan \frac{-\sqrt{2}/2}{-\sqrt{2}/2}$$

$$= \arctan \left(\frac{-1}{-1}\right) = 225° = -135°$$

$$= \frac{5\pi}{4} = -\frac{3\pi}{4}$$

Fig. 1.11
Example 1.2-2(i)

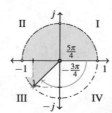

(j)   Given $z = -j$, we write
$$\left.\begin{array}{l}\Re(z) = 0 \\ \Im(z) = -1 < 0\end{array}\right\} \text{(III)},$$
then (Fig. 1.12), the module is

$$|x| = \sqrt{(0)^2 + (-1)^2} = 1$$

$$\angle z = \arctan \frac{\Im(z)}{\Re(z)} = \arctan \frac{-1}{0}$$

$$= 270° = -90° = \frac{3\pi}{2} = -\frac{\pi}{2}$$

Fig. 1.12
Example 1.2-2(j)

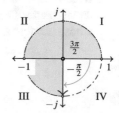

(k)   Given $z = \sqrt{2}/2 - j\sqrt{2}/2$, we write
$$\left.\begin{array}{l}\Re(z) = \sqrt{2}/2 > 0 \\ \Im(z) = -\sqrt{2}/2 < 0\end{array}\right\} \text{(IV)},$$
then (Fig. 1.13), the module is

$$|x| = \sqrt{\left(\sqrt{2}/2\right)^2 + \left(-\sqrt{2}/2\right)^2} = 1$$

$$\angle z = \arctan\frac{\Im(z)}{\Re(z)} = \arctan\frac{-\frac{\sqrt{2}}{2}}{\frac{\sqrt{2}}{2}}$$

$$= \arctan\left(\frac{-1}{1}\right) = 315° = -45°$$

$$= \frac{7\pi}{4} = -\frac{\pi}{4}$$

(l)   Given $z = \sqrt{3}/2 - j1/2$, we write
$$\left.\begin{array}{l}\Re(z) = \sqrt{3}/2 > 0 \\ \Im(z) = -1/2 < 0\end{array}\right\} \text{(IV)},$$
then (Fig. 1.14), the module is

$$|x| = \sqrt{\left(\sqrt{3}/2\right)^2 + (-1/2)^2} = 1$$

$$\angle z = \arctan\frac{\Im(z)}{\Re(z)} = \arctan\frac{-1/2}{\sqrt{3}/2}$$

$$= \arctan\frac{-1}{\sqrt{3}} = 330° = -30°$$

$$= \frac{11\pi}{6} = -\frac{\pi}{6}$$

Fig. 1.13
Example 1.2-2(k)

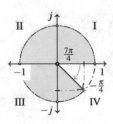

Fig. 1.14
Example 1.2-2(l)

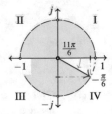

3. Complex numbers given here in the exponential forms are identical to those in Example 1.2-2. First, the phase is determined by comparison with (1.12), then for real and imaginary parts, as well as the module can be calculated.

However, aside from Pythagorean theorem used to calculate $|z|$, it is important to know how to calculate $|z|$ in the exponential form,

$$|z|^2 = z\,z^* = e^{j\frac{\pi}{6}}\,e^{-j\frac{\pi}{6}} = e^{j\frac{\pi}{6} - j\frac{\pi}{6}}$$

$$= e^0 = 1 \quad \therefore \quad |z| = 1,$$

which illustrates that the module of any complex number found on the unity circle is therefore equal to one. That is to say, the complex exponent by itself keeps only the phase, and the numbers whose module $|z| = A \neq 1$ are written in the form

$$z = A\,e^{j\phi} \quad \therefore \quad |z| = \left|A\,e^{j\phi}\right|$$

$$= |A|\,\underbrace{\left|e^{j\phi}\right|}_{1},$$

where A is any real number.

(a) $z = e^{j\,0}$; therefore, $\phi = 0\text{rad}$, and its vector graph is in Example 1.2-2(a):

$$\Re(z) = \cos 0 = 1$$

$$\Im(z) = \sin 0 = 0$$

$$|z| = \sqrt{\Re(z)^2 + \Im(z)^2} = 1$$

(b) $z = e^{j\,\frac{\pi}{6}}$; therefore, $\phi = \pi/6$, and its vector graph is in Example 1.2-2(b):

$$\Re(z) = \cos \pi/6 = \sqrt{3}/2$$

$$\Im(z) = \sin \pi/6 = 1/2$$

$$|z| = \sqrt{\Re(z)^2 + \Im(z)^2} = 1$$

(c) $z = e^{j\,\frac{\pi}{4}}$; therefore, $\phi = \pi/4$, and its vector graph is in Example 1.2-2(c):

$$\Re(z) = \cos \pi/4 = \sqrt{2}/2$$

$$\Im(z) = \sin \pi/4 = \sqrt{2}/2$$

$$|z|^2 = z\,z^* = e^{j\,\frac{\pi}{4}}\,e^{-j\,\frac{\pi}{4}} = e^0 = 1$$

(d) $z = e^{j\,\frac{\pi}{3}}$; therefore, $\phi = \pi/3$, and its vector graph is in Example 1.2-2(d):

$$\Re(z) = \cos \pi/3 = 1/2$$

$$\Im(z) = \sin \pi/3 = \sqrt{3}/2$$

$$|z|^2 = z\,z^* = e^{j\,(\frac{\pi}{3} - \frac{\pi}{3})} = e^0 = 1$$

(e) $z = e^{j\,\frac{\pi}{2}}$; therefore, $\phi = \pi/2$, and its vector graph is in Example 1.2-2(e):

$$\Re(z) = \cos \pi/2 = 0$$

$$\Im(z) = \sin \pi/3 = 1$$

$$|z|^2 = z\,z^* = e^{j\,(\frac{\pi}{2} - \frac{\pi}{2})} = e^0 = 1$$

(f) $z = e^{j\,\frac{3\pi}{4}}$; therefore, $\phi = 3\pi/4$, and its vector graph is in Example 1.2-2(f):

$$\Re(z) = \cos 3\pi/4 = -\sqrt{2}/2$$

$$\Im(z) = \sin 3\pi/4 = \sqrt{2}/2$$

$$|z|^2 = z\,z^* = e^{j\,(\frac{3\pi}{4} - \frac{3\pi}{4})} = e^0 = 1$$

(g) $z = e^{j\,\frac{5\pi}{6}}$; therefore, $\phi = 5\pi/6$, and its vector graph is in Example 1.2-2(g):

$$\Re(z) = \cos 5\pi/6 = -\sqrt{3}/2$$

$$\Im(z) = \sin 5\pi/6 = 1/2$$

$$|z|^2 = z\,z^* = e^{j\,(\frac{5\pi}{6} - \frac{5\pi}{6})} = e^0 = 1$$

(h) $z = e^{j\,\pi}$; therefore, $\phi = \pi$, and its vector graph is in Example 1.2-2(h):

$$\Re(z) = \cos \pi = -1$$

$$\Im(z) = \sin \pi = 0$$

$$|z|^2 = z\,z^* = e^{j\,(\pi - \pi)} = e^0 = 1$$

(i) $z = e^{j\frac{5\pi}{4}}$; therefore, $\phi = {}^{5\pi}/_4 = -{}^{3\pi}/_4$,   (j) $z = e^{j\frac{3\pi}{2}}$; therefore, $\phi = {}^{3\pi}/_2 = -{}^{\pi}/_2$,
   and its vector graph is in Example 1.2-   and its vector graph is in Example 1.2-
   2(i):   2(j):

$$\Re(z) = \cos {}^{5\pi}/_4 = -\sqrt{2}/_2$$

$$\Im(z) = \sin {}^{5\pi}/_4 = -\sqrt{2}/_2$$

$$|z|^2 = z\,z^* = e^{j\,(\frac{5\pi}{4} - \frac{5\pi}{4})} = e^0 = 1$$

$$\Re(z) = \cos {}^{3\pi}/_2 = 0$$

$$\Im(z) = \sin {}^{3\pi}/_2 = -1$$

$$|z|^2 = z\,z^* = e^{j\,(\frac{3\pi}{2} - \frac{3\pi}{2})} = e^0 = 1$$

(k) $z = e^{j\frac{7\pi}{4}}$; therefore, $\phi = {}^{7\pi}/_4 = -{}^{\pi}/_4$,   (l) $z = e^{j\frac{11\pi}{6}}$; therefore, $\phi = {}^{11\pi}/_6 = -{}^{\pi}/_6$,
   and its vector graph is in Example 1.2-   and its vector graph is in Example 1.2-
   2(k):   2(l):

$$\Re(z) = \cos {}^{7\pi}/_4 = \sqrt{2}/_2$$

$$\Im(z) = \sin {}^{7\pi}/_4 = -\sqrt{2}/_2$$

$$|z|^2 = z\,z^* = e^{j\,(\frac{7\pi}{4} - \frac{7\pi}{4})} = e^0 = 1$$

$$\Re(z) = \cos {}^{11\pi}/_6 = \sqrt{3}/_2$$

$$\Im(z) = \sin {}^{11\pi}/_6 = -{}^1/_2$$

$$|z|^2 = z\,z^* = e^{j\,(\frac{11\pi}{6} - \frac{11\pi}{6})} = e^0 = 1$$

4. By comparison with (1.12), the phase and the module are found as

(a) $z = 2\,e^{j\frac{\pi}{4}}$; therefore, $\phi = {}^{\pi}/_4$ and   (b) $z = {}^1/_2\,e^{j\frac{\pi}{3}}$; therefore,
                                                                 $\phi = {}^{\pi}/_3$ and $|z| = {}^1/_2$

$$|z|^2 = z\,z^* = 2e^{j\frac{\pi}{4}}\,2e^{-j\frac{\pi}{4}} = 4e^0$$

$$|z| = 2$$

(c) $z = -2\,e^{-j\frac{5\pi}{6}}$; therefore,   (d) $z = -{}^1/_2\,e^{j\frac{7\pi}{4}}$; therefore,
   $\phi = -{}^{5\pi}/_6$ and $|z| = 2$ $\phi = {}^{7\pi}/_4 = -{}^{\pi}/_4$ and $|z| = {}^1/_2$

Here, it was not even necessary to calculate anything; both the phase and the module are found simply by comparison with (1.12). In the following examples, we note that powers of a complex number, in effect, aside from increasing its module also *rotate* the associated vector by multiples of the original phase.

(e) First, we must calculate the exponential form of z as

$$z = \left(\sqrt{2}\,e^{j\frac{\pi}{4}}\right)^2$$

$$= \left(\sqrt{2}\right)^2 \left(e^{j\frac{\pi}{4}}\right)^2$$

$$= 2\,e^{j\frac{\pi}{4}\times 2} = 2\,e^{j\frac{\pi}{2}};$$

therefore, $\phi = \pi/2$ and $|z| = 2$.

(f) First, we must calculate the exponential form of z as

$$z = \left(3\,e^{j\frac{\pi}{3}}\right)^3$$

$$= 9\,e^{j\pi} = 9\,(-1) = -9;$$

therefore, $\phi = \pi$ and $|z| = 9$. We note that this complex number rotated into horizontal position, and thus into a real number.

(g) First, we must calculate the exponential form of z as

$$z = \left(4\,e^{j\frac{5\pi}{6}}\right)^5 = 4^5\,e^{j\frac{25\pi}{6}}$$

$$= 2^{10}\,e^{j\frac{\pi}{6}} = 1024\,e^{j\frac{\pi}{6}};$$

therefore, after making four turns around the unity circle $\phi = \pi/6$, and $|z| = 1024$.

(h) First, we must calculate the exponential form of z as

$$z = \left(0.2\,e^{j\frac{7\pi}{4}}\right)^4 = 0.2^4\,e^{j7\pi}$$

$$= 5^{-4}\,e^{j\pi} = 1.6 \times 10^{-3}\,(-1)$$

$$= -1.6 \times 10^{-3};$$

therefore, after making two turns around the unity circle $\phi = \pi$, the complex number rotated into its position, and thus into a real number, and $|z| = 1.6 \times 10^{-3}$ (in engineering units).

(i) First, we must calculate the exponential form of z as

$$z = \left(-3\,e^{-j\frac{\pi}{6}}\right)^3 = (-3)^3\,e^{-j\frac{\pi}{2}}$$

$$= -27\,(-j) = 27j;$$

therefore, $\phi = -\pi/2$ and $|z| = 27$.

Exercise 1.3, page 6

1. Main idea is to decompose given angles into the basic angles found in the unity circle, whose sin and cos (therefore, also tan) values are already known. In addition, understanding of relations among similar angles, for example $180° \pm \alpha$ and $90° \pm \alpha$ is very important (Fig. 1.15).

Fig. 1.15 Example 1.3-1

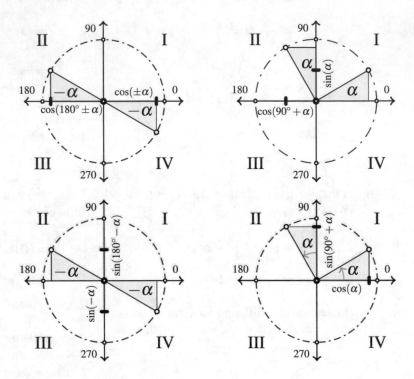

(a) Important to know values of sin and cos are found on the unity circle,

$$\cos\left(\frac{\pi}{3}\right) = \frac{1}{2}$$

(b) Function tan is found by sin and cos ratio

$$\tan\left(\frac{\pi}{4}\right) = \frac{\sin(\pi/4)}{\cos(\pi/4)} = \frac{\sqrt{2}/2}{\sqrt{2}/2} = 1$$

(c) Difference between 180 and 30 gives

$$\cos(150°) = \cos(180° - 30°) = -\cos(30°) = -\frac{\sqrt{3}}{2}$$

(d) Sum of 90 and 30 gives

$$\sin(120°) = \sin(90° + 30°) = \cos(30°) = \frac{\sqrt{3}}{2}$$

(e) Functions sin and cos of negative angles result in

$$\tan(-45°) = \frac{\sin(-45°)}{\cos(-45°)} = \frac{-\sin(45°)}{\cos(45°)} = -\frac{\sin(45°)}{\cos(45°)} = -\tan(45°) = -1$$

(f) Functions sin and cos of negative angles result in

$$\tan(-60°) = \frac{\sin(-60°)}{\cos(-60°)} = \frac{-\sin(60°)}{\cos(60°)} = -\frac{\sin(60°)}{\cos(60°)} = -\frac{\sqrt{3}/2}{1/2} = -\sqrt{3}$$

(g) Difference between 360 and 60 gives

$$\tan(300°) = \tan(360° - 60°) = \tan(-60°) = \frac{\sin(-60°)}{\cos(-60°)} = \frac{-\sin(60°)}{\cos(60°)}$$

$$= -\frac{\sqrt{3}/2}{1/2} = -\sqrt{3}$$

(h) Sum of 180 and 45 gives

$$\tan(225°) = \tan(180° + 45°) = \frac{\sin(180° + 45°)}{\cos(180° + 45°)} = \frac{-\sin(+45°)}{-\cos(+45°)}$$

$$= \tan 45° = 1$$

(i) Difference between 360, 120, and 60 gives

$$\tan(480°) = \tan(360° + 120°) = \tan(180° - 60°) = \tan(-60°)$$

$$= -\tan(60°) = -\sqrt{3}$$

(j) Multiple revolutions around the 2π circle result in

$$\tan(840°) = \tan(2 \times 360° + 120°) = \tan(180° - 60°) = \tan(-60°)$$

$$= -\tan(60°) = -\sqrt{3}$$

(k) Multiple revolutions around the 2π circle result in

$$\cos(1320°) = \cos(3 \times 360° + 240°) = \cos(180° + 60°) = -\cos(60°) = -1/2$$

(l) More complicated problems may also be reduced as, for example

$$\sin\left(\frac{\pi}{6}\right) \sin^2\left(\frac{\pi}{3}\right) = \frac{1}{2}\left(\frac{\sqrt{3}}{2}\right)^2 = \frac{1}{2}\frac{3}{4} = \frac{3}{8}$$

2. The idea is to use trigonometric identities for products of sin and cos functions,

(a) Product of two sin functions is identical to

$$-4\sin\left(\frac{7\pi}{12}\right)\sin\left(\frac{13\pi}{12}\right) = -4\frac{1}{2}\left[\cos\left(\frac{7\pi}{12} - \frac{13\pi}{12}\right) - \cos\left(\frac{7\pi}{12} + \frac{13\pi}{12}\right)\right]$$

$$= -2\left[\cos\left(-\frac{6\pi}{12}\right) - \cos\left(\frac{20\pi}{12}\right)\right]$$

$$= -2\left[\cos\left(\frac{\pi}{2}\right)^0 - \cos\left(\frac{5\pi}{3}\right)\right]$$

$$= -2\left[-\cos\left(-\frac{\pi}{3}\right)\right] = 1$$

(b) Inter-product of sin and cos functions is identical to

$$4 \sin\left(\frac{5\pi}{8}\right) \cos\left(\frac{\pi}{8}\right) = 4 \frac{1}{2}\left[\sin\left(\frac{5\pi}{8} + \frac{\pi}{8}\right) + \sin\left(\frac{5\pi}{8} - \frac{\pi}{8}\right)\right]$$

$$= 4 \frac{1}{2}\left[\sin\left(\frac{3\pi}{4}\right) + \sin\left(\frac{\pi}{2}\right)\right] = 4 \frac{1}{2}\left[\frac{\sqrt{2}}{2} + 1\right]$$

$$= \sqrt{2} + 2$$

(c) Product of cos functions results in

$$4 \cos\left(\frac{5\pi}{8}\right) \cos\left(\frac{3\pi}{8}\right) = 4 \frac{1}{2}\left[\cos\left(\frac{5\pi}{8} + \frac{3\pi}{8}\right) + \cos\left(\frac{5\pi}{8} - \frac{3\pi}{8}\right)\right]$$

$$= 4 \frac{1}{2}\left[\cos(\pi) + \cos\left(\frac{\pi}{4}\right)\right] = 4 \frac{1}{2}\left[-1 + \frac{\sqrt{2}}{2}\right]$$

$$= \sqrt{2} - 2$$

(d) This product is found first, the product of numerators, and then the product of denominators,

$$\tan 20° \; \tan 40° \; \tan 80° = \frac{\sin 20° \; \sin 40° \; \sin 80°}{\cos 20° \; \cos 40° \; \cos 80°}$$

$$\therefore$$

$$\sin 20° \sin 40° \sin 80° = \frac{1}{2}[\cos(20° - 40°) - \cos(20° + 40°)] \sin 80°$$

$$= \frac{1}{2}\left[\underset{\overset{\displaystyle\cos 20°}{\nearrow}}{\cos(-20°)} - \cos(60°)\right] \sin 80°$$

$$= \frac{1}{2}\left[\frac{1}{2}(\sin 100° + \sin 60°) - \underset{\overset{\displaystyle 1/2}{\nearrow}}{\cos 60°} \sin 80°\right]$$

$$= \frac{1}{2}\frac{1}{2}\left(\sin 100° + \underset{\overset{\displaystyle \sqrt{3}/2}{\nearrow}}{\sin 60°} - \sin 80°\right)$$

$$= \frac{\sqrt{3}}{8}$$

Here, we note that $\sin 100° = \sin(90° + 10°)$ and $\sin(80°) = \sin(90° - 10°)$; in other words, these two values are equal and thus cancelled. Similarly, we find that $\cos 100° = -\cos 80°$, and the product of denominators is calculated as

$$\therefore$$

$$\cos 20° \cos 40° \cos 80° = \frac{1}{2}[\cos(20° + 40°) + \cos(20° - 40°)]\cos 80°$$

$$= \frac{1}{2}\left[\underbrace{\cos 60°}_{1/2} + \underbrace{\cos(-20°)}_{\cos 20°}\right]\cos 80°$$

$$= \frac{1}{2}\left[\frac{1}{2}\cos 80° + \cos 20° \cos 80°\right]$$

$$= \frac{1}{2}\left[\frac{1}{2}\underbrace{\cos 80°}_{} + \frac{1}{2}\underbrace{\cos 100°}_{} + \frac{1}{2}\underbrace{\cos 60°}_{1/2}\right]$$

$$= \frac{1}{8}$$

Therefore, we conclude that

$$\tan 20° \tan 40° \tan 80° = \frac{\sqrt{3}/8}{1/8} = \sqrt{3}$$

Exercise 1.4, page 6

1. The three integrals in this exercise are typical forms encountered, and for that reason, we review practical integration techniques:

   (a) the change of variable technique in case of $\sin(2x)$ or $\cos(2x)$ functions. Here, the new variable is

$$t = 2x \quad \therefore \quad \frac{dt}{dx} = 2 \quad \therefore \quad dx = \frac{1}{2}dt$$

$$\therefore$$

$$\int \sin(2x)\,dx = \frac{1}{2}\int \sin t\,dt = -\frac{1}{2}\cos t = -\frac{1}{2}\cos(2x) + C$$

   (b) demonstration of how trigonometric identities can be used to decompose more a complicated function. Here,

$$\sin x \sin x = \frac{1}{2}[\cos(x - x) - \cos(x + x)]$$

$$\sin^2(x) = \frac{1}{2}[\cos(0) - \cos(2x)] = \frac{1}{2}[1 - \cos(2x)]$$

Now, integral in this problem is question to the integral in (a), as

$$\int \sin^2(x)\,dx = \int \frac{1}{2}[1 - \cos(2x)]\,dx = \frac{1}{2}\int dx - \frac{1}{2}\int \cos(2x)\,dx$$
$$= \frac{1}{2}\left(x - \frac{1}{2}\sin(2x)\right) + C$$

(c) functions defined by linear pieces are among most often used approximations in engineering. Given their graphical representation, it is necessary to derive analytical forms for *each* of the linear sections separately. Main idea is that the operation of "integration" itself is fundamentally an operation of addition. Also, a linear function is the easiest function to integrate. Therefore, knowing a priori the analytical form of each linear piece, the overall integral is reduced to the sum of simple integrals.

A linear section $y = ax + b$ is determined by two points in the plane, that is to say that knowing coordinates (x, y) of these two points that are both found at the linear, two constants (a, b) are determined with a simple algebra.

Here, we derive three linear sections, as emphasized by different colours in graph and points $(\mathscr{A}, \mathscr{B}, \mathscr{C}, \mathscr{D})$, see Fig. 1.16.

1. $\overline{\mathscr{A}\mathscr{B}}$: at coordinates of the two end points (x, y), we write $y = ax + b$ as

$$(0, 0) \quad \therefore \quad 0 = a \times 0 + b \Rightarrow b = 0$$

$$(0.25, 1) \quad \therefore \quad 1 = a \times \frac{1}{4} + b \Rightarrow a = 4$$

Therefore, the analytical form of $\overline{\mathscr{A}\mathscr{B}}$ linear segment is $f(x) = 4x$, which is valid in the interval $(0, 0.25)$. Therefore, in this interval, we calculate

$$I_1 = \int_0^{\frac{1}{4}} f(x)^2\,dx = \int_0^{\frac{1}{4}} (4x)^2\,dx = 16\left.\frac{x^3}{3}\right|_0^{\frac{1}{4}} = \frac{16}{3}\left(\frac{1}{4^3} - 0\right) = \frac{1}{12}$$

2. $\overline{\mathscr{B}\mathscr{C}}$: at coordinates of the two end points (x, y), we write $y = ax + b$ as

$$(0.25, 1) \quad \therefore \quad 1 = a \times \frac{1}{4} + b$$

$$(0.75, -1) \quad \therefore \quad -1 = a \times \frac{3}{4} + b$$

Solution of this system of two linear equation is $(a, b) = (-4, 2)$. Therefore, the analytical form of $\overline{\mathscr{C}\mathscr{D}}$ linear segment is $f(x) = -4x + 2$, which is valid in the interval $(0.25, 0.75)$. Therefore, in this interval, we calculate

$$I_2 = \int_{\frac{1}{4}}^{\frac{3}{4}} f(x)^2 \, dx = \int_{\frac{1}{4}}^{\frac{3}{4}} (-4x + 2)^2 \, dx = \int_{\frac{1}{4}}^{\frac{3}{4}} (16x^2 - 16x + 4) \, dx$$

$$= 4 \left(4 \frac{x^3}{3} \Big|_{\frac{1}{4}}^{\frac{3}{4}} - 4 \frac{x^2}{2} \Big|_{\frac{1}{4}}^{\frac{3}{4}} + x \Big|_{\frac{1}{4}}^{\frac{3}{4}} \right)$$

$$= \left(16 \frac{1}{3} \frac{26}{64} - 16 \frac{1}{2} \frac{8}{16} + 4 \frac{2}{4} \right) = \left(\frac{13}{6} - 2 \right) = \frac{1}{6}$$

3. $\overline{\mathscr{CD}}$: at coordinates of the two end points (x, y), we write $y = ax + b$ as

$$(0.75, -1) \quad \therefore \quad -1 = a \times \frac{3}{4} + b$$

$$(1, 0) \quad \therefore \quad 0 = a \times 1 + b$$

Solution of this system of two linear equation is $(a, b) = (4, -4)$. Therefore, the analytical form of $\overline{\mathscr{CD}}$ linear segment is $f(x) = 4x - 4$, which is valid in the interval $(0.75, 1)$. Therefore, in this interval, we calculate

$$I_3 = \int_{\frac{3}{4}}^{1} f(x)^2 \, dx = \int_{\frac{3}{4}}^{1} (4x - 4)^2 \, dx = \int_{\frac{3}{4}}^{1} (16x^2 - 32x + 16) \, dx$$

$$= 16 \left(\frac{x^3}{3} \Big|_{\frac{3}{4}}^{1} - 2 \frac{x^2}{2} \Big|_{\frac{3}{4}}^{1} + x \Big|_{\frac{3}{4}}^{1} \right)$$

$$= \left(16 \frac{1}{3} \frac{37}{64} - 16 \frac{7}{16} + 16 \frac{1}{4} \right) = \left(\frac{37}{12} - 3 \right) = \frac{1}{12}$$

After finding these three integrals, we return to the original questions to conclude

$$I = \int_0^1 f(x)^2 \, dx = I_1 + I_2 + I_3 = \frac{1}{12} + \frac{1}{6} + \frac{1}{12} = \frac{1}{3}$$

Geometrical interpretation is that integral I equals to the total area underneath quadratic function $f(x)^2$ that consists of three distinct regions. These three quadratic sub-functions and their respective areas are shown by coloured sections in the solution graph.
As an additional exercise, a simple integral,

$$I = \int_0^1 f(x) = \int_0^{\frac{1}{4}} 4x \, dx + \int_{\frac{1}{4}}^{\frac{3}{4}} (-4x + 2) \, dx + \int_{\frac{3}{4}}^1 (4x - 4) \, dx = 0$$

which may also be concluded by graph inspection; the area of triangle above zero (i.e. positive) equals the area of triangle below zero (i.e. negative); therefore, their sum equals zero. Another way to say is that the average of this function equals zero.

Fig. 1.16 Example 1.4-1

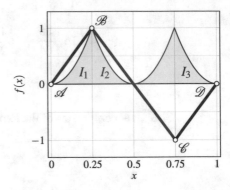

Exercise 1.5, page 7

1. After using the identities (1.1)–(1.8), we write

(a)   Given $z(x) = j^{x}/2$, then $x_0 = 2$ and

$$|z(x)| = \left| j\frac{x}{2} \right| = \frac{x}{2} \quad \therefore$$

$$H(x) = 20 \log |z(x)| = 20 \log \left(\frac{x}{2} \right) = 20 \log(x) - 20 \log(2)$$

$$\lim_{x \to 0} H(x) = \lim_{x \to 0} 20 \log(x) - 20 \log(2) = -\infty$$

$$\lim_{x \to 2} H(x) = \lim_{x \to 2} 20 \log(x) - 20 \log(2) = 0$$

$$\lim_{x \to +\infty} H(x) = \lim_{x \to +\infty} 20 \log(x) - 20 \log(2) = +\infty,$$

and the phase function $P(x)$ is found by definition as

$$P(x) = \arctan \left[\frac{\Im(z(x))}{\Re(z(x))} \right] = \arctan \left(\frac{x/2}{0} \right) = \arctan(+\infty) = \frac{\pi}{2} = \text{const.};$$

therefore, all limits of $P(x) = \pi/2$, i.e. not function of x.

We calculate $H(x)$ at points $x = 0.2$, i.e. ten times smaller, and $x = 20$, i.e. ten times greater than $x_0 = 2$ as

$$H(x = 0.2) = 20 \log \left(\frac{0.2}{2} \right) = 20 \log \left(\frac{1}{10} \right) = -20$$

$$H(x = 20) = 20 \log \left(\frac{20}{2} \right) = 20 \log(10) = +20$$

(b) Given $z(x) = j\dfrac{x}{10}$, then $x_0 = 10$ and

$$|z(x)| = \left| j\frac{x}{10} \right| = \frac{x}{10} \quad \therefore$$

$$H(x) = 20\log|z(x)| = 20\log\left(\frac{x}{10}\right) = 20\log(x) - \cancel{20\log(10)}^{20}$$

$$\lim_{x\to+\infty} H(x) = \lim_{x\to+\infty} 20\log(x) - 20 = +\infty$$

$$\lim_{x\to 10} H(x) = \lim_{x\to 10} 20\log(x) - 20 = 0$$

$$\lim_{x\to 0} H(x) = \lim_{x\to 0} 20\log(x) - 20 = -\infty,$$

and the phase function $P(x)$ is found by definition as

$$P(x) = \arctan\left[\frac{\Im(z(x))}{\Re(z(x))}\right] = \arctan\left(\frac{x/10}{0}\right) = \arctan(+\infty) = \frac{\pi}{2} = \text{const.};$$

therefore, all limits of $P(x) = \pi/2$, i.e. not function of x.

We calculate $H(x)$ at points $x = 1$, i.e. ten times smaller, and $x = 100$, i.e. ten times greater than $x_0 = 10$ as

$$H(x = 1) = 20\log\left(\frac{1}{10}\right) = -20$$

$$H(x = 100) = 20\log\left(\frac{10\cancel{0}}{1\cancel{0}}\right) = +20$$

(c) Given $z(x) = 1 + j\dfrac{x}{2}$, then $x_0 = 2$ and

$$|z(x)| = \left| 1 + j\frac{x}{2} \right| = \sqrt{1^2 + \left(\frac{x}{2}\right)^2} \quad \therefore$$

$$H(x) = 20\log|z(x)| = 20\log\sqrt{1 + \left(\frac{x}{2}\right)^2}$$

$$\lim_{x\to 0} H(x) = 20\log\left[\lim_{x\to 0}\sqrt{1 + \left(\frac{x}{2}\right)^2}\right] = 20\log(1) = 0$$

$$\lim_{x\to 2} H(x) = 20\log\left[\lim_{x\to 2}\sqrt{1 + \left(\frac{x}{2}\right)^2}\right] = 20\log\sqrt{2} = 3$$

$$\lim_{x\gg x_0} H(x) = 20\log\left[\lim_{x\gg x_0}\sqrt{1 + \left(\frac{x}{2}\right)^2}\right] \approx 20\log\left[\lim_{x\gg x_0}\sqrt{\left(\frac{x}{2}\right)^2}\right]$$

$$\approx 20\log\left[\lim_{x\gg x_0}\left(\frac{x}{2}\right)\right] \approx 20\log|x|,$$

and the phase function $P(x)$ is found by definition as

$$P(x) = \arctan\left[\frac{\Im(z(x))}{\Re(z(x))}\right] = \arctan\left(\frac{x/2}{1}\right) = \arctan\left(\frac{x}{2}\right)$$

$$\lim_{x \to 0} P(x) = \lim_{x \to 0} \arctan\left(\frac{x}{2}\right) = \arctan(0) = 0°$$

$$\lim_{x \to 2} P(x) = \lim_{x \to 2} \arctan\left(\frac{x}{2}\right) = \arctan\left(\frac{1}{1}\right) = \frac{\pi}{4} = 45°$$

$$\lim_{x \to +\infty} P(x) = \lim_{x \to +\infty} \arctan\left(\frac{x}{2}\right) = \arctan(\infty) = \frac{\pi}{2} = 90°$$

We calculate $H(x)$ and $P(x)$ at points $x = 0.2$, i.e. ten times smaller, and $x = 20$, i.e. ten times greater than $x_0 = 2$ as

$$H(x = 0.2) = 20\log\sqrt{1 + \left(\frac{0.2}{2}\right)^2} \approx 20\log(1) = 0$$

$$H(x = 20) = 20\log\sqrt{1 + \left(\frac{20}{2}\right)^2} \approx 20\log(10) = +20$$

$$P(x = 0.2) = \arctan\left(\frac{0.2}{2}\right) = \arctan(0.1) = 5.7° \approx 0$$

$$P(x = 20) = \arctan\left(\frac{20}{2}\right) = \arctan(10) = 84.3° \approx -90° = \frac{\pi}{2}$$

(d)   Given $z(x) = 1 + j\dfrac{x}{10}$, then $x_0 = 10$ and

$$|z(x)| = \left|1 + j\frac{x}{10}\right| = \sqrt{1^2 + \left(\frac{x}{10}\right)^2} \quad \therefore$$

$$H(x) = 20\log|z(x)| = 20\log\sqrt{1 + \left(\frac{x}{10}\right)^2}$$

$$\lim_{x \to 0} H(x) = 20\log\left[\lim_{x \to 0}\sqrt{1 + \left(\frac{x}{10}\right)^2}\right] = 20\log(1) = 0$$

$$\lim_{x \to 10} H(x) = 20\log\left[\lim_{x \to 10}\sqrt{1 + \left(\frac{x}{10}\right)^2}\right] = 20\log\sqrt{2} = 3$$

$$\lim_{x \gg x_0} H(x) = 20\log\left[\lim_{x \gg x_0}\sqrt{1 + \left(\frac{x}{10}\right)^2}\right] \approx 20\log\left[\lim_{x \gg x_0}\sqrt{\left(\frac{x}{10}\right)^2}\right]$$

$$\approx 20\log\left[\lim_{x \gg x_0}\left(\frac{x}{10}\right)\right] \approx 20\log|x|,$$

and the phase function $P(x)$ is found by definition as

$$P(x) = \arctan\left[\frac{\Im(z(x))}{\Re(z(x))}\right] = \arctan\left(\frac{x/10}{1}\right) = \arctan\left(\frac{x}{10}\right)$$

$$\lim_{x\to 0} P(x) = \lim_{x\to 0} \arctan\left(\frac{x}{10}\right) = \arctan(0) = 0°$$

$$\lim_{x\to 10} P(x) = \lim_{x\to 10} \arctan\left(\frac{x}{10}\right) = \arctan\left(\frac{1}{1}\right) = \frac{\pi}{4} = 45°$$

$$\lim_{x\to +\infty} P(x) = \lim_{x\to +\infty} \arctan\left(\frac{x}{10}\right) = \arctan(\infty) = \frac{\pi}{2} = 90°$$

We calculate $H(x)$ and $P(x)$ at points $x = 1$, i.e. ten times smaller, and $x = 100$, i.e. ten times greater than $x_0 = 10$ as

$$H(x = 1) = 20\log\sqrt{1 + \left(\frac{0.1}{10}\right)^2} \approx 20\log(1) = 0$$

$$H(x = 100) = 20\log\sqrt{1 + \left(\frac{100}{10}\right)^2} \approx 20\log(10) = +20$$

$$P(x = 1) = \arctan\left(\frac{1}{10}\right) = \arctan(0.1) = 5.7° \approx 0$$

$$P(x = 100) = \arctan\left(\frac{100}{10}\right) = \arctan(10) = 84.3° \approx 90° = \frac{\pi}{2}$$

(e) Given $z(x) = 1 - j\frac{x}{2}$, then $x_0 = 2$ and

$$|z(x)| = \left|1 - j\frac{x}{2}\right| = \sqrt{1^2 + \left(\frac{x}{2}\right)^2} \quad \therefore$$

$$H(x) = 20\log|z(x)| = 20\log\sqrt{1 + \left(\frac{x}{2}\right)^2}$$

$$\lim_{x\to 0} H(x) = 20\log\left[\lim_{x\to 0}\sqrt{1 + \left(\frac{x}{2}\right)^2}\right] = 20\log(1) = 0$$

$$\lim_{x\to 2} H(x) = 20\log\left[\lim_{x\to 2}\sqrt{1 + \left(\frac{x}{2}\right)^2}\right] = 20\log\sqrt{2} = 3$$

$$\lim_{x\gg x_0} H(x) = 20\log\left[\lim_{x\gg x_0}\sqrt{1 + \left(\frac{x}{2}\right)^2}\right] \approx 20\log\left[\lim_{x\gg x_0}\sqrt{\left(\frac{x}{2}\right)^2}\right]$$

$$\approx 20\log\left[\lim_{x\gg x_0}\left(\frac{x}{2}\right)\right] \approx 20\log|x|,$$

and the phase function $P(x)$ is found by definition as

$$P(x) = \arctan\left[\frac{\Im(z(x))}{\Re(z(x))}\right] = \arctan\left(\frac{-x/2}{1}\right) = \arctan\left(\frac{-x}{2}\right)$$

$$\lim_{x\to 0} P(x) = \lim_{x\to 0} \arctan\left(\frac{-x}{2}\right) = \arctan(0) = 0°$$

$$\lim_{x\to 2} P(x) = \lim_{x\to 2} \arctan\left(\frac{-x}{2}\right) = \arctan\left(\frac{-1}{1}\right) = -\frac{\pi}{4} = -45°$$

$$\lim_{x\to +\infty} P(x) = \lim_{x\to +\infty} \arctan\left(\frac{-x}{2}\right) = \arctan(-\infty) = -\frac{\pi}{2} = -90°$$

We calculate $H(x)$ and $P(x)$ at points $x = 0.2$, i.e. ten times smaller, and $x = 20$, i.e. ten times greater than $x_0 = 2$ as

$$H(x = 0.2) = 20 \log\sqrt{1 + \left(\frac{0.2}{2}\right)^2} \approx 20 \log(1) = 0$$

$$H(x = 20) = 20 \log\sqrt{1 + \left(\frac{20}{2}\right)^2} \approx 20 \log(10) = +20$$

$$P(x = 0.2) = \arctan\left(\frac{-0.2}{2}\right) = \arctan(-0.1) = -5.7° \approx 0$$

$$P(x = 20) = \arctan\left(\frac{-20}{2}\right) = \arctan(-10) = -84.3° \approx -90° = -\frac{\pi}{2}$$

(f)   Given $z(x) = 1 - j\dfrac{x}{10}$, then $x_0 = 10$ and

$$|z(x)| = \left|1 - j\frac{x}{10}\right| = \sqrt{1 + \left(\frac{x}{10}\right)^2} \quad\therefore$$

$$H(x) = 20 \log|z(x)| = 20 \log\sqrt{1 + \left(\frac{x}{10}\right)^2}$$

$$\lim_{x\to 0} H(x) = 20 \log\left[\lim_{x\to 0}\sqrt{1 + \left(\frac{x}{10}\right)^2}\right] = 20 \log(1) = 0$$

$$\lim_{x\to 10} H(x) = 20 \log\left[\lim_{x\to 10}\sqrt{1 + \left(\frac{x}{10}\right)^2}\right] = 20 \log\sqrt{2} = 3$$

$$\lim_{x\gg x_0} H(x) = 20 \log\left[\lim_{x\gg x_0}\sqrt{1 + \left(\frac{x}{10}\right)^2}\right] \approx 20 \log\left[\lim_{x\gg x_0}\sqrt{\left(\frac{x}{10}\right)^2}\right]$$

$$\approx 20 \log\left[\lim_{x\gg x_0}\left(\frac{x}{10}\right)\right] \approx 20 \log|x|,$$

and the phase function $P(x)$ is found by definition as

$$P(x) = \arctan \left[\frac{\Im(z(x))}{\Re(z(x))} \right] = \arctan \left(\frac{-x/10}{1} \right) = \arctan \left(\frac{-x}{10} \right)$$

$$\lim_{x \to 0} P(x) = \lim_{x \to 0} \arctan \left(\frac{-x}{10} \right) = \arctan(0) = 0°$$

$$\lim_{x \to 10} P(x) = \lim_{x \to 10} \arctan \left(\frac{-x}{10} \right) = \arctan \left(\frac{-1}{1} \right) = -\frac{\pi}{4} = -45°$$

$$\lim_{x \to +\infty} P(x) = \lim_{x \to +\infty} \arctan \left(\frac{-x}{10} \right) = \arctan(-\infty) = -\frac{\pi}{2} = -90°$$

We calculate $H(x)$ and $P(x)$ at points $x = 0.2$, i.e. ten times smaller, and $x = 20$, i.e. ten times greater than $x_0 = 2$ as

$$H(x = 1) = 20 \log \sqrt{1 + \left(\frac{1}{10} \right)^2} \approx 20 \log(1) = 0$$

$$H(x = 100) = 20 \log \sqrt{1 + \left(\frac{100}{10} \right)^2} \approx 20 \log(10) = +20$$

$$P(x = 1) = \arctan \left(\frac{-1}{10} \right) = \arctan(-0.1) = -5.7° \approx 0$$

$$P(x = 100) = \arctan \left(\frac{-100}{10} \right) = \arctan(-10) = -84.3° \approx -90° = -\frac{\pi}{2}$$

(g) Given $z(x) = \dfrac{1}{1 + j\frac{x}{2}}$, then $x_0 = 2$ and

$$|z(x)| = \left| \frac{1}{1 + j\frac{x}{2}} \right| = \frac{1}{\sqrt{1 + \left(\frac{x}{2} \right)^2}} \quad \therefore$$

$$H(x) = 20 \log |z(x)| = \underbrace{20 \log(1)}_{0} - 20 \log \sqrt{1 + \left(\frac{x}{2} \right)^2}$$

$$\lim_{x \to 0} H(x) = -20 \log \left[\lim_{x \to 0} \sqrt{1 + \left(\frac{x}{2} \right)^2} \right] = -20 \log(1) = 0$$

$$\lim_{x \to 2} H(x) = -20 \log \left[\lim_{x \to 2} \sqrt{1 + \left(\frac{x}{2} \right)^2} \right] = -20 \log \sqrt{2} = -3$$

$$\lim_{x \to +\infty} H(x) = -20 \log \left[\lim_{x \to +\infty} \sqrt{1 + \left(\frac{x}{2} \right)^2} \right] \approx -20 \log |x|,$$

and the phase function $P(x)$ is found by definition, after explicitly deriving the real and imaginary parts of $z(x)$, as

$$z(x) = \frac{1}{1 + j^{x}/2} \frac{1 - j^{x}/2}{1 - j^{x}/2} = \frac{1}{\sqrt{1 + (x/2)^2}} - j\frac{x/2}{\sqrt{1 + (x/2)^2}} \quad \therefore$$

$$\Re(z(x)) = \frac{1}{\sqrt{1 + (x/2)^2}} \quad \text{and} \quad \Im(z(x)) = -\frac{x/2}{\sqrt{1 + (x/2)^2}}$$

which again is reduced to

$$P(x) = \arctan\left[\frac{\Im(z(x))}{\Re(z(x))}\right] = \arctan\left(\frac{-x/2}{1}\right) = \arctan\left(\frac{-x}{2}\right)$$

$$\lim_{x \to 0} P(x) = \lim_{x \to 0} \arctan\left(\frac{-x}{2}\right) = \arctan(0) = 0°$$

$$\lim_{x \to 2} P(x) = \lim_{x \to 2} \arctan\left(\frac{-x}{2}\right) = \arctan(-1) = -\frac{\pi}{4} = -45°$$

$$\lim_{x \to +\infty} P(x) = \lim_{x \to +\infty} \arctan\left(\frac{-x}{2}\right) = \arctan(-\infty) = -\frac{\pi}{2} = -90°$$

We calculate $H(x)$ and $P(x)$ at points $x = 0.2$, i.e. ten times smaller, and $x = 20$, i.e. ten times greater than $x_0 = 2$ as

$$H(x = 0.2) = -20\log\sqrt{1 + \left(\frac{0.2}{2}\right)^2} \approx -20\log(1) = 0$$

$$H(x = 20) = -20\log\sqrt{1 + \left(\frac{20}{2}\right)^2} \approx -20\log(10) = -20$$

$$P(x = 0.2) = \arctan\left(\frac{-0.2}{2}\right) = \arctan(-0.1) = -5.7° \approx 0$$

$$P(x = 20) = \arctan\left(\frac{-20}{2}\right) = \arctan(-10) = -84.3° \approx -90° = -\frac{\pi}{2}$$

(h)   Given $z(x) = \frac{1}{1 + j\frac{x}{10}}$, then $x_0 = 10$ and

$$|z(x)| = \left|\frac{1}{1 + j\frac{x}{10}}\right| = \frac{1}{\sqrt{1^2 + \left(\frac{x}{10}\right)^2}} \quad \therefore$$

$$H(x) = 20\log|z(x)| = \underbrace{20\log(1)}_{0} - 20\log\sqrt{1 + \left(\frac{x}{10}\right)^2}$$

$$\lim_{x \to 0} H(x) = -20\log\left[\lim_{x \to 0}\sqrt{1 + \left(\frac{x}{10}\right)^2}\right] = -20\log(1) = 0$$

$$\lim_{x \to 10} H(x) = -20\log\left[\lim_{x \to 10}\sqrt{1 + \left(\frac{x}{10}\right)^2}\right] = -20\log\sqrt{2} = -3$$

$$\lim_{x \gg x_0} H(x) = -20\log\left[\lim_{x \gg x_0}\sqrt{1 + \left(\frac{x}{10}\right)^2}\right] \approx -20\log|x|,$$

and the phase function $P(x)$ is found by definition, after explicitly deriving the real and imaginary parts of $z(x)$, as

$$z(x) = \frac{1}{1 + jx/10} \frac{1 - jx/10}{1 - jx/10} = \frac{1}{\sqrt{1 + (x/10)^2}} - j\frac{x/10}{\sqrt{1 + (x/10)^2}} \quad \therefore$$

$$\Re(z(x)) = \frac{1}{\sqrt{1 + (x/10)^2}} \quad \text{and} \quad \Im(z(x)) = -\frac{x/10}{\sqrt{1 + (x/10)^2}}$$

which again is reduced to

$$P(x) = \arctan\left[\frac{\Im(z(x))}{\Re(z(x))}\right] = \arctan\left(\frac{-x/10}{1}\right) = \arctan\left(\frac{-x}{10}\right); \quad \text{therefore,}$$

$$\lim_{x \to 0} P(x) = \lim_{x \to 0} \arctan\left(\frac{-x}{10}\right) = \arctan(0) = 0°$$

$$\lim_{x \to 10} P(x) = \lim_{x \to 10} \arctan\left(\frac{-x}{10}\right) = \arctan(-1) = -\frac{\pi}{4} = -45°$$

$$\lim_{x \to +\infty} P(x) = \lim_{x \to +\infty} \arctan\left(\frac{-x}{10}\right) = \arctan(-\infty) = -\frac{\pi}{2} = -90°$$

We calculate $H(x)$ and $P(x)$ at points $x = 1$, i.e. ten times smaller, and $x = 100$, i.e. ten times greater than $x_0 = 10$ as

$$H(x = 1) = -20\log\sqrt{1 + \left(\frac{1}{10}\right)^2} \approx -20\log(1) = 0$$

$$H(x = 100) = -20\log\sqrt{1 + \left(\frac{100}{10}\right)^2} \approx -20\log(10) = -20$$

$$P(x = 1) = \arctan\left(\frac{-1}{10}\right) = \arctan(-0.1) = -5.7° \approx 0$$

$$P(x = 100) = \arctan\left(\frac{-100}{10}\right) = \arctan(-10) = -84.3° \approx -90° = -\frac{\pi}{2}$$

(i) Given $z(x) = \dfrac{1}{1 - j\frac{x}{10}}$, then $x_0 = 10$ and

$$|z(x)| = \left|\frac{1}{1 - j\frac{x}{10}}\right| = \frac{1}{\sqrt{1^2 + \left(\frac{x}{10}\right)^2}} \quad \therefore$$

$$H(x) = 20\log(1)^{\,0} - 20\log\sqrt{1 + \left(\frac{x}{10}\right)^2}$$

$$\lim_{x \to 0} H(x) = -20\log\left[\lim_{x \to 0}\sqrt{1 + \left(\frac{x}{10}\right)^2}\right] = -20\log(1) = 0$$

$$\lim_{x \to 10} H(x) = -20\log\left[\lim_{x \to 10}\sqrt{1 + \left(\frac{x}{10}\right)^2}\right] = -20\log\sqrt{2} = -3$$

$$\lim_{x \gg x_0} H(x) = -20\log\left[\lim_{x \gg x_0}\sqrt{1 + \left(\frac{x}{10}\right)^2}\right] \approx -20\log|x|,$$

and the phase function $P(x)$ is found by definition, after explicitly deriving the real and imaginary parts of $z(x)$, as

$$z(x) = \frac{1}{1 - j^{x}/10} \frac{1 + j^{x}/10}{1 + j^{x}/10} = \frac{1}{\sqrt{1 + (^{x}/10)^2}} + j \frac{^{x}/10}{\sqrt{1 + (^{x}/10)^2}} \quad \therefore$$

$$\Re(z(x)) = \frac{1}{\sqrt{1 + (^{x}/10)^2}} \quad \text{and} \quad \Im(z(x)) = \frac{^{x}/10}{\sqrt{1 + (^{x}/10)^2}}$$

which again is reduced to

$$P(x) = \arctan\left(\frac{^{x}/10}{1}\right) = \arctan\left(\frac{x}{10}\right) \quad \text{therefore,}$$

$$\lim_{x \to 0} P(x) = \lim_{x \to 0} \arctan\left(\frac{x}{10}\right) = \arctan(0) = 0°$$

$$\lim_{x \to 10} P(x) = \lim_{x \to 10} \arctan\left(\frac{x}{10}\right) = \arctan(1) = \frac{\pi}{4} = 45°$$

$$\lim_{x \to +\infty} P(x) = \lim_{x \to +\infty} \arctan\left(\frac{x}{10}\right) = \arctan(+\infty) = \frac{\pi}{2} = 90°$$

We calculate $H(x)$ and $P(x)$ at points $x = 1$, i.e. ten times smaller, and $x = 100$, i.e. ten times greater than $x_0 = 10$ as

$$H(x = 1) = -20 \log \sqrt{1 + \left(\frac{1}{10}\right)^2} \approx -20 \log(1) = 0$$

$$H(x = 100) = -20 \log \sqrt{1 + \left(\frac{100}{10}\right)^2} \approx -20 \log(10) = -20$$

$$P(x = 1) = \arctan\left(\frac{1}{10}\right) = \arctan(0.1) = 5.7° \approx 0$$

$$P(x = 100) = \arctan\left(\frac{100}{10}\right) = \arctan(10) = 84.3° \approx 90° = \frac{\pi}{2}$$

(j) Given $z(x) = \dfrac{1}{1 - j\frac{x}{2}}$, then $x_0 = 2$ and

$$|z(x)| = \left| \frac{1}{1 - j\frac{x}{2}} \right| = \frac{1}{\sqrt{1^2 + \left(\frac{x}{2}\right)^2}} \quad \therefore$$

$$H(x) = \underline{20 \log(1)}^{\,0} - 20 \log \sqrt{1 + \left(\frac{x}{2}\right)^2}$$

$$\lim_{x \to 0} H(x) = -20 \log \left[\lim_{x \to 0} \sqrt{1 + \left(\frac{x}{2}\right)^2} \right] = -20 \log(1) = 0$$

$$\lim_{x \to 2} H(x) = -20 \log \left[\lim_{x \to 2} \sqrt{1 + \left(\frac{x}{2}\right)^2} \right] = -20 \log \sqrt{2} = -3$$

$$\lim_{x \gg x_0} H(x) = -20 \log \left[\lim_{x \gg x_0} \sqrt{1 + \left(\frac{x}{2}\right)^2} \right] \approx -20 \log |x|,$$

and the phase function $P(x)$ is found by definition, after explicitly deriving the real and imaginary parts of $z(x)$, as

$$z(x) = \frac{1}{1 - jx/2} \frac{1 + jx/2}{1 + jx/2} = \frac{1}{\sqrt{1 + (x/2)^2}} + j \frac{x/2}{\sqrt{1 + (x/2)^2}} \quad \therefore$$

$$\Re(z(x)) = \frac{1}{\sqrt{1 + (x/2)^2}} \quad \text{and} \quad \Im(z(x)) = \frac{x/2}{\sqrt{1 + (x/2)^2}}$$

which again is reduced to

$$P(x) = \arctan\left(\frac{x/2}{1}\right) = \arctan\left(\frac{x}{2}\right) \quad \text{therefore,}$$

$$\lim_{x \to 0} P(x) = \lim_{x \to 0} \arctan\left(\frac{x}{2}\right) = \arctan(0) = 0°$$

$$\lim_{x \to 2} P(x) = \lim_{x \to 2} \arctan\left(\frac{x}{2}\right) = \arctan(1) = \frac{\pi}{4} = 45°$$

$$\lim_{x \to +\infty} P(x) = \lim_{x \to +\infty} \arctan\left(\frac{x}{2}\right) = \arctan(+\infty) = \frac{\pi}{2} = 90°$$

We calculate $H(x)$ and $P(x)$ at points $x = 0.2$, i.e. ten times smaller, and $x = 20$, i.e. ten times greater than $x_0 = 2$ as

$$H(x = 0.2) = -20 \log \sqrt{1 + \left(\frac{0.2}{2}\right)^2} \approx -20 \log(1) = 0$$

$$H(x = 20) = -20 \log \sqrt{1 + \left(\frac{20}{2}\right)^2} \approx -20 \log(10) = -20$$

$$P(x = 0.2) = \arctan\left(\frac{0.2}{2}\right) = \arctan(0.1) = 5.7° \approx 0$$

$$P(x = 20) = \arctan\left(\frac{20}{2}\right) = \arctan(10) = 84.3° \approx 90° = \frac{\pi}{2}$$

Exercise 1.6, page 7

1.

(a) For $z(x) = j^x/2$, Example 1.5-1(a), we found that when $x = 2$, the value of $H(x = 2) = 0$, and at the same time, angle is constant $\angle P(x) = +\pi/2$. In addition, every tenfold increase of x results in $+20$ increase of $H(x)$ (see Fig. 1.17).

(b) For $z(x) = j^x/10$, Example 1.5-1(b), we found that when $x = 10$, the value of $H(x = 10) = 0$, and at the same time, angle is constant $\angle P(x) = +\pi/2$. In addition, every tenfold increase of x results in $+20$ increase of $H(x)$ (see Fig. 1.18).

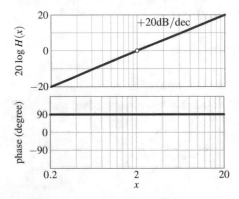

Fig. 1.17 Example 1.6-1(a)

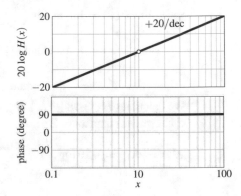

Fig. 1.18 Example 1.6-1(b)

(c) For $z(x) = 1 + j^x/2$, Example 1.5-1(c), we found that when $x = 2$, the value of $H(x = 2) = 3$, and at the same time, angle is $\angle P(x = 2) = +\pi/4$. In addition, angle limits are $0°$ and $+90°$, while every tenfold increase of x results in $+20$ increase of $H(x)$. After accounting for all limits of $\lim H(x)$, see Fig. 1.19.

(d) For $z(x) = 1 + j^x/10$, Example 1.5-1(d), we found that when $x = 10$, the value of $H(x = 10) = 3$, and at the same time, angle is $\angle P(x = 10) = +\pi/4$. In addition, angle limits are $0°$ and $+90°$, while every tenfold increase of x results in $+20$ increase of $H(x)$. After accounting for all limits of $\lim H(x)$, see Fig. 1.20.

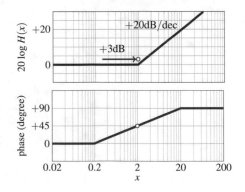

Fig. 1.19 Example 1.6-1(c)

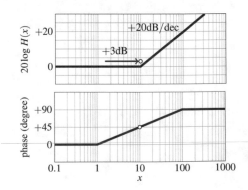

Fig. 1.20 Example 1.6-1(d)

(e) For $z(x) = 1 - j^x/2$, Example 1.5-1(e), we found that when $x = 2$, the value of $H(x = 2) = 3$, and at the same time, angle is $\angle P(x = 2) = -\pi/4$. In addition, angle limits are $0°$ and $-90°$, while every tenfold increase of x results in $+20$ increase of $H(x)$. After accounting for all limits of $\lim H(x)$, see Fig. 1.21.

(f) For $z(x) = 1 - j^x/10$, Example 1.5-1(f), we found that when $x = 10$, the value of $H(x = 10) = 3$, and at the same time, angle is $\angle P(x = 10) = -\pi/4$. In addition, angle limits are $0°$ and $-90°$, while every tenfold increase of x results in $+20$ increase of $H(x)$. After accounting for all limits of $\lim H(x)$, see Fig. 1.22.

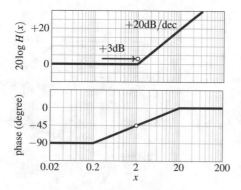

Fig. 1.21 Example 1.6-1(e)

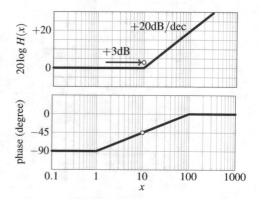

Fig. 1.22 Example 1.6-1(f)

(g) For $z(x) = \dfrac{1}{1 + j\frac{x}{2}}$, Example 1.5-1(g), we found that when $x = 2$, the value of $H(x = 2) = -3$, and at the same time, angle is $\angle P(x = 2) = -\pi/4$. In addition, angle limits are $0°$ and $-90°$, while every tenfold increase of x results in -20 increase of $H(x)$. After accounting for all limits of $\lim H(x)$, see Fig. 1.23.

(h) For $z(x) = \dfrac{1}{1 + j\frac{x}{10}}$, Example 1.5-1(h), we found that when $x = 10$, the value of $H(x = 10) = -3$, and at the same time, angle is $\angle P(x = 10) = -\pi/4$. In addition, angle limits are $0°$ and $-90°$, while every tenfold increase of x results in -20 increase of $H(x)$. After accounting for all limits of $\lim H(x)$, see Fig. 1.24.

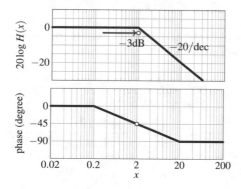

Fig. 1.23 Example 1.6-1(g)

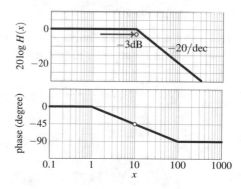

Fig. 1.24 Example 1.6-1(h)

(i) For $z(x) = \dfrac{1}{1 - j\frac{x}{10}}$, Example 1.5-1(i), we found that when $x = 10$, the value of $H(x = 10) = -3$, and at the same time, angle is $\angle P(x = 10) = \pi/4$. In addition, angle limits are $0°$ and $+90°$, while every tenfold increase of x results in -20 increase of $H(x)$. After accounting for all limits of $\lim H(x)$, see Fig. 1.25.

(j) For $z(x) = \dfrac{1}{1 - j\frac{x}{2}}$, Example 1.5-1(j), we found that when $x = 2$, the value of $H(x = 2) = -3$, and at the same time, angle is $\angle P(x = 2) = \pi/4$. In addition, angle limits are $0°$ and $+90°$, while every tenfold increase of x results in -20 increase of $H(x)$. After accounting for all limits of $\lim H(x)$, see Fig. 1.26.

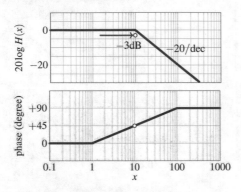

Fig. 1.25 Example 1.6-1(i)   **Fig. 1.26** Example 1.6-1(j)

It is important to note that, as long as the form of analytical function $z(x)$ is same, where the only change is the value of x_0, then all respective graphs have the same shapes. That is to say, the graphs are only shifted to $x = x_0$. This property greatly simplifies the analysis of more complicated $z(x)$ functions. General idea is to factorize complicated $z(x)$ functions into these basic simpler "building blocks" and then "assemble" the final response by adding these basic functions.

Exercise 1.7, page 7

1. This is second order function; thus, in order to find out if roots of its denominator are reel or complex, it is necessary to factorize its denominator.

$$H(j\omega) = 2\,000\,\frac{2 + jx}{220jx + 4\,000 - x^2} = 2\,000\,\frac{2 + jx}{j^2x^2 + 220jx + 4\,000}$$

$$= 2\,000\,\frac{2 + jx}{j^2x^2 + 20jx + 200jx + 4\,000} = 2\,000\,\frac{2 + jx}{jx\,(20 + jx) + 200\,(20 + jx)}$$

$$= 2\,000\,\frac{2 + jx}{(20 + jx)(200 + jx)}$$

$$= 2\,000\,\frac{\cancel{2}\left(1 + j\dfrac{x}{2}\right)}{\cancel{20}\left(1 + j\dfrac{x}{20}\right)\cancel{200}\left(1 + j\dfrac{x}{200}\right)}$$

$$= \frac{1 + j\,\dfrac{x}{2}}{\left(1 + j\,\dfrac{x}{20}\right)\left(1 + j\,\dfrac{x}{200}\right)}$$

This factorized form is suitable for conversion into sum of the basic form simply by rewriting it in its logarithmic form,

$$20\log z(x) = 20\log\left[\frac{1 + j\,\dfrac{x}{2}}{\left(1 + j\,\dfrac{x}{20}\right)\left(1 + j\,\dfrac{x}{200}\right)}\right]$$

$$= \underbrace{+20\log\left(1 + j\,\frac{x}{2}\right)}_{\textcircled{1}\ x_0=2}\underbrace{-20\log\left(1 + j\,\frac{x}{20}\right)}_{\textcircled{2}\ x_0=20}\underbrace{-20\log\left(1 + j\,\frac{x}{200}\right)}_{\textcircled{3}\ x_0=200}$$

Obviously, this example is a second order function whose poles are real; thus, it is possible to decompose it into the basic first order functions. Each summing term in the logarithmic form is in effect as one of the already studied basic forms, as annotated. Similarly, the phase plot is created as the sum of the corresponding linear terms,

$$\angle z(x) = \underbrace{+\arctan\frac{x}{2}}_{\textcircled{1}\ x_0=2}\underbrace{-\arctan\frac{x}{20}}_{\textcircled{2}\ x_0=20}\underbrace{-\arctan\frac{x}{200}}_{\textcircled{3}\ x_0=200}$$

Once the gain and the phase logarithmic forms are factorized, first we plot graphs all four terms (both amplitude and phase), and then we create plots of the total sums, Fig. 1.27.

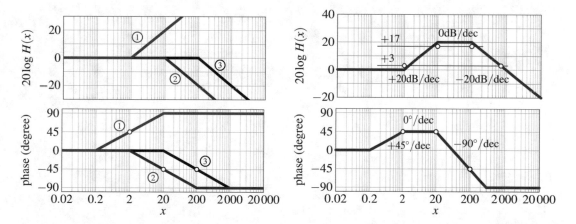

Fig. 1.27 Example 1.7-1

Introduction

2

Wireless transmission of information over vast distances is one of the finest examples of Clarke's third law, which states that "any sufficiently advanced technology is indistinguishable from magic". Even though a radio represents one of the most ingenious achievements of humankind and is now taken for granted; for the majority of the modern human population (including some of its highly educated members), this phenomenon still appears to be magical. In order to understand this magic, it is important to review some of fundamental concepts and definitions in physics, mathematics, and engineering.

2.1 Important to Know

1. Sine wave definitions,

$$f \stackrel{\text{def}}{=} \frac{1}{T} \ [\text{Hz}]; \quad \omega \stackrel{\text{def}}{=} 2\pi f = \frac{2\pi}{T} \ \left[\frac{\text{rad}}{\text{s}}\right]; \quad \lambda = v\,T = \frac{v}{f} \ [\text{m}] \tag{2.1}$$

2. Average of a periodic function

$$\langle f(x) \rangle \stackrel{\text{def}}{=} \frac{1}{T} \int_0^T f(x)\,dx \tag{2.2}$$

3. Power

$$P \stackrel{\text{def}}{=} \frac{dW}{dt} = \frac{dW}{dQ}\frac{dQ}{dt} = V\,I \ [\text{W}] \tag{2.3}$$

4. RMS of sine function

$$\langle P(t) \rangle = \langle i(t)^2 \rangle \, R \stackrel{\text{def}}{=} I_{\text{rms}}^2 \, R$$

$$= \left[\frac{I_m}{\sqrt{2}}\right]^2 R \tag{2.4}$$

© Springer Nature Switzerland AG 2021
R. Sobot, *Wireless Communication Electronics by Example*,
https://doi.org/10.1007/978-3-030-59498-5_2

5. Decibel scale

$$G_{dB} \stackrel{\text{def}}{=} 10 \log \frac{P_2}{P_1} \tag{2.5}$$

$$A_{dB} \stackrel{\text{def}}{=} 10 \log \frac{P_2}{P_1} = 10 \log \frac{v_2^2/\cancel{Z}}{v_1^2/\cancel{Z}} = 10 \log \left(\frac{v_2}{v_1} \right)^2 = 20 \log \frac{v_2}{v_1} \tag{2.6}$$

6. dBm definition

$$G_{dBm} \stackrel{\text{def}}{=} 10 \log \frac{P_2}{1\,\text{mW}} \tag{2.7}$$

7. First order complex transfer functions

$$H_1(j\omega) = a_0 \quad (a_0 \in \mathfrak{R}) \tag{2.8}$$

$$H_2(j\omega) = j\frac{\omega}{\omega_0} \quad (j^2 = -1,\ \omega_0 = \text{const.}) \tag{2.9}$$

$$H_3(j\omega) = 1 + j\frac{\omega}{\omega_0} \tag{2.10}$$

$$H_4(j\omega) = \frac{1}{1 + j\frac{\omega}{\omega_0}} \tag{2.11}$$

8. Gain and phase margins

$$GM = 0 - |H(j\omega_M)| \quad [\text{dB}] \quad \text{and,} \quad PM = \phi(\omega_0) - (-180°) \tag{2.12}$$

2.2   Exercises

2.1 *   Basic Physics

1. An average sized snowflake consists of approximately $n = 6.68559 \times 10^{19}$ molecules. Assuming the complete matter of the snowflake is converted into energy, estimate for how long a laptop computer whose average power consumption is $P = 25\,\text{W}$ could be powered?
   Data: atomic weight $H = 1.00794\,\text{g/mol}$, atomic weight $O = 15.9994\,\text{g/mol}$.

2. Estimate the distance of lightning if approximately $t = 9\,\text{s}$ pass between the time the lightning flash is registered and the time the thunder is heard.
   Data: speed of sound, $343\,\text{m/s}$, speed of light, $3 \times 10^8\,\text{m/s}$.

3. Given that light crossing the boundary from an optical fibre to air changes its wavelength from $\lambda_1 = 452\,\text{nm}$ to $\lambda_0 = 633\,\text{nm}$, calculate:

   (a)   index of refraction of fibre,   (b)   speed of light in fibre,

   (c)   frequency of light in fibre,   (d)   frequency of light in air.

4. Historically, two best known experimental estimates of velocity of light are:

(a) in 1676 Rømer noticed approximately 22 min difference in times when eclipses of Jupiter's moon Io are detected, as measured at points of min and max distance from Earth (while Earth is orbiting around the Sun). At that time, the diameter of Earth's orbit was known to be approximately $d = 2.98 \times 10^{11}$ m. What was his estimated speed of light ?

(b) between in 1922 and 1924, in the series of experiments based on geodetic measurements of distances between two mountains and the rotating-mirror method, Michelson established that speed of light is $c = 299\,796$ km/s. In his setup the rotating mirrors were placed at 35 km distance from each other. How much time it takes light to travel that distance?

2.2 *   Basic Definitions

1. For a voltage disturbance wave traveling at the speed of light and described as

$$v_1(t) = \sin(20\pi \times 10^6\, t)$$

find:

(a) its maximum amplitude;   (b) its frequency;

(c) its period;   (d) its phase at time $t = 0$ s.

2. A sinusoidal wave is defined as $v(t) = 10\,\text{V}\,\sin(100\,t + 45°)$. Determine its:

(a) frequency,   (b) period,   (c) max amplitude,

(d) $\phi(t = (\pi/100)\text{s})$,   (e) equivalent cos form,   (f) average value.

3. Calculate wavelengths of EM waves propagating in air with the following frequencies:

(a) $f_1 = 3$ kHz (i.e. in the audio range),

(b) $f_2 = 3$ MHz (i.e. frequency of a simple LC oscillator), and

(c) $f_3 = 3$ GHz (i.e. close to the operational frequency of cell phones).

4. Calculate period, wavelength, the propagation constant, and phase velocity an electromagnetic wave in free space for the following frequencies:

(a) $f_1 = 10$ MHz,   (b) $f_2 = 100$ MHz,   (c) $f_3 = 10$ GHz.

5. Instantaneous voltage of a waveform is described as $v(t) = V_m \cos(\omega t + \phi)$ where, $\omega = (2\pi \times 10^3)$ rad/s and $\phi = \pi/4$. Calculate its frequency f as well as its instantaneous phase $\varphi(t)$ at $t = 125\,\mu$s.

6. By comparing, for example, maximum amplitude, calculate differences in the arrival times Δt (which is absolute measure) for EM wave pairs with phase difference of $\Delta\phi = \pi/2$ (which is relative measure) at each of the following frequencies:

(a) $f_1 = 1$ kHz,   (b) $f_2 = 1$ MHz,   (c) $f_3 = 1$ GHz.

7. Given EM wave described as $v_1(t) = \sin(20\pi \times 10^6 t)$

(a)   calculate wavelength $\lambda_1(v_1)$.

(b)   given a second wave $v_2(t)$ with the same amplitude and frequency as $v_1(t)$, however, there is the phase difference $\Delta\phi = +45°$. Calculate amplitude $v_2(t = 0)$ and the space distance d between one of its peaks and the first following peak of $v_1(t)$.

8. Two cosine waveforms with the same frequency $f = 10\,\text{MHz}$ and amplitude of $V_{pp} = 2\,\text{V}$ travel along a conductive wire, however, the second wave's phase is increased by $\phi = \pi/2$. At the end of the wire, there is a $R = 1\,\text{k}\Omega$ loading resistor connected to the ground. Assuming that $v_1(t = 0) = V_p$, calculate:

(a)   time Δt and spatial Δx differences,   (b) I_R when $t = 0$ and $t = {}^{T}/_8$,

(c) I_R along the wire at $x = 0$ and $x = \lambda/8$,   (d) I_R along the wire at $x = 3\lambda/8$ and $x = 7\lambda/8$,
when $t = 0$,   when $t = 0$.

2.3 ***   Signals

1. Two single-tone signals, $v_1 = f(\omega_1, t)$ and $v_2 = f(\omega_2, t)$, are applied at the input nodes of a passive network, Fig. 2.1. Assuming that the two signals are drawn in scale, and $v_1(t)$ has period of $T_1 = 1\,\mu\text{s}$:

(a)   derive $v_o(t)$, if $R_1 = R_2$;

(b)   derive $v_o(t)$, if $R_2 = 2\,R_1$;

(c)   sketch the frequency spectrum of V_o if $R_1 = R_2$;

(d)   sketch the frequency spectrum of V_o if $R_2 = 2\,R_1$.

Fig. 2.1 Example 2.3-1

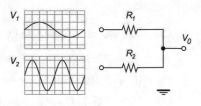

2. Given three waveforms,

$$v_1 = 1 + 0.5\,\sin(\omega t)$$

$$v_2 = 1 + 0.5\,\sin(\omega t - \pi)$$

$$v_3 = 0.5 + 0.5\,\sin(\omega t - \pi)$$

sketch by hands in correct scale and comment on the following plots:

(a) v_1, v_2, and $v_1(t) - v_2(t)$;   (b) v_1, v_2, and $v_1(t) + v_2(t)$;

(c) v_1, v_3, and $v_1(t) - v_3(t)$;   (d) v_1, v_3, and $v_1(t) + v_3(t)$.

3. The instantaneous voltage of EM wave is given as

$$v(t) = V_m \sin(2\pi f\, t + \phi_0)$$

where, $\omega = (2\pi\ 100)\ \mathrm{rad/s}$ and $\phi_0 = \pi/4$. Calculate:

(a) wavelength λ;

(b) at $t = 15\,\mathrm{ms}$: phase and traveled distance ;

(c) voltage amplitudes V at $t = 0\,\mathrm{s}$ and $t = 15\,\mathrm{ms}$.

2.4 *   Signal's Average, rms, and Power

1. A Thévenin generator $v(t)$ consists of two voltage sources, a simple sinusoidal source v_{AM} and a DC battery V_{DC}, see Fig. 2.2.

Given data, write expression for $v(t)$, then calculate its average $\langle v(t) \rangle$.

Data: (a) $V_{DC} = -1\,\mathrm{V}$, (b) $V_{DC} = 0\,\mathrm{V}$, and (c) $V_{DC} = 1\,\mathrm{V}$

Fig. 2.2 Example 2.4-1

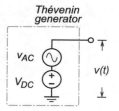

2. Derive the average and rms values of periodic waveforms, see Fig. 2.3:

    (a) sine,
    (b) square,
    (c) triangle, and
    (d) sawtooth.

Fig. 2.3 Example 2.4-2

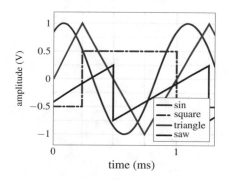

3. Assuming current $i(t)$, see Fig. 2.4, calculate:

    (a)   period and frequency of $i(t)$,

    (b)   min and max amplitude of $i(t)$,

    (c)   AC amplitude I_m,

    (d)   average in $(0\,\text{s}, 20\,\text{s})$ interval,

    (e)   common mode I_{CM} of $i(t)$,

    (f)   RMS of AC component only,

    (g)   RMS of $i(t)$ signal.

Fig. 2.4 Example 2.4-3

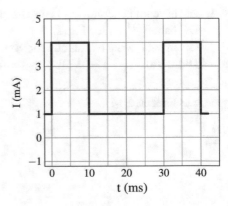

4. Given AM modulated waveform in Fig. 2.5, estimate frequencies of carrier f_C (i.e. high-frequency) and modulating f_M (i.e. the embedded low-frequency) waveforms if horizontal time axes are shown in the units of:

    (a)   [ms];   (b)   [μs];   (c)   [ns].

Fig. 2.5 Example 2.4-4

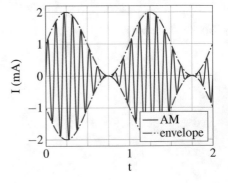

5. How much energy is used by a 60 W amplifier turned on over $t = 8\,\text{h}$? Compare with a 1000 W amplifier turned on over $t = 60\,\text{s}$.

6. The voltage and current values of a 50 Hz sine waveforms are given as: $v(t) = 220 \sin(\omega t + 5\pi/6)$ V and $i(t) = 10 \sin(\omega t - 2\pi/3)$ A. Calculate the values of the instantaneous power $p(t)$ and the average power $\langle P \rangle$ generated by this voltage and current.

7. Calculate the total power gain of the two amplifiers, see block diagram in Fig. 2.6. Comment on the calculations when using the two different gain units shown in the illustration.

Fig. 2.6 Example 2.4-7

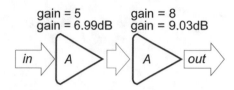

gain = 5　　gain = 8
gain = 6.99dB　gain = 9.03dB

8. A cell phone transmits $P_1 = +30\,\text{dBm}$ of signal power from its antenna. At the receiving unit the detected signal power is measured as $P_2 = 5\,\text{pW}$. Calculate the propagation loss in the transmitting medium.

2.5 *** Transfer Function

1. Show Bode plot of the following transfer function,

$$H(\omega) = \frac{100}{j\omega + 10}$$

2. Show Bode plot of the following transfer function,

$$H(\omega) = \frac{10\left(1 + j\frac{\omega}{10}\right)}{j\omega + 10}$$

3. Show Bode plot of the following transfer function ($s = j\omega$),

$$H(s) = s^2 + 22s + 40$$

4. Show Bode plot of the following transfer function ($s = j\omega$),

$$H(s) = \frac{100}{(s + 10)(s + 100)}$$

5. Show Bode plot of the following transfer function ($s = j\omega$),

$$H(\omega) = \frac{100\,(s + 1)}{s^2 + 110\,s + 1000}$$

6. Show Bode plot of the following transfer function, a case of second order function with complex roots ($s = j\omega$),

$$H(s) = 10^2 \frac{s + 10}{s^2 + 40\,s + 10^4}$$

7. Given the Bode plot in Fig. 2.7, write the analytical form or $H(\omega)$.

Fig. 2.7 Example 2.5-7

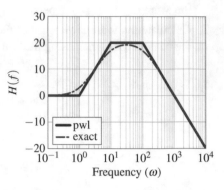

8. Time domain form of a waveform is given as

$$v(t) = \frac{4}{\pi} \left[\sum_{n=1,3,5,...} \frac{1}{n} \sin\left(\frac{n}{2}t\right) \right]$$

Using plotting software or SPICE simulation, create plot of $v(t)$ when first ten terms are used. What waveform is being created? Intentionally change amplitude and/or sin argument of one or more individual terms in the sum, and observe the consequence on the shape of this waveform. What happens if some terms are completely removed? What if starting with the last term, one by one is systematically removed? What if starting with the first term, one by one is systematically removed?

9. Repeat exercises in Question 8 with waveform whose amplitude is given as

$$v(t) = \frac{2}{\pi} \left[\sum_{n=1,2,3,...} \frac{1}{n} \sin(nt) \right]$$

Solutions

Exercise 2.1, page 38

1. Einstein formalized the mass–energy equivalence as

$$E = \sqrt{(mc^2)^2 + (pc)^2} \tag{2.13}$$

where m is the mass of object whose equivalent energy is calculated, c is the velocity of light in vacuum, and p is momentum of the mass m in motion. However, if the mass is not moving, then $p = 0$ and (2.13) reduces into its simple well known form

$$E = mc^2 \tag{2.14}$$

A snowflake consists of water molecules (H_2O), thus, first we find its total mass (m_S) by adding masses of all n individual molecules that constitute the snowflake. Each water molecule consists of two hydrogen atoms (atomic weight $H = 1.00794$ g/mol) and one atom of oxygen (atomic weight $O = 15.9994$ g/mol). Thus, a single water molecule has atomic weight of:

$$H_2O = 2 \times H + O = 18.01528 \text{ g/mol}$$

or, the molecule mass in units of g is

$$H_2O = \frac{H_2O}{N_A} = \frac{18.01528 \text{ g/mol}}{6.022141 \times 10^{23} \text{ g/mol}}$$
$$= 2.99151 \times 10^{-23} \text{ g}$$

thus, the total snowflake mass is

$$m_S = n \times H_2O$$
$$= 6.68 \times 10^{19} \times 2.99 \times 10^{-23} \text{ g}$$
$$= 2 \text{ mg} = 2 \times 10^{-6} \text{ kg}$$

Velocity of the snowflake is negligible relative to the speed of light, thus we calculate the equivalent energy as:

$$E = m c^2 = 2 \text{ mg} \times 3 \times 10^8 \text{ m/s}$$
$$= 1.8 \times 10^{11} \text{ J}$$

Average power is defined as the energy transfer rate $E = P/t$, therefore in order to provide the average power of $P = 25$ W the total energy E must be distributed over the following time

$$t = \frac{E}{P} = \frac{1.8 \times 10^{11} \text{ J}}{25 \text{ W}} = 7.2 \times 10^9 \text{ s}$$
$$\approx 228 \text{ years}$$

We conclude that if we were able to completely convert, for instance, only 2 mg of any matter (including a snowflake) into energy we would be able to provide power to our hand-held electronic equipment for many years. However, until we develop such a source of energy we must obey the limits imposed by the capacity of our modern batteries and design our circuits accordingly.

2. In $t = 9$ s sound travels approximately $s = v t = 343$ m/s $\times 9$ s ≈ 3 km. At the same time, light travels this distance in $t = s v = 3$ km/3×10^8 m/s ≈ 10 μs which is negligible relative to 9 s. Therefore, we can ignore the travel time of light and simply estimate that the lightning happened at approximately 3 km away.

3. We recall most basic relationship in physics where traveled distance s is calculated as product of velocity v and time t, i.e. $s = v t$. In the context of waves, λ is distance traveled over period of time T, where the time period is defined as the inverse of frequency, i.e. $T = 1/f$. Therefore, we write $\lambda = c T$, where c is the velocity of light. Refraction index n is defined as the ratio of wave velocities in vacuum (vacuum variables are usually indexed with "0") and the given media. By

experiment it was found that the refraction index of dry air is as same as for vacuum. In addition, light frequency stays constant while propagating in various media. Therefore, it follows

$$n = \frac{c}{v} = \frac{\lambda_0/T}{\lambda_1/T} = \frac{\lambda_0 \, \cancel{f_0}}{\lambda_1 \, \cancel{f_0}} \tag{2.15}$$

(a) by definition (2.15), refraction index of fibre is

$$n = \frac{\lambda_0}{\lambda_1} = \frac{633\,\text{nm}}{452\,\text{nm}} = 1.400$$

(b) knowing n and (2.15), velocity of light in fibre

$$v = \frac{c}{n} \approx \frac{3 \times 10^8 \; \text{m/s}}{1.4}$$

$$\approx 214 \times 10^6 \; \text{m/s}$$

(c) Light frequency in the fibre can be found from its speed and wavelength:

$$v = \frac{v}{\lambda_1} = \frac{214 \times 10^6 \; \text{m/s}}{452\,\text{nm}}$$

$$= 474 \times 10^{12}\,\text{Hz}$$

(d) Light frequency in the air can be found from its speed and wavelength:

$$v = \frac{c}{\lambda_0} = \frac{3 \times 10^8 \; \text{m/s}}{633\,\text{nm}}$$

$$= 474 \times 10^{12}\,\text{Hz}$$

4. Given the technologies of their respective days, the two experiments not only proved that velocity of light is not infinite but also gave very good numerical estimates, as

(a) velocity is ratio of distance and time, thus

$$v = \frac{d}{t} = \frac{2.98 \times 10^{11}\,\text{m}}{22 \times 60\,\text{s}}$$

$$= 2.26 \times 10^8 \; \text{m/s}$$

(b) time is ratio of distance and velocity,

$$t = \frac{d}{v} = \frac{35 \times 10^3\,\text{m}}{299\,796 \times 10^3 \; \text{m/s}}$$

$$= 117\,\mu\text{s}$$

Exercise 2.2, page 39

1. By inspection of the given wave $v_1(t)$ equation, we rewrite its equation in general form to explicitly include its amplitude and phase as:

$$v_1(t) = A_m \sin(\omega t + \phi) = 1\text{V} \times \sin(2\pi \times 10 \times 10^6\,t + 0)$$

where radial frequency ω is by definition $\omega = 2\pi f$, thus,

(a) being a voltage waveform, its maximum amplitude is in the units of volts, i.e. $A_m = 1\,\text{V}$,

(b) from definition of radial frequency, $\omega = 2\pi f = 2\pi \times 10 \times 10^6\,\text{Hz}$, i.e. $f = 10 \times 10^6\,\text{Hz} = 10\,\text{MHz}$,

(c) by definition period is equal to inverse of frequency, i.e. $T = 1/f = 1/10 \times 10^6\,\mathrm{Hz} = 100\,\mathrm{ns}$,

(d) by comparison with general form of sine, the initial phase (i.e. at $t = 0$) is $\phi = 0°$.

2. By inspection of the given waveform definition, we write

(a) given $\omega t = 2\pi f t = 100\,t$, we write $\omega = 100\,\mathrm{rad/s}$ $\therefore$ $f = (100/2\pi)\,\mathrm{Hz}$,

(b) given $f = (100/2\pi)\,\mathrm{Hz}$ $\therefore$ $T = 1/T = (2\pi/100)\,\mathrm{s}$,

(c) maximum value of a sin (or cos) function is measured either as its positive amplitude ("peak") or difference between its minimum and maximum levels ("peak to peak") $A_p = 10\,\mathrm{V}$, or $A_{pp} = 20\,\mathrm{V}$,

(d) given $T = (2\pi/100)\,\mathrm{s}$, then $T/2 = (\pi/100)\,\mathrm{s}$. One full period is equivalent to phase of 2π, thus half-period equals π. The initial phase is $\pi/4$, hence the total phase is $\phi(t = (\pi/100)\,\mathrm{s}) = \pi + \pi/4 = 5\pi/4$.

(e) a cos function equivalent to sin function delayed by $\pi/2$. Therefore we write,
$$v(t) = 10\,\mathrm{V}\,\sin{(100\,t + 45°)}$$
$$= 10\,\mathrm{V}\,\cos{(100\,t + 45° - 90°)}$$
$$= 10\,\mathrm{V}\,\cos{(100\,t - 45°)}$$

(f) by definition of average value for a periodic function, average of a sin function is shown to be
$$\langle f(x) \rangle \overset{\mathrm{def}}{=} \frac{1}{2\pi} \int_0^{2\pi} \sin{(x)}\,dx$$
$$= -\frac{1}{2\pi}\,\cos(x)|_0^{2\pi} = 0$$

3. While propagating in air, EM waves have phase velocity of $v_p = c_0 \approx 3 \times 10^8$ [m/s], hence, we write phase velocity and the propagation constant as

$$\omega = 2\pi f, \quad v_p = \frac{\omega}{\beta}, \quad \beta = \frac{2\pi}{\lambda} \quad \therefore \quad \lambda = \frac{2\pi}{\beta} = \frac{2\pi\,v_p}{\omega} = \frac{v_p}{f} = \frac{c_0}{f}$$

which results in:

(a) audio range:
$$\lambda_1 \approx \frac{3 \times 10^8\,\mathrm{m/s}}{3\,\mathrm{kHz}}$$
$$= 100 \times 10^3\,\mathrm{m}$$
$$= 100\,\mathrm{km}$$

(b) LF wave:
$$\lambda_2 \approx \frac{3 \times 10^8\,\mathrm{m/s}}{3\,\mathrm{MHz}}$$
$$= 100\,\mathrm{m}$$

(c) HF wave:
$$\lambda_3 \approx \frac{3 \times 10^8\,\mathrm{m/s}}{3\,\mathrm{GHz}}$$
$$= 100 \times 10^{-3}\,\mathrm{m}$$
$$= 100\,\mathrm{mm}$$

These lengths are interpreted as the physical distance over which the wave's amplitude undergoes one full cycle of changes: from zero to positive maximum, back to zero, down to the negative minimum, and finally back to zero amplitude again. In order to apply LF approximations, the media length d must be much shorter than the signal wavelength, i.e. $d \ll \lambda$. Only in that case, the media can be assumed to be equipotential, i.e. to have constant signal amplitude at any point in *space* but necessarily in *time*. Otherwise, the transmission line model and exact Maxwell's equations must be used.

4. By textbook definitions, we calculate

(a) $T_1 = \dfrac{1}{f_1} = \dfrac{1}{10\,\text{MHz}} = 100\,\text{ns},\ \lambda \approx 30\,\text{m},\ \beta = 0.21\,\text{m}^{-1},\ v_p \approx 3 \times 10^8\,\text{m/s}$

(b) $T_1 = \dfrac{1}{f_1} = \dfrac{1}{100\,\text{MHz}} = 10\,\text{ns},\ \lambda \approx 3\,\text{m},\ \beta = 2.1\,\text{m}^{-1},\ v_p \approx 3 \times 10^8\,\text{m/s}$

(c) $T_1 = \dfrac{1}{f_1} = \dfrac{1}{1\,\text{GHz}} = 1\,\text{ns},\ \lambda \approx 0.3\,\text{m},\ \beta = 21\,\text{m}^{-1},\ v_p \approx 3 \times 10^8\,\text{m/s}$

This example gives us medium lengths where frequencies of the respective signals are considered to have "high frequency" because the signal amplitudes are not approximately constant along their traveling paths.

5. By inspection, we write,

$$\omega = (2\pi \times 10^3)\ \text{rad/s}$$

$$\therefore$$

$$f = \omega/2\pi = 1\,\text{kHz}$$

where the frequency $f = 1\,\text{kHz}$ and the initial phase $\phi_0 = \pi/4$ are constant.

Given $f = 1\,\text{kHz}$, therefore $T = 1/f = 1\,\text{ms}$. In addition, one period in units of time is equivalent to $360° = 2\pi$ in units of angle. Therefore, time $t = 125\,\mu\text{s}$ represents $1\,\text{ms}/125\,\mu\text{s} = 8$, in other words a $1\,\text{kHz}$ signal advances only $T/8 = 2\pi/8 = \pi/4$ angle during the first $125\,\mu\text{s}$, see Fig. 2.8.

Therefore, we conclude that at $t = 125\,\mu\text{s}$ the instantaneous phase is $\phi(125\,\mu\text{s}) = \phi_0 + \pi/4 = \pi/2$. As a consequence, $v_1(t = 125\,\mu\text{s}) = 0\,\text{V}$.

Fig. 2.8 Example 2.2-5

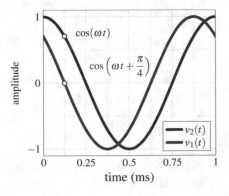

6. Conversion of the given frequencies into their equivalent periods and by knowing that period $T \overset{\text{def}}{=} 2\pi$ angle (i.e. one full cycle), we write

(a) $T_1 = \dfrac{1}{f_1} = \dfrac{1}{1\,\text{kHz}} = 1\,\text{ms}\ \therefore\ \Delta t_1 = \dfrac{T_1}{4} = \dfrac{1\,\text{ms}}{4} = 250\,\mu\text{s}$

(b) $T_2 = \dfrac{1}{f_2} = \dfrac{1}{1\,\text{MHz}} = 1\,\mu s \quad \therefore \quad \Delta t_2 = \dfrac{T_2}{4} = \dfrac{1\,\mu s}{4} = 250\,\text{ns}$

(c) $T_3 = \dfrac{1}{f_3} = \dfrac{1}{1\,\text{GHz}} = 1\,\text{ns} \quad \therefore \quad \Delta t_3 = \dfrac{T_3}{4} = \dfrac{1\,\text{ns}}{4} = 250\,\text{ps}$

which illustrates the equivalence between phase and the arrival times.

7. By inspection of wave $v_1(t)$ equation, we write:

(a) by definition its wavelength is

$$f = 10\,\text{MHz}, \quad \therefore \quad T = 100\,\text{ns}$$
$$\lambda = cT = 3 \times 10^8 \ \text{m/s} \times 100 \times 10^{-9}\,\text{s} \approx 30\,\text{m}$$

(b) the second wave is leading with phase difference of

$$\Delta\phi = 45° = \pi/4 = 2\pi/8 = T/8$$

and its amplitude at $t = 0\,\text{s}$ is therefore

$$v_2 = 1\,\text{V} \, \sin(\omega \times 0 + \pi/4) = 1\,\text{V} \, \sin(\pi/4) = 1/\sqrt{2}\,\text{V} \approx 0.707\,\text{V}$$

the phase difference is $T/8$, therefore the distance in space has to be

$$d = \frac{\lambda}{8} = \frac{30\,\text{m}}{8} = 3.75\,\text{m}$$

8. Amplitude equations relative to period and wavelength are conveniently correlated so that the two equations are symmetric: signal quantities (i.e. voltage and current) calculated at a given moment (space independent) are equal to the quantities calculated at their respective point in space (time independent). From Maxwell's equations for EM wave that give wave propagation in time as well as in space, as in[1]

$$E_x = E_{0x} \sin(\omega t - \beta z)$$

by setting $t = 0$ we have the equation for the wave propagation in space, and by setting $z = 0$ we have the equation for the wave propagation in time. In addition, from definitions of the propagation constant $\beta = 2\pi/\lambda$ and phase velocity $v_p = \omega/\beta$, for EM wave in vacuum (or air) we easily show direct proportionality between T and λ as $\lambda = cT$. Alternatively, we write,

$$\Delta\phi = \frac{\pi}{2} \stackrel{\text{def}}{=} \frac{T}{4} = \frac{\lambda}{4}$$

Declaring $x = 0$ as the reference point we write

---

[1] We note that sin and cos functions differ only in the initial condition (i.e. phase).

$$(t = 0, x = 0); \quad v_1(0) = V_p \cos(0) = 1V$$
$$v_2(0) = V_p \cos(\pi/2) = V_p \sin(0) = 0\,V$$
$$(t = {}^T/8, x = 0); \quad v_1({}^T/8) = V_p \cos(\pi/4) = +\sqrt{2}/2V$$
$$v_2({}^T/8) = V_p \cos(\pi/4 + \pi/2) = V_p \cos(3\pi/4) = -\sqrt{2}/2V$$
$$(t = {}^{3T}/8, x = 0); \quad v_1(3T/8) = V_p \cos(3\pi/4) = -\sqrt{2}/2V$$
$$v_2(3T/8) = V_p \cos(3\pi/4 + \pi/2) = V_p \cos(5\pi/4) = -\sqrt{2}/2V$$
$$(t = {}^{7T}/8, x = 0); \quad v_1(7T/8) = V_p \cos(-\pi/4) = +\sqrt{2}/2V$$
$$v_2(7T/8) = V_p \cos(-\pi/4 + \pi/2) = V_p \cos(\pi/4) = +\sqrt{2}/2V$$

and similarly, declaring any of time points $t = nT = 0, (n = 1, 2, \ldots)$ as the time measuring reference point $t = 0$, we write

$$(t = 0, x = 0); \quad v_1(0) = V_p \cos(0) = 1\,V$$
$$v_2(0) = V_p \cos(\pi/2) = V_p \sin(0) = 0\,V$$
$$(t = 0, x = \lambda/8); \quad v_1(\lambda/8) = V_p \cos(\pi/4) = +\sqrt{2}/2V$$
$$v_2(\lambda/8) = V_p \cos(3\pi/4) = -\sqrt{2}/2V$$
$$(t = 0, x = 3\lambda/8); \quad v_1(3\lambda/8) = V_p \cos(3\pi/4) = -\sqrt{2}/2V$$
$$v_2(3\lambda/8) = V_p \cos(5\pi/4) = -\sqrt{2}/2V$$
$$(t = 0, x = 7\lambda/8); \quad v_1(7\lambda/8) = V_p \cos(-\pi/4) = +\sqrt{2}/2V$$
$$v_2(7\lambda/8) = V_p \cos(\pi/4) = +\sqrt{2}/2V$$

The load resistor current at any point in time and space is

$$i(t, x) = \frac{v_R(t, x)}{R} = \frac{v_1(t, x) + v_2(t, x)}{R}$$

therefore, we calculate:

(a) by definition period and wavelength are

$$f = 10\,\text{MHz}, \quad \therefore \quad T = 100\,\text{ns}$$

$$\lambda = cT = 3 \times 10^8 \text{ m/s} \times 100 \times 10^{-9}\,\text{s} \approx 30\,\text{m}$$

and, relative to the phase of $v_1(t)$, phase difference of the second wave is

$$\Delta\phi = 90° = \pi/2 = {}^{2\pi}/4 = {}^T/4$$

therefore, $\Delta t = {}^T/4 = 25\,\text{ns}$ and $\Delta x = \lambda/4 = 7.5\,\text{m}$

(b) at $x = 0\,\text{mm}$, we calculate instantaneous currents in two time points, i.e. $t = 0\,\text{s}$ and $t = T/8 = 12.5\,\text{ns}$ as

$$v(0\text{s}) = v_1(0) + v_2(0) = 1V \quad \therefore \quad I_R(0\,\text{s}) = \frac{v(t)}{R} = \frac{1\,V}{1\,\text{k}\Omega} = 1\,\text{mA}$$

$$v(12.5\,\text{ns}) = v_1(T/8) + v_2(T/8) = \left(+\sqrt{2}/2 - \sqrt{2}/2\right)V = 0V \quad \therefore \quad I_R(12.5\,\text{ns}) = 0\,\text{A}$$

(c) at time reference $t = 0\,\text{s}$, we calculate instantaneous currents in two space points, i.e. $x = 0\,\text{m}$ and $x = \lambda/8 = 3.75\,\text{m}$. To travel distance of $x = 3.75\,\text{m}$ EM wave needs $t = x/c = 12.5\,\text{ns} \equiv T/8$, therefore we write

$$v(0) = v_1(0) + v_2(0) = 1\,\text{V}$$

$$\therefore\ I_R(0) = \frac{v(x)}{R} = \frac{1\,\text{V}}{1\,\text{k}\Omega} = 1\,\text{mA}$$

$$v(3.75\,\text{m}) = v_1(3.75\,\text{m}) + v_2(3.75\,\text{m}) = \left(+\sqrt{2}/2 - \sqrt{2}/2\right)\,\text{V} = 0\,\text{V}$$

$$\therefore\ I_R(3.75\,\text{m}) = 0\,\text{A}$$

(d) at time reference $t = 0\,\text{s}$, we calculate instantaneous currents in two space points, i.e. $x = 3\lambda/8 = 11.25\,\text{m}$ and $x = 7\lambda/8 = 26.25\,\text{m}$. To travel these two distances, EM wave needs $t = 3T/8$, and $t = 7T/8$, respectively, therefore we write

$$v(11.25\,\text{m}) = \left(-\sqrt{2}/2 - \sqrt{2}/2\right)\,\text{V} = -\sqrt{2}\,\text{V}$$

$$\therefore\ I_R(11.25\,\text{m}) = \frac{v(3\lambda/8)}{R} = \frac{-\sqrt{2}}{1\,\text{k}\Omega} = -\sqrt{2}\,\text{mA}$$

$$v(26.25\,\text{m}) = \left(+\sqrt{2}/2 + \sqrt{2}/2\right)\,\text{V} = \sqrt{2}\,\text{V}$$

$$\therefore\ I_R(26.25\,\text{m}) = +\sqrt{2}\,\text{mA}$$

Exercise 2.3, page 40

1. The two voltages are added by means of a resistive voltage divider. By inspection of Fig. 2.1, given data, we write expressions for the two input signals as

$$v_1(t) = 1\ \sin[(2\pi \times 1\,\text{MHz})\,t]$$

$$v_2(t) = 2\ \sin[(2\pi \times 2\,\text{MHz})\,t] \tag{2.16}$$

(a) Because the two resistors are equal $R = R_1 = R_2$ by inspection of Fig. 2.9 we write two expressions for $i(t)$ as

$$i(t) = \frac{v_1 - v_2}{2R} = \frac{v_1 - v_o}{R}$$

$$\therefore$$

$$v_1 - v_2 = 2(v_1 - v_o) \quad \therefore\quad v_o = \frac{1}{2}(v_1 + v_2) \tag{2.17}$$

Side effect of this addition is that there is $1/2$ scaling factor, which means that correct implementation of analog voltage adder requires one additional amplifier whose gain is $G = 2$ so that it delivers true sum as $v_{\text{sum}} = v_1 + v_2$, see Fig. 2.9. That being the case, we write

$$v_{\text{sum}}(t) = 1 \, \sin[(2\pi \times 1\,\text{MHz})\, t] + 2 \, \sin[(2\pi \times 2\,\text{MHz})\, t] \qquad (2.18)$$

Fig. 2.9 Example 2.3-1(a)

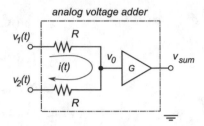

analog voltage adder

(b) Given that $R_2 = 2\, R_1 = 2R$, we write

$$i(t) = \frac{v_1 - v_2}{3R} = \frac{v_1 - v_o}{R}$$

$$\therefore$$

$$v_1 - v_2 = 3(v_1 - v_o) \quad \therefore \quad v_o = \frac{1}{3}\,(2\,v_1 + v_2) \qquad (2.19)$$

In this case the output voltage v_o is not just a simple sum of the two input signals, instead voltage v_1 is multiplied by factor two *before* it is added to v_2, which is then altogether multiplied by $1/3$ factor. Therefore, after setting $G = 3$, we write

$$v_{\text{sum}}(t) = 2 \, \sin[(2\pi\, 1\,\text{MHz})\, t] + 2 \, \sin[(2\pi\, 2\,\text{MHz})\, t] \qquad (2.20)$$

(c) From (2.18), we conclude that the linear addition of two single-tone signals does not produce any new frequency tones in the output spectrum (we still have only the 1 MHz and 2 MHz single tones at the output), while that the output amplitude of v_2 is still twice the amplitude of v_1.

Knowing the two voltages $(v_{1\text{m}} = 1)$, $(v_{2\text{m}} = 2)$ and resistance R, we calculate the two powers (P_1, P_2) as

$$P_{1\text{avg}} = \frac{v_{1\text{rms}}^2}{R} = \frac{v_{1\text{m}}^2}{2R} = \frac{1}{2R} \quad \text{and,} \quad P_{2\text{avg}} = \frac{v_{2\text{rms}}^2}{R} = \frac{v_{2\text{m}}^2}{2R} = \frac{4}{2R}$$

$$\therefore$$

$$\frac{P_{1\text{avg}}}{P_{2\text{avg}}} = \frac{1}{4} \qquad (2.21)$$

which is to say that $1 : 4$ is *relative* power ratio between the two signals, both at the input and the output nodes. Frequency spectrum graph is shown in Fig. 2.10 (left).

Fig. 2.10
Example 2.3-1(c)

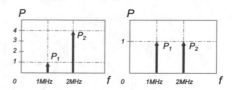

(d) From (2.20), we conclude that the two output powers are now equal, as shown in Fig. 2.10 (right).

This example illustrates *linear* operation of addition, which may change relative powers of the internal tones in the output signal, however, it does not change the frequency content. This is very important observation, because design of wireless RF circuits is based on using *non-linear* operations, such as multiplication, to produce output tones whose frequency is different relative to the input tones. This "frequency shifting" operation is fundamental for wireless RF communication systems and must be recognized as different than linear addition.

2. In this example illustrates a special and very important case of "constructive" addition of two waveforms, that is, we find out how to create "differential" signal out of two single-ended signals.

(a) Before we proceed, let us first take a closer look at the given signals

$$v_1(t) = 1 + 0.5 \sin(\omega t)$$

$$v_2(t) = 1 + 0.5 \sin(\omega t - \pi)$$

We observe:

a. The two constant terms (i.e. DC=1) are equal,
b. The two amplitudes are equal,
c. The two waveforms are inverted, i.e. phase difference $\Delta\phi = \pi$.

Subtraction of $v_2(t)$ from $v_1(t)$ gives

$$
\begin{aligned}
v(t) = v_1(t) - v_2(t) &= [1 + 0.5 \sin(\omega t)] - [1 + 0.5 \sin(\omega t - \pi)] \\
&= 0.5\,[\sin(\omega t) - \sin(\omega t - \pi)] \\
&= 0.5\,[\sin(\omega t) + \sin(\omega t)] \\
&= \sin(\omega t) \quad\quad\quad\quad\quad\quad\quad\quad\quad (2.22)
\end{aligned}
$$

The resulting sinusoidal waveform $v(t)$ has average $DC = 1 - 1 = 0$. Maximum amplitude of the resulting sinusoid is the sum $V_m = V_{m1} - (-V_{m2}) = 0.5 - (-0.5) = 1$.

To summarize, when two sinusoids have same DC, frequency, and amplitude, while the phase difference is π (i.e. they are inverted relatively to each other), their difference is a sinusoid with the same frequency, double amplitude, and zero DC, Fig. 2.11 (left).

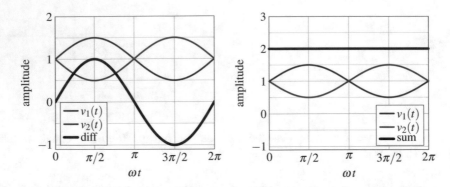

Fig. 2.11 Example 2.3-2(a)

(b) The sum of $v_1(t)$ and $v_2(t)$ gives

$$v(t) = v_1(t) + v_2(t) = [1 + 0.5 \sin(\omega t)] + [1 + 0.5 \sin(\omega t - \pi)]$$

$$= 2 + 0.5 [\sin(\omega t) + \sin(\omega t - \pi)]$$

$$= 2 + 0.5 [\sin(\omega t) - \sin(\omega t)]$$

$$= 2 \tag{2.23}$$

This time, two time varying components (AC) cancelled each other while the two DC levels added, Fig. 2.11 (right)

(c) When two waveforms are not perfectly symmetrical we find

$$v(t) = v_1(t) - v_3(t) = [1 + 0.5 \sin(\omega t)] - [0.5 + 0.5 \sin(\omega t - \pi)]$$

$$= 0.5 + 0.5 [\sin(\omega t) - \sin(\omega t - \pi)]$$

$$= 0.5 + 0.5 [\sin(\omega t) + \sin(\omega t)]$$

$$= 0.5 + \sin(\omega t) \tag{2.24}$$

that is, the two DC levels did not cancel, as a consequence, the resulting differential signal does not have zero CM and the sinusoid takes asymmetrically positive and negative values, Fig. 2.12 (left).

(d) Addition of two inverted sinusoids with misaligned DC levels results in

$$v(t) = v_1(t) + v_3(t) = [1 + 0.5 \sin(\omega t)] + [0.5 + 0.5 \sin(\omega t - \pi)]$$

$$= 1.5 + 0.5 [\sin(\omega t) + \sin(\omega t - \pi)]$$

$$= 1.5 + 0.5 [\sin(\omega t) - \sin(\omega t)]$$

$$= 1.5 \tag{2.25}$$

which means that two inverted sinusoids cancel even if their CM are different, while the resulting DC level equals to the sum of two initial DC levels, Fig. 2.12 (right).

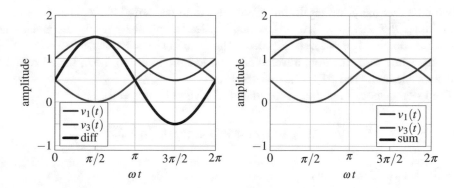

Fig. 2.12 Example 2.3-2(d)

Differential signal processing is extremely important in modern signal processing, and it is, therefore, very important to quantify all possible imperfections due to the signal misalignments, or departures of any kind from the ideal symmetrical case.

3. Inquiring, for example, about a wave's instantaneous phase in radians is equivalent to asking about location of the wave's front end in time and space.

   (a) By inspection of the given waveform

$$v(t) = V_m \sin(\omega t + \phi_0) \tag{2.26}$$

where radial frequency is $\omega = (2\pi\ 100)\ \mathrm{rad/s}$ and the initial phase $\phi_0 = \pi/4$, then by definition it follows that

$$\omega = (2\pi\ 100)\ \mathrm{rad/s} \quad \Rightarrow \quad f = 100\,\mathrm{Hz}$$

$$\therefore$$

$$T = \frac{1}{f} = 10\,\mathrm{ms}$$

$$\therefore$$

$$\lambda = c\,T \approx 3 \times 10^8\,\mathrm{m/s} \times 10\,\mathrm{ms} \approx 3000\,\mathrm{km}$$

where velocity of light is c and the wave period is T.

   (b) Similarly, at $t = 15\,\mathrm{ms}$ we find

$$T = 2\pi \quad \therefore \quad 2\pi \equiv 10\,\mathrm{ms} \quad \therefore \quad 15\,\mathrm{ms} \equiv 3\pi = \frac{12\pi}{4}$$

which is phase of a waveform whose initial phase is $\phi_0 = 0$, for example, dashed line sinusoidal waveform in Fig. 2.13. However, the initial phase is $\phi_0 = \pi/4$. Thus, at the moment $t = 15\,\mathrm{ms}$ the phase is shifted to

$$\phi = \frac{12\pi}{4} + \frac{\pi}{4} = \frac{13\pi}{4} \equiv \frac{5\pi}{4}$$

We note that the phase results are given relative to the period T, i.e. after removing integer multiples $n \times 2\pi$, $(n = 2, 3, \ldots)$, which implies that all phase results are given within $0 \le \phi \le 2\pi$ range (i.e. $n = 1$), see Fig. 2.13. That is to say, $\phi = {}^{13\pi}/_4 \equiv {}^{5\pi}/_4$.

At the same time, the time traveled is equal to $t = 15\,\text{ms} = 1.5T$, therefore the traveled distance is found as $1.5T \equiv 1.5\lambda = 4500\,\text{km}$.

(c) The instantaneous amplitude is calculated relative to V_m at $t = 0\,\text{s}$ as

$$V = V_m \sin(2\pi f t + \phi_0) = V_m \sin\left(0 + \frac{\pi}{4}\right) = \frac{V_m}{\sqrt{2}}$$

while at $t = 15\,\text{ms}$ we find,

$$V = V_m \sin\left(\frac{12\pi}{4} + \frac{\pi}{4}\right) = -\frac{V_m}{\sqrt{2}}$$

In conclusion, the equivalence between time, phase, wavelength is fundamental property of traveling EM waves. Various frequencies of EM waves are essentially the corresponding colours of light, thus they travel at the speed of light. This equivalence is summarized in Fig. 2.13.

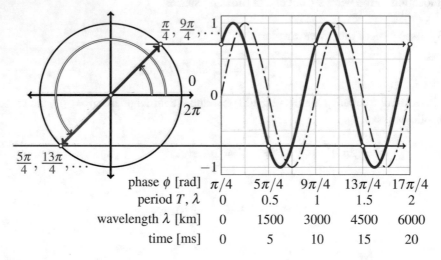

phase ϕ [rad]	$\pi/4$	$5\pi/4$	$9\pi/4$	$13\pi/4$	$17\pi/4$
period T, λ	0	0.5	1	1.5	2
wavelength λ [km]	0	1500	3000	4500	6000
time [ms]	0	5	10	15	20

Fig. 2.13 Example 2.3-3

Exercise 2.4, page 41

1. By definition a sine waveform is periodic with $T = 2\pi$, while the given $v(t)$ is simply the sum $v(t) = V_{DC} + v_{AM} = V_{DC} + \sin \omega t$, thus average value is found by definition:

$$\langle v(t) \rangle = \frac{1}{T} \int_0^T v(t)\, dt = \frac{1}{2\pi} \int_0^{2\pi} [V_{DC} + \sin(\omega t)]\, dt$$

$$= \frac{V_{DC}}{2\pi} \int_0^{2\pi} dt + \frac{1}{2\pi} \int_0^{2\pi} \sin(\omega t)\, dt$$

$$= \left. \frac{V_{DC}}{2\pi} t \right|_0^{2\pi} + \frac{1}{2\pi} \frac{1}{\omega} \int_0^{2\pi} \overset{\nearrow 0}{\sin(x)} \, dx$$

$$\therefore$$

$$\langle v(t) \rangle = V_{DC}$$

where the $\sin \omega t$ integral is solved by change of variables $x = \omega t$. This example illustrates the definition of common-mode (CM), or average value of a sine waveform. Therefore: (a) $\langle v(t) \rangle = -1\,\text{V}$, (b) $\langle v(t) \rangle = 0\,\text{V}$, and (c) $\langle v(t) \rangle = +1\,\text{V}$, we note that this value is found simply by inspection of $v(t)$ expression.

2.

(a) given sin waveform, by inspection we write: $T = 1\,\text{ms}$, $f = 1/T = 1\,\text{kHz}$, $\omega = 2\pi\ \text{krad/s}$, $V_p = 0.5\,\text{V}$, $V_{pp} = 1\,\text{V}$, and average $V_{CM} = -0.5\,\text{V}$. It RMS value is calculated by definition

$$V_{\text{rms}} \overset{\text{def}}{=} \sqrt{\frac{1}{T} \int_0^T |v(t)|^2 \, dt} = \sqrt{\frac{1}{2\pi} \int_0^{2\pi} (V_{CM} + V_p \sin \omega t)^2 \, dt}$$

$$= \sqrt{\frac{1}{2\pi} \int_0^{2\pi} \left[V_{CM}^2 + 2 V_{CM} V_p \sin \omega t + (V_p \sin \omega t)^2 \right] \, dt}$$

$$= \sqrt{\frac{1}{2\pi} \left[\int_0^{2\pi} V_{CM}^2 \, dt + \overset{\nearrow 0}{\int_0^{2\pi} A \sin \omega t \, dt} + \int_0^{2\pi} V_p^2 \sin^2 \omega t \, dt \right]}$$

$$= \sqrt{\frac{1}{2\pi} \left[2\pi\, V_{CM}^2 + \int_0^{2\pi} V_p^2 \sin^2 \omega t \, dt \right]}$$

$$= \sqrt{\frac{1}{2\pi} \left[2\pi\, V_{CM}^2 + V_p^2 \int_0^{2\pi} \sin^2 \omega t \, dt \right]}$$

$$= \sqrt{\frac{1}{2\pi} \left[2\pi\, V_{CM}^2 + V_p^2 \int_0^{2\pi} \frac{1 - \cos(2\omega t)}{2} \, dt \right]}$$

$$= \sqrt{\frac{1}{2\pi} \left[2\pi\, V_{CM}^2 + V_p^2 \left(\frac{t}{2} - \frac{\sin(2\omega t)}{4\omega} \right) \right]_0^{2\pi}}$$

$$= \sqrt{\frac{1}{2\pi} \left[2\pi\, V_{CM}^2 + V_p^2 \frac{2\pi}{2} - \frac{V_p^2}{4\omega} \overset{\nearrow 0}{\sin(2\omega t)} \Big|_0^{2\pi} \right]}$$

$$= |V_{CM}| + \frac{V_p}{\sqrt{2}} = \left(0.5 + \frac{0.5}{\sqrt{2}} \right) \text{V} = 0.25(\sqrt{2} + 2)\text{V}$$

(b) given square waveform, by inspection we write: $T = 1.5\,\text{ms}$, $f = 1/T = 667\,\text{Hz}$, $\omega = 2\pi \times 667\ \text{rad/s}$, $V_p = 0.5\,\text{V}$, $V(min) = -V_p = -0.5\,\text{V}$, $V(max) = +V_p = +0.5\,\text{V}$, and average $V_{CM} = 0$V (this is 50% "on" 50% "off" wave). It RMS value of piecewise linear function is calculated by definition

$$V_{rms} \overset{\text{def}}{=} \sqrt{\frac{1}{T} \int_0^T |v(t)|^2 \, dt} = \sqrt{\frac{1}{T} \left(\int_0^{T/2} V_{max}^2 \, dt + \int_{T/2}^T V_{min}^2 \, dt \right)}$$

$$= \sqrt{\frac{V_p^2}{T} \left(\int_0^{T/2} dt + \int_{T/2}^T dt \right)} = \sqrt{\frac{V_p^2}{T} \left(\frac{T}{2} + T - \frac{T}{2} \right)}$$

$$= V_p = 0.5\text{V}$$

(c) given triangle waveform, by inspection we write: $T = 1$ ms, $f = 1/T = 1$ kHz, $\omega = 2\pi$ krad/s, $V_p = 1$ V, $V(min) = -V_p = -1$ V, $V(max) = +V_p = +1$ V, and average $V_{CM} = 0$ V (this is 50% "on" 50% "off" wave). This piecewise linear function consists of three linear functions:

$$(0, ^T/_4) \qquad (^T/_4, \, ^{3T}/_4) \qquad \qquad (^{3T}/_4, T)$$
$$v_1(t) = V_p \frac{4}{T} t \quad v_2(t) = V_p \left(-\frac{4}{T} t + 2 \right) \quad v_3(t) = V_p \left(\frac{4}{T} t - 4 \right)$$

First, we derive the following three integrals separately,

$$I_1 = V_p^2 \int_0^{T/4} \left(\frac{4}{T} t \right)^2 dt = V_p^2 \frac{16}{T^2} \frac{t^3}{3} \Big|_0^{T/4} = V_p^2 \frac{16}{3 T^2} \frac{T^3}{64} = V_p^2 \frac{T}{12}$$

$$I_2 = V_p^2 \int_{T/4}^{3T/4} \left(-\frac{4}{T} t + 2 \right)^2 dt = V_p^2 \int_{T/4}^{3T/4} \left(\frac{16}{T^2} t^2 - \frac{16}{T} t + 4 \right) dt$$

$$= V_p^2 \left[\frac{16}{T^2} \frac{t^3}{3} \Big|_{T/4}^{3T/4} - \frac{16}{T} \frac{t^2}{2} \Big|_{T/4}^{3T/4} + 4t \Big|_{T/4}^{3T/4} \right]$$

$$= V_p^2 \left[\frac{16}{3 T^2} \frac{1326 T^3}{664} - \frac{16}{2T} \frac{84 T^2}{16} + 4 \frac{2T}{2} \right]$$

$$= V_p^2 \frac{T}{6}$$

$$I_3 = V_p^2 \int_{3T/4}^T \left(\frac{4}{T} t - 4 \right)^2 dt = V_p^2 \int_{3T/4}^T \left(\frac{16}{T^2} t^2 - \frac{32}{T} t + 16 \right) dt$$

$$= V_p^2 \left[\frac{16}{T^2} \frac{t^3}{3} \Big|_{3T/4}^T - \frac{32}{T} \frac{t^2}{2} \Big|_{3T/4}^T + 16t \Big|_{3T/4}^T \right]$$

$$= V_p^2 \left[\frac{16}{3T^2} \left(T^3 - \frac{27 T^3}{64} \right) - \frac{16}{T} \left(T^2 - \frac{9T^2}{16} \right) + 16 \left(T - \frac{3T}{4} \right) \right]$$

$$= V_p^2 \frac{T}{6}$$

Now, RMS value of piecewise linear function is calculated by definition

$$V_{rms} \overset{def}{=} \sqrt{\frac{1}{T} \int_0^T |v(t)|^2 \, dt} = \sqrt{\frac{V_p^2}{T} (I_1 + I_2 + I_3)} = \sqrt{\frac{V_p^2}{T} \left(\frac{T}{12} + \frac{T}{6} + \frac{T}{12} \right)}$$

$$= \sqrt{\frac{V_p^2}{\not{T}} \frac{\not{T}}{3}} = \frac{V_p}{\sqrt{3}} = \frac{1}{\sqrt{3}} \, V$$

(d) given sawtooth waveform, by inspection we write: $T = 0.75\,\text{ms}$, $f = 1/T = 1.33\,\text{kHz}$, $\omega = 2\pi \times 1.33\,\text{krad/s}$, $V(min) = -0.75\,\text{V}$, $V(max) = +0.25\,\text{V}$, and average $V_{CM} = -0.25\,\text{V}$ (this is 50% "on" 50% "off" wave). AC component of this piecewise linear function consists of only one linear functions whose average is $V_{CM} = -0.25\,\text{V}$, therefore relative to its average AC component's amplitude is $V_p = \pm 0.5\,\text{V}$, i.e.

$$(0, T); \quad v_{AM}(t) = V_p \left(\frac{2}{T} t - 1 \right)$$

First, we can derive RMS for the AC component only, as the following integral,

$$I_1 = V_p^2 \int_0^T \left(\frac{2}{T} t - 1 \right)^2 dt = V_p^2 \int_0^T \left(\frac{4}{T^2} t^2 - \frac{4}{T} t + 1 \right)$$

$$= V_p^2 \left(\frac{4}{3T^2} t^3 - \frac{4}{2T} t^2 + t \right) \Big|_0^T = V_p^2 \frac{T}{3}$$

Now, RMS value of piecewise linear function is calculated by definition

$$V_{rms} \overset{def}{=} \sqrt{\frac{1}{T} \int_0^T |v(t)|^2 \, dt} = \sqrt{\frac{1}{\not{T}} \left(V_{CM}^2 \not{T} + V_p^2 \frac{\not{T}}{3} \right)}$$

$$= \sqrt{V_{CM}^2 + \frac{V_p^2}{3}} = \sqrt{\frac{1}{4 \times 4} + \frac{1}{4 \times 3}} = \frac{\sqrt{7}}{4\sqrt{3}} \, V$$

We note that, in the case of $V_{CM} = 0$ the RMS of sawtooth waveform is also $V_p/\sqrt{3}$ as for the equivalent triangle waveform.

3. We recall that definite integral of a function is numerically equal to the surface area under that function. For this reason, definite integral of a piecewise linear function is found simply by adding areas of rectangular shapes under the function. Thus we write,

   (a) by inspection of $i(t)$, we write period $T = 30\,\text{ms}$ and frequency $f = 1/T = 33.33\,\text{Hz}$,
   (b) by inspection of $i(t)$, we write minimal $i(t)_{min} = \text{mA}$ and maximal $i(t)_{max} = 4\,\text{mA}$,
   (c) "AC" refers to pulse amplitude, i.e. $I_m = i(t)_{max} - i(t)_{min} = 3\,\text{mA}$,
   (d) the geometrical interpretation of an average is the height of rectangle spread over the full interval and whose area equals to the area of the function within the same interval, see Fig. 2.14.

By adding square units in (0 s, 20 ms) interval, the total surface is

$$\underbrace{(4\,\text{mA} \times 10\,\text{ms})}_{A} + \underbrace{(1\,\text{mA} \times 10\,\text{ms})}_{B} = \underbrace{50\ \text{square units}}_{C}$$

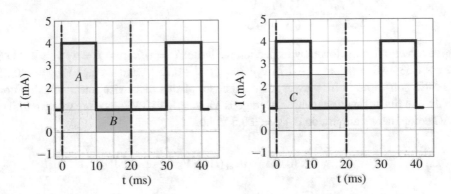

Fig. 2.14   Example 2.4-3(d)

When the same rectangle area is equally spread over the full interval, i.e. area C, its height is calculated as

$$\frac{50\ \text{square units}}{20\,\text{ms}} = 2.5\,\text{mA}$$

which is therefore the current average in (0 s, 20 ms) interval only.

(e) "common mode" I_{CM} of $i(t)$ is as another way of saying "average" of the full $i(t)$. This is a periodic signal with $T = 30\,\text{ms}$, therefore it is sufficient to calculate its average over one period only. By using the geometrical addition of square unit areas in (0 s, 30 ms) interval, the total surface is

$$(4\,\text{mA} \times 10\,\text{ms}) + (1\,\text{mA} \times 20\,\text{ms}) = 60\ \text{square units}$$

When the same rectangle area is equally spread over the full interval, its height is calculated as

$$\frac{60\ \text{square units}}{30\,\text{ms}} = 2\,\text{mA}$$

which is by definition average value of this square signal. As an exercise, let us use the integral definition to again calculate this average as

$$\langle i(t) \rangle = \frac{1}{T} \int_0^T i(t)\, dt = \frac{1}{30\,\text{ms}} \int_0^{30\,\text{ms}} i(t)\, dt$$

$$= \frac{1}{30\,\text{ms}} \left[\int_0^{10\,\text{ms}} 4\,\text{mA}\, dt + \int_{10\,\text{ms}}^{30\,\text{ms}} 1\,\text{mA}\, dt \right]$$

$$= \frac{1}{30 \, \text{ms}} \, (10 \, \text{ms} \times 4 \, \text{mA} + 20 \, \text{ms} \times 1 \, \text{mA})$$

$$= \frac{60 \, \text{ms} \times \text{mA}}{30 \, \text{ms}} = 2 \, \text{mA}$$

(f) a time domain signal is the sum of its DC (i.e. average) and AC signal components. Graphical interpretation of this sum is (Fig. 2.15)

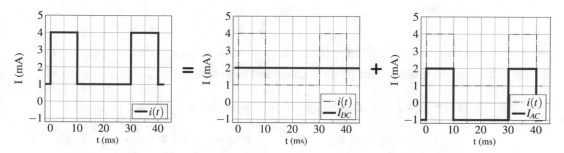

Fig. 2.15 Example 2.4-3(f)

where, after removing DC component $i_{ADC} = 2 \, \text{mA}$, the remaining i_{AC} signal component is found between $-1 \, \text{mA}$ and $2 \, \text{mA}$ levels. As a consequence, RMS (i.e. "root-mean-square") of AC component only is found by definition as

$$I_{\text{rms}} \stackrel{\text{def}}{=} \sqrt{\frac{1}{T} \int_0^T |i(t)|^2 \, dt} = \sqrt{\frac{(2 \, \text{mA})^2 \times 10 \, \text{ms} + (-1 \, \text{mA})^2 \times 20 \, \text{ms}}{30 \, \text{ms}}}$$

$$= \sqrt{2} \, \text{mA}$$

(g) similarly, RMS of the complete square waveform $i(t)$ is calculated as

$$I_{\text{rms}} = \sqrt{\frac{(4 \, \text{mA})^2 \times 10 \, \text{ms} + (1 \, \text{mA})^2 \times 20 \, \text{ms}}{30 \, \text{ms}}} = \sqrt{6} \, \text{mA}$$

Alternatively, the total RMS value could be calculated as the sum of the RMS of the DC and AC components, as

$$I_{\text{rms}} = \sqrt{I_{DC}^2 + I_{\text{rms}}^2 (AC)} = \sqrt{2^2 + \left(\sqrt{2} \right)^2} \, \text{mA} = \sqrt{6} \, \text{mA}$$

which gives the same result because the RMS value of a DC level is a constant number, which is to say that its RMS is the same number.

4. By inspection of the given graph, we count that high frequency carrier waveform takes one time unit to finish 10 full periods. For that reason, its frequency is $f_C = 1/10$ relative to the time unit. At the same time, the embedded low frequency sinusoid takes one time unit to finish one period, thus its frequency is $f_M = 1/1$ relative to the time unit.

(a) one time unit equals 1 ms, therefore $T_C = 100\,\mu s$ and $T_M = 1\,ms$, i.e. $f_C = 10\,kHz$ and $f_M = 1\,kHz$.

(b) one time unit equals $1\,\mu s$, therefore $T_C = 100\,ns$ and $T_M = 1\,\mu s$, i.e. $f_C = 10\,MHz$ and $f_M = 1\,MHz$.

(c) one time unit equals 1 ns, therefore $T_C = 100\,ps$ and $T_M = 1\,ns$, i.e. $f_C = 10\,GHz$ and $f_M = 1\,GHz$

5. $E = P \times t$, therefore $P_1 = 1.728 \times 10^6\,J$, and $P_1 = 60 \times 10^3\,J$

6. By definition, the instantaneous power is calculated as product,

$$p(t) = v(t) \times i(t) = 220 \sin(\omega t + {}^{5\pi}/_6)V \times 10 \sin(\omega t - {}^{2\pi}/_3)A$$

$$= 2200 \times \frac{1}{2}(\cos(\omega t + {}^{5\pi}/_6 - \omega t + {}^{2\pi}/_3) - \cos(\omega t + {}^{5\pi}/_6 + \omega t - {}^{2\pi}/_3))$$

$$= 1100 \times \left[\cos {}^{3\pi}/_2{}^{\,0} - \cos(2\omega t + {}^{\pi}/_6) \right] = -1\,100 \cos(2\omega t + {}^{\pi}/_6)$$

Average power is calculated by definition as

$$\langle P \rangle = \frac{1}{T}\int_0^T p(t)\,dt = -1\,100\,\frac{1}{T}\int_0^T \cos(2\omega t + {}^{\pi}/_6)\,dt {}^{\nearrow 0} = 0$$

and we conclude that because $\langle P \rangle = 0$ this must be purely reactive circuit.

7. Calculated in gain units, the total power gain is then the product of two gains, $G = 5 \times 8 = 40$. However, if gain of each stage is given in dB, then the total power gain is calculated as the sum $G_{dB} = 6.99 + 9.03 = 16\,dB$. We can easily verify that indeed $10 \log(40) = 16\,dB$.

8. We convert the received power into dBm units as

$$P_2 = 10 \log \frac{P_2}{1\,mW} = 10 \log \frac{5\,pW}{1\,mW} = -83\,dBm$$

Therefore the total signal power loss is the difference between the power levels at the end $P_2 = -83\,dBm$ and at the beginning, i.e. $P_1 = +30\,dBm$, that is,

$$A = P_2 - P_1 = -83\,dBm - 30\,dBm = -113\,dBm$$

Exercise 2.5, page 43

1. First, we factorize transfer given function to the product of its basic forms.

$$H(\omega) = \frac{100}{j\omega + 10} = 100\frac{1}{j\omega + 10} = \frac{100}{10}\frac{1}{1 + j\dfrac{\omega}{10}} = 10\frac{1}{1 + j\dfrac{\omega}{10}}$$

By inspection, we find one pole at $\omega_0 = 10$ and "DC gain" $A = 10$. Logarithmic form of gain function is

$$20 \log H(\omega) = 20 \log \left(10 \, \frac{1}{1 + j \dfrac{\omega}{10}} \right)$$

$$= \underbrace{20 \log 10}_{\text{DC gain}} + \underbrace{20 \log(1)}^{0} \underbrace{- 20 \log \left(1 + j \, \frac{\omega}{10} \right)}_{\text{pole, } \omega_0 = 10}$$

$$\underbrace{}_{\textcircled{1}} \qquad \underbrace{}_{\textcircled{4}}$$

The first term is a constant term whose value is 20 dB (DC gain, i.e. when $\omega = 0$), the second term is logarithm of denominator (i.e. of 1) and it equals zero. Third term sets the function's pole. In the math review section (see Chap. 1) we reviewed basic techniques to calculate limits and angles for each term of gain and phase functions, thus here we reuse already calculated results. Then, its corresponding phase function is

$$\angle H(\omega) = \underbrace{0°}_{\substack{\text{DC, real} \\ \textcircled{1}}} \underbrace{- \arctan \frac{\omega}{10}}_{\substack{\text{pole, } \omega_0 = 10 \\ \textcircled{4}}}$$

With the prior knowledge of the basic functions and their piecewise linear forms, gain and phase Bode plots are created as simple superposition of linear sections both in the gain and phase graphs, see Fig. 2.16.

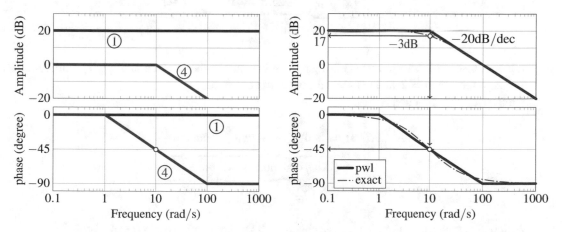

Fig. 2.16 Example 2.5-1

With piecewise linear approximation we are able to perform rapid hand analysis of complicated transfer functions. Error made by this approximation is no more than 3 dB for gain function. However, the overall transfer function form is correct, and here shows a classic first order (i.e. only one pole) LPF response.

2. Factorization of the given function shows that

$$H(\omega) = \frac{100\left(1 + j\frac{\omega}{10}\right)}{j\omega + 10} = \frac{100\left(1 + j\frac{\omega}{10}\right)}{10\left(1 + j\frac{\omega}{10}\right)} = 10 = 20\,\text{dB}$$

This case is interesting because it demonstrates "pole-zero" cancelling technique. Even though there is one pole and one zero in the transfer function, they happen to be set at the same frequency ω_0, therefore they cancel, and in this case we are left with the frequency independent transfer function whose gain is, theoretically, constant from DC to infinite frequency. Phase of a real number is zero, again independent of frequency.

3. Factorization of the given function shows its basic terms as

$$H(s) = s^2 + 22s + 40 = s^2 + 20s + 2s + 40 = s(s + 20) + 2(s + 20)$$
$$= (s + 20)(s + 2)$$

$$\therefore$$

$$H(\omega) = (j\omega + 20)(j\omega + 2) = 20\left(1 + j\frac{\omega}{20}\right)2\left(1 + j\frac{\omega}{2}\right)$$
$$= 40\left(1 + j\frac{\omega}{2}\right)\left(1 + j\frac{\omega}{20}\right)$$

It is now straightforward to conclude there is DC gain of 40 and there are two "zeros", first at $\omega_1 = 2$ and the second at one decade higher $\omega_2 = 20$.
Logarithmic form of the gain function is

$$20\log H(\omega) = \underbrace{20\log 40}_{\substack{\text{DC gain} \\ \textcircled{1}}} + \underbrace{20\log\left(1 + j\,\frac{\omega}{2}\right)}_{\substack{\text{zero, } \omega_0=2 \\ \textcircled{2}}} + \underbrace{20\log\left(1 + j\,\frac{\omega}{20}\right)}_{\substack{\text{zero, } \omega_0=20 \\ \textcircled{2}}}$$

Which consists of three terms: (1) DC gain that is result of the factored multiplying number, (2) first zero term at $\omega = 2$, and (3) second zero term at $\omega = 20$.
Similarly, phase function is

$$\angle H(\omega) = \underbrace{0°}_{\substack{\text{DC, real} \\ \textcircled{1}}} + \underbrace{\arctan\frac{\omega}{2}}_{\substack{\text{pole, } \omega_0=2 \\ \textcircled{2}}} + \underbrace{\arctan\frac{\omega}{20}}_{\substack{\text{pole, } \omega_0=20 \\ \textcircled{2}}}$$

Which also consists of three terms associated with DC gain and two zero terns.
Overall gain and phase functions are produced by summing piecewise linear sections of gain and phase functions, as shown in these two plots. We note that the gain slope starts as $+20\,\text{dB/dec}$ due to the first zero, then it increases to $+40\,\text{dB/dec}$ after slope of the second zero is added. Similarly, each zero term introduces linear phase shift from $0°$ to $90°$ spread over two decades and centred at its zero frequency. Because there are two zero terms, at frequency one decade above the second

zero the total phase shift is therefore doubled, that is to say, the total phase shifts from 0° to 180°, see Fig. 2.17.

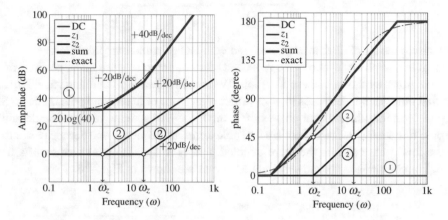

Fig. 2.17 Example 2.5-3

4. Factorization of the given function shows its basic terms as

$$H(s) = \frac{100}{(s + 10)(s + 100)}$$

$$\therefore$$

$$H(\omega) = \frac{100}{(j\omega + 10)(j\omega + 100)} = \frac{100}{10(1 + j\frac{\omega}{10})100(1 + j\frac{\omega}{100})}$$

$$= \frac{100}{1000(1 + j\frac{\omega}{10})(1 + j\frac{\omega}{100})} = \frac{1}{10}\frac{1}{(1 + j\frac{\omega}{10})(1 + j\frac{\omega}{100})}$$

It is now straightforward to conclude there is DC gain of 0.1 and there are two "poles", first at $\omega_1 = 10$ and the second at one decade higher $\omega_2 = 100$.
Logarithmic form of the gain function is

$$20 \log H(\omega) = \underbrace{20 \log 0.1}_{\substack{\text{DC gain} \\ \textcircled{1}}} + \underbrace{20 \log \left(1 + j\frac{\omega}{10}\right)}_{\substack{\text{pole, } \omega_0 = 10 \\ \textcircled{4}}} + \underbrace{20 \log \left(1 + j\frac{\omega}{100}\right)}_{\substack{\text{pole, } \omega_0 = 100 \\ \textcircled{4}}}$$

Which consists of three terms: (1) DC gain that is result of the factored multiplying number, (2) first pole term at $\omega = 10$, and (3) second pole term at $\omega = 100$.
Similarly, phase function is

$$\angle H(\omega) = \underbrace{0°}_{\substack{\text{DC, real} \\ \textcircled{1}}} \underbrace{- \arctan \frac{\omega}{10}}_{\substack{\text{pole, } \omega_0=10 \\ \textcircled{4}}} \underbrace{- \arctan \frac{\omega}{100}}_{\substack{\text{pole, } \omega_0=100 \\ \textcircled{4}}}$$

Which also consists of three terms associated with DC gain and two pole terns.

Overall gain and phase functions are produced by summing piecewise linear sections of gain and phase functions, as shown in these two plots. We note that the gain slope starts as -20 dB/dec due to the first pole, then it decreases to -40 dB/dec after slope of the second pole is added. Similarly, each pole term introduces linear phase shift from $0°$ to $-90°$ spread over two decades and centred at its pole frequency. Because there are two pole terms, at frequency one decade above the second pole the total phase shift is therefore doubled, that is to say, the total phase shifts from $0°$ to $-180°$. For comparison, the exact simulated curves are shown in Fig. 2.18.

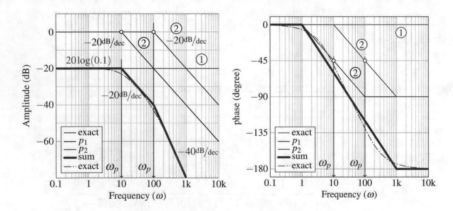

Fig. 2.18 Example 2.5-4

5. We note that $H(s)$ is a rational polynomial function, where numerator contains one $j\omega$ term, thus there is one zero in the transfer function. The denominator is a second order polynomial, and as a consequence there are two poles of $H(s)$. Factorization of the given function gives

$$H(j\omega) = 100 \frac{s+1}{s^2 + 110\,s + 1000} = 100 \frac{s+1}{s^2 + 10\,s + 100\,s + 1000}$$

$$= 100 \frac{s+1}{s\,(s+10) + 100\,(s+10)} = 100 \frac{s+1}{(s+10)(s+100)}$$

$$= \cancel{100} \frac{1 + \dfrac{s}{1}}{10 \left(1 + \dfrac{s}{10}\right) \cancel{100} \left(1 + \dfrac{s}{100}\right)} = \frac{1}{10} \frac{1 + j\dfrac{\omega}{1}}{\left(1 + j\dfrac{\omega}{10}\right)\left(1 + j\dfrac{\omega}{100}\right)}$$

where its logarithmic form is

$$20 \log H(j\omega) = 20 \log \left[\frac{1}{10} \frac{1 + j\dfrac{\omega}{1}}{\left(1 + j\dfrac{\omega}{10}\right)\left(1 + j\dfrac{\omega}{100}\right)} \right]$$

$$= \underbrace{20\log\frac{1}{10}}_{\substack{\text{DC gain} \\ \text{①}}} + \underbrace{20\log\left(1+j\frac{\omega}{1}\right)}_{\substack{\text{zero},\omega_0=1 \\ \text{②}}} - \underbrace{20\log\left(1+j\frac{\omega}{10}\right)}_{\substack{\text{pole},\omega_0=10 \\ \text{③}}} - \underbrace{20\log\left(1+j\frac{\omega}{100}\right)}_{\substack{\text{pole},\omega_0=100 \\ \text{④}}}$$

Similarly, the phase plot is created as the sum of the corresponding linear terms,

$$\angle H(\omega) = \underbrace{0°}_{\substack{\text{DC gain} \\ \text{①}}} + \underbrace{\arctan\frac{\omega}{1}}_{\substack{\text{zero},\omega_0=1 \\ \text{②}}} - \underbrace{\arctan\frac{\omega}{10}}_{\substack{\text{pole},\omega_0=10 \\ \text{③}}} - \underbrace{\arctan\frac{\omega}{100}}_{\substack{\text{pole},\omega_0=100 \\ \text{④}}}$$

Once the gain and phase logarithmic forms are factorized, first we plot graphs all four terms (both amplitude and phase), then we create plots of the total sums, see Fig. 2.19.

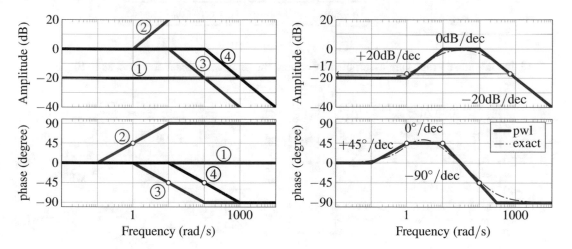

Fig. 2.19 Example 2.5-5

6. Given transfer function is a second order function with complex poles. It can be rewritten as

$$H(s) = 100\frac{(s+10)}{s^2+40s+10^4} = 100\frac{10}{10^4}\frac{1+\dfrac{s}{10}}{\left(\dfrac{s}{10^2}\right)^2+\left(\dfrac{40}{10^4}\right)s+1}$$

Discriminant of the quadratic equation is

$$\Delta = 40^2 - 4 \times 1 \times 10^4 = -38.4 \times 10^3 < 0$$

therefore, indeed this second order function has complex poles. In accordance to with its theoretical form of rational functions with complex roots found in textbook, by inspection we write

$$\omega_0^2 = 10^4 \quad \therefore \quad \omega_0 = 100$$

$$2\xi\,\omega_0 = 40 \quad \therefore \quad \xi = \frac{40}{2 \times 100} = 0.2$$

and conclude that $H(\omega_0) = 8\,\text{dB}$ at $\omega_m = 95.9\,\text{rad/s}$. Logarithmic form is

$$20\log H(j\omega) = 20\log\left[\frac{1}{10}\left(1 + \frac{s}{10}\right)\frac{10^4}{s^2 + 40\,s + 10^4}\right]$$

$$= \underbrace{-20\log 10}_{\substack{\text{DC gain}\\ \text{①}}} + \underbrace{20\log\left(1 + j\,\frac{\omega}{10}\right)}_{\substack{\text{zero},\omega_0=10\\ \text{②}}} + \underbrace{20\log\left(\frac{10^4}{s^2 + 40\,s + 10^4}\right)}_{\substack{\text{two zeros},\omega_0=100,\xi=0.2\\ \text{③}}}$$

Similarly, the phase plot is created as the sum of the corresponding linear terms,

$$\angle H(\omega) = \underbrace{0°}_{\substack{\text{DC gain}\\ \text{①}}} + \underbrace{\arctan\frac{\omega}{10}}_{\substack{\text{zero},\omega_0=10\\ \text{②}}} - \underbrace{\arctan\frac{-0.4\,(\omega/100)}{1 - (\omega/100)^2}}_{\substack{\text{two zeros},\omega_0=100,\xi=0.2\\ \text{③}}}$$

Gain and phase graphs are assembled by adding the piecewise linear sections, as in Fig. 2.20.

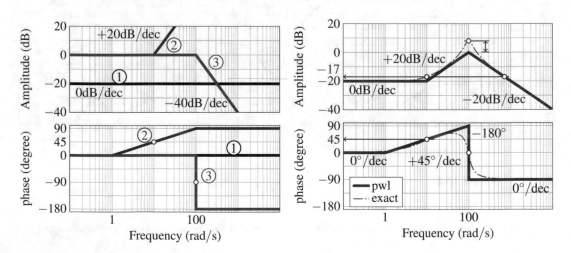

Fig. 2.20 Example 2.5-6

7. By inspection of the graph, we write

(a) DC gain: for frequencies lower than 1 Hz, the gain is 0 dB, i.e. 1, which is the function's multiplication factor, we write

$$A_0 = 1$$

(b) Zero: after $f = 1\,\text{Hz}$ there is a positive slope, $+20\,\text{dB/dec}$, which the consequence of the transfer function's zero at $f = 1\,\text{Hz}$, we write the zero term as

$$\left(1 + \frac{s}{1}\right)$$

(c) Pole: at $f = 10$ Hz the initial positive slope is cancelled and the gain became constant. We conclude that at $f = 10$ Hz there must be a pole to give -20 dB/dec slope and therefore cancel the further gain increase. We write this pole term as

$$\left(1 + \frac{s}{10}\right)$$

(d) Pole: at $f = 100$ Hz and after the transfer function starts its negative slope, -20 dB/dec, an indication of pole at $f = 100$ Hz. We write this pole term as

$$\left(1 + \frac{s}{100}\right)$$

The total transfer function is assembled by writing the zero terms as factors of numerator, and pole terms as factors of denominator as,

$$H(s) = \frac{A_0^{\,1}\left(1 + \frac{s}{1}\right)}{\left(1 + \frac{s}{10}\right)\left(1 + \frac{s}{100}\right)} = \frac{1000}{1000}\frac{(1 + s)}{\left(1 + \frac{s}{10}\right)\left(1 + \frac{s}{100}\right)}$$

$$= \frac{1000\,(1 + s)}{10\left(1 + \frac{s}{10}\right)100\left(1 + \frac{s}{100}\right)} = \frac{1000\,(1 + s)}{(10 + s)\,(100 + s)}$$

where we note that DC gain is $A_0 = 1$, not the 1000 term in the numerator that is consequence of algebraic forms. We keep in mind that to make plot with horizontal axis in Hz instead of rad, each 's' term is replaced by '$s/(2\pi)$', as shown in the problem figure.

8. First few terms of this waveform are added as

$$v(t) = \frac{4}{\pi}\left[\sum_{n=1,3,5,\ldots} \frac{1}{n} \sin\left(\frac{n}{2}t\right)\right]$$

$$= \frac{4}{\pi}\left[\frac{1}{1}\sin\left(\frac{1}{2}t\right) + \frac{1}{3}\sin\left(\frac{3}{2}t\right) + \frac{1}{5}\sin\left(\frac{5}{2}t\right) + \frac{1}{7}\sin\left(\frac{7}{2}t\right) + \cdots\right]$$

In accordance with Fourier theorem, a sum of single sinusoidal waveforms ("tones") can be used to approximate an arbitrary function, for example, the waveforms of speech. In this simple exercise, it is a square wave being created by adding more and more (in theory infinitely) terms in the sum, Fig. 2.21. The lowest frequency is called fundamental frequency while, strictly speaking, "harmonics" are whole number multiples of the fundamental frequency.

Disturbing any of terms results in distortion of the original waveform. Signal distortions are consequence of the external interference that affects one or more tones in the signal spectrum. Systematic removal of a group of tones is called filtering, done by LPF or HPF, whose main function is in effect to "shape frequency spectrum".

Fig. 2.21 Example 2.5-8

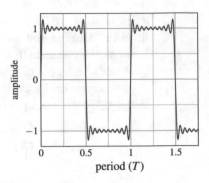

9. First few terms of sawtooth waveform are added as

$$v(t) = \frac{2}{\pi} \left[\sum_{n=1,2,3,..} \frac{1}{n} \sin(n\,t) \right]$$

$$= \frac{2}{\pi} \left[\frac{1}{1} \sin(t) + \frac{1}{2} \sin(2\,t) + \frac{1}{3} \sin(3\,t) + \frac{1}{4} \sin(4\,t) + \cdots \right]$$

Disturbing any of the sum terms results in distortion of the original waveform. For example, removing only one of the high end tones is visible as if HF noise were added ("disto1"), while removing the fundamental tone produces violent distortion ("disto2"), see Fig. 2.22. Understanding frequency spectrum and signal structure is essential for RF circuit design and signal processing.

Fig. 2.22 Example 2.5-9

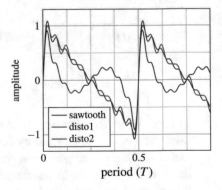

Basic Behavioural and Device Models

3

At the system level, analysis and design of electronic circuits are based on the set of fundamental building blocks represented by their behavioural models. In the initial phase of the design, it is important to validate the intended functionality of the overall circuit at the level of mathematics, without regard for the implementation details. Therefore, the knowledge of functionality of fundamental devices, namely, a switch, voltage and current sources, and RLC devices as well as their respective impedances is the prerequisite for the next phase in the design process.

3.1 Important to Know

1. Fundamental I/O definitions

$$v_{out} = A_v \times v_{in} \quad \therefore \quad A_v \overset{\text{def}}{=} \frac{v_{out}}{v_{in}} \quad \left[\frac{\text{V}}{\text{V}}\right] \tag{3.1}$$

$$i_{out} = A_i \times i_{in} \quad \therefore \quad A_i \overset{\text{def}}{=} \frac{i_{out}}{i_{in}} \quad \left[\frac{\text{A}}{\text{A}}\right] \tag{3.2}$$

$$v_{out} = R \times i_{in} \quad \therefore \quad R \overset{\text{def}}{=} \frac{v_{out}}{i_{in}} \quad \left[\frac{\text{V}}{\text{A}} \overset{\text{def}}{=} \Omega\right] \tag{3.3}$$

$$i_{out} = g_m \times v_{in} \quad \therefore \quad g_m \overset{\text{def}}{=} \frac{1}{R} = \frac{i_{out}}{v_{in}} \quad \left[\frac{\text{A}}{\text{V}} \overset{\text{def}}{=} \text{S}\right] \tag{3.4}$$

2. Physical RLC component sizing

$$R_{DC} = \rho \frac{l}{S} \quad ; \quad C = \epsilon \frac{S}{d} \quad ; \quad L = \mu \frac{AN^2}{l} \tag{3.5}$$

3. DC and AC resistances

$$R_{DC}\Big|_{\mathscr{A}} = \frac{V_A}{I_A} \qquad R_{AC}\Big|_{\mathscr{A}} = \frac{dV_0}{dI_0}\bigg|_{\mathscr{A}} \approx \frac{\Delta V_0}{\Delta I_0}\bigg|_{\mathscr{A}} \tag{3.6}$$

© Springer Nature Switzerland AG 2021
R. Sobot, *Wireless Communication Electronics by Example*,
https://doi.org/10.1007/978-3-030-59498-5_3

4. Impedances

$$Z_R \stackrel{\text{def}}{=} \frac{V_R}{I_R} = R \neq \overset{\cdot}{f}(\omega) \Rightarrow R \in \mathbb{R}, \angle Z_R = 0° \tag{3.7}$$

$$Z_C(j\omega) \stackrel{\text{def}}{=} \frac{V_C}{I_C} = \frac{1}{j\omega C} = -j\frac{1}{\omega C} \Rightarrow Z_C \in \mathbb{C}, \angle Z_C = -90° \tag{3.8}$$

$$Z_L(j\omega) \stackrel{\text{def}}{=} \frac{V_L}{I_L} = j\omega L \Rightarrow Z_L \in \mathbb{C}, \angle Z_L = +90° \tag{3.9}$$

5. Time constants,

$$\tau = RC \quad \text{and}, \quad \tau = \frac{L}{R} \tag{3.10}$$

3.2   Exercises

3.1 *   Linear Resistor

1. A typical bonding wire has a diameter of $d = 75\,\mu\text{m}$. Calculate the wire length if its resistance is $R_{DC} = 1\,\Omega$, and it is made of:

   (a) copper, $\rho_{Cu} = 1.68 \times 10^{-8}\,\Omega\text{m}$,

   (b) Al, $\rho_{Al} = 2.65 \times 10^{-8}\,\Omega\text{m}$,

   (c) iron, $\rho_{Fe} = 9.70 \times 10^{-8}\,\Omega\text{m}$,

   (d) silicon, $\rho_{Si} = 6.4 \times 10^{2}\,\Omega\text{m}$.

2. Given the V/I transfer function, see Fig. 3.1, what basic electrical component functionality it represents? Assume that the vertical axis is scaled in steps of 1 V, and the horizontal axis in steps of:

   (a) 1 mA,   (b) 10 μA,   (c) 1 μA,

Fig. 3.1 Example 3.1-2

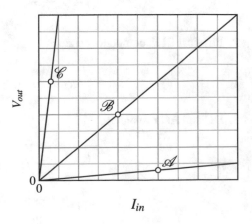

3. Given a 1 mA DC current flowing through a resistor R, calculate its dissipated power P if voltage V_R equals

(a) 1 V,   (b) 10 V,

(c) 100 V,   (d) 1 kV.

4. Given a 1 mA DC current flowing through a resistor, calculate its dissipated power P if R equals:

(a) 1 kΩ,   (b) 10 kΩ,

(c) 100 kΩ,   (d) 1 MΩ.

5. Given a 1 kΩ resistor, calculate the DC voltage at its terminals if its dissipated power P equals:

(a) 1 mW,   (b) 10 mW,

(c) 100 mW,   (d) 1 W.

6. Given a 1 kΩ resistor, calculate the DC current at its terminals if its dissipated power P equals:

(a) 1 mW,   (b) 10 mW,

(c) 100 mW,   (d) 1 W.

7. Given voltage/current waveforms

$$v(t) = 2\text{V} \sin(\omega t)$$

$$i(t) = 1.5\,\text{mA} \sin(\omega t)$$

at resistor terminals, calculate:

(a) average power dissipated,

(b) its resistance,

(c) equivalent DC voltage that would generate the same power level as the average power calculated in Question (a).

3.2 *   Linear Capacitor

1. Calculate the impedance of a capacitor in the following two limiting cases:

(a) $\lim_{f \to DC} Z_C$,   (b) $\lim_{f \to \infty} Z_C$.

2. Given a 159nF capacitor, calculate its impedance Z_C at the following frequencies:

(a) 1 GHz,   (b) 100 MHz,

(c) 1 MHz,   (d) 1 kHz.

3. A rotary plate capacitor consists of five pairs of semicircular metallic plates, five static plates, and five rotating plates, see Fig. 3.2. All static plates are connected to the same potential V_1, and all rotating plates are connected to the same potential V_2.
Assuming the initial position of the capacitor to be when the plates are fully overlapped, calculate the total capacitance when the plates are:

(a) at the initial position,

(b) when the rotating plates are rotated by $\alpha = \pi/4$, and

(c) when the rotating plates are rotated by $\alpha = \pi/2$.

Data: The plate radius is $r = 10\,\mathrm{mm}$, the distance between each two neighbouring plates is $d = 1\,\mathrm{mm}$, and they are separated by air, vacuum permittivity $\epsilon_0 = 8.854 \times 10^{-12}\,\mathrm{F/m}$.

Fig. 3.2 Example 3.2-3

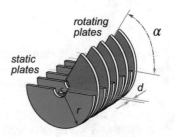

4. Given a $f = 10\,\mathrm{MHz}$ operating frequency, calculate the capacitance to create the following impedances:

(a) $1\,\Omega$,   (b) $100\,\Omega$,   (c) $1\,\mathrm{k}\Omega$.

5. In reference to graph in Fig. 3.3 showing impedance versus frequency of a realistic capacitor, estimate the values of the capacitance C and its parasitic inductance L.
    Write complex impedance Z_C in two regions, when

(a) $f < 20\,\mathrm{MHz}$,

(b) $f > 200\,\mathrm{MHz}$,

and calculate its module and phase.

Fig. 3.3 Example 3.2-5

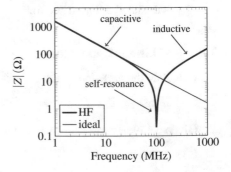

6. For the given voltage waveform, sketch the graph of its corresponding current $i(t)$ across, for example, a $C = 3\,\mu\mathrm{F}$ capacitor (Fig. 3.4).

Fig. 3.4 Example 3.2-6

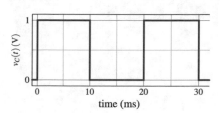

3.3 *   Linear Inductor

1. Calculate the impedance of an inductor in the following two limiting cases:

    (a) $\lim_{f \to DC} Z_L$,   (b) $\lim_{f \to \infty} Z_L$.

2. Given a $(1/2\pi)\,\mathrm{mH} \approx 159\,\mu\mathrm{H}$ inductor, calculate its impedance Z_L at the following frequencies:

    (a) 1 kHz,   (b) 1 MHz,

    (c) 100 MHz,   (d) 1 GHz.

3. Design a coil inductor, see Fig. 3.5, with air core whose inductance is Lr. Available wire has a radius a, and in the given space, we can accommodate a coil radius of r.
   Data: $L = 3.3\,\mu\mathrm{m}$, $a = 100\,\mu\mathrm{m}$, $r = 2.5\,\mathrm{mm}$, $\mu_0 = 4\pi \times 10^{-7}\,\mathrm{H/m}$.

Fig. 3.5 Example 3.3-3

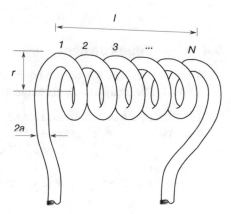

4. Given a 10 MHz operating frequency, calculate the inductance L to create the following impedances:

    (a) 1 Ω,   (b) 100 Ω,   (c) 1 kΩ.

5. In reference to graph in Fig. 3.6 showing impedance versus frequency of a realistic inductor, estimate the values of the inductance C and its parasitic capacitance C.
   Write complex impedance Z_L in two regions, when (a) $f < 100\,\mathrm{MHz}$, (b) $f > 1\,\mathrm{GHz}$, and calculate its module and phase.

Fig. 3.6 Example 3.3-5

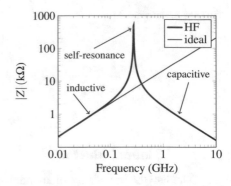

6. For the given current waveform, see Fig. 3.7, sketch the graph of its corresponding voltage $v(t)$ across an inductor L. **Data:** $L = 3\,\mu\text{H}$.

Fig. 3.7 Example 3.3-6

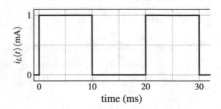

3.4 **   Time Constant

1. Derive an expression for the equivalent impedance Z_{RL} if a resistor $R = 1\,\text{k}\Omega$ and an inductor $L = (1/20\pi)\,\text{mH}$ are connected in series.

   (a) Calculate the absolute value of the impedance $|Z_{RL}(\omega)|$ and the associated phase $\varphi_{RL}(\omega)$ at the following two frequencies:
   (1) $\omega_1 = 2\pi\,f_1 = 2\pi \times 1\,\text{MHz}$,
   (2) $\omega_2 = 2\pi\,f_2 = 2\pi \times 10\,\text{MHz}$,
   (3) $\omega_3 = 2\pi\,f_2 = 2\pi \times 100\,\text{MHz}$.

   (b) What are limits of $|Z_{RL}(\omega)|$ and $\varphi_{RL}(\omega)$ when:
   (1) $\omega \to 0$ (i.e. at zero frequency, a.k.a. *DC*),
   (2) $\omega \to \infty$ (i.e. at very high frequencies).

2. Repeat the calculations from Question 1, but this time in the case of parallel RL network.

3. Derive an expression for the equivalent impedance Z_{RC} if a resistor $R = 100\,\Omega$ and a capacitor $C = (1/2\pi)\,\mu\text{F}$ are connected in series.

   (a) Calculate the absolute value of the impedance $|Z_{RC}(\omega)|$ and the associated phase $\varphi_{RC}(\omega)$ at the following two frequencies:
   (1) $\omega_1 = 2\pi\,f_1 = 2\pi \times 1\,\text{kHz}$,
   (2) $\omega_2 = 2\pi\,f_2 = 2\pi \times 10\,\text{kHz}$,
   (3) $\omega_3 = 2\pi\,f_3 = 2\pi \times 100\,\text{kHz}$.

(b) What are the limits of $|Z_{RC}(\omega)|$ and $\varphi_{RC}(\omega)$ when:
(1) $\omega \to 0$ (i.e. at zero frequency, a.k.a. DC),
(2) $\omega \to \infty$ (i.e. at very high frequencies).

4. Repeat the calculations from Question 3, but this time in the case of parallel RC network.

5. Capacitor C and resistor R are connected in series as in Fig. 3.8. Consider the following two scenarios, given the initial conditions:

(a) consider the following:
a. at time $t_0 \le 0$ s, the capacitor is completely discharged, i.e. $v_C(t \le 0) = 0$ V,
b. the voltage source function is $v(t < 0) = 0$ and $v(t \ge 0) = V_0$, i.e. when there is $OFF \to ON$ voltage change at $t = 0$, derive the expression for $v_C(t \ge 0)$ and sketch its graph.
(b) consider the following:
a. at time $t_0 \le 0$ s, the capacitor is completely charged, i.e. $v_C(t \le 0) = V_0$, and
b. the voltage source function is $v(t < 0) = V_0$ and $v(t \ge 0) = 0$, i.e. when there is $ON \to OFF$ voltage change at $t = 0$, derive the expression for $v_C(t \ge 0)$ and sketch its graph.

Fig. 3.8 Example 3.4-5

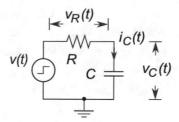

6. Inductor L and resistor R are connected in series as in Fig. 3.9. Consider the following two scenarios, given the initial conditions:

(a) the voltage source function is $v(t < 0) = 0$ and $v(t \ge 0) = V_0$, i.e. when there is $OFF \to ON$ voltage change at $t = 0$, derive the expression for $i_L(t \ge 0)$ and sketch its graph.

(b) the voltage source function is $v(t < 0) = V_0$ and $v(t \ge 0) = 0$, i.e. when there is $ON \to OFF$ voltage change at $t = 0$, derive the expression for $i_L(t \ge 0)$ and sketch its graph.

Fig. 3.9 Example 3.4-6

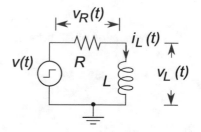

7. Illustrate the consequence of having a repetitive piecewise linear signal whose $ON \rightarrow OFF$ and back to ON cycle period is either too short or too long relative to the given circuit's time constant τ_0. Then, explain the resulting fundamental limit on the minimum clock frequency that can be used for a square-wave clock signal in our modern electronics.

3.5 **   **Time Constant, RC** As a reminder, the equivalent serial/parallel branches of multiple components are:

$$(n \text{ resistors in series}): R_{eq} = R_1 + R_2 + \cdots + R_n \tag{3.11}$$

$$(n \text{ resistors in parallel}): \frac{1}{R_{eq}} = \frac{1}{R_1} + \frac{1}{R_2} + \cdots + \frac{1}{R_n} \tag{3.12}$$

$$(n \text{ capacitors in parallel}): C_{eq} = C_1 + C_2 + \cdots + C_n \tag{3.13}$$

$$(n \text{ capacitors in series}): \frac{1}{C_{eq}} = \frac{1}{C_1} + \frac{1}{C_2} + \cdots + \frac{1}{C_n} \tag{3.14}$$

$$(n \text{ inductors in series}): L_{eq} = L_1 + L_2 + \cdots + L_n \tag{3.15}$$

$$(n \text{ inductors in parallel}): \frac{1}{L_{eq}} = \frac{1}{L_1} + \frac{1}{L_2} + \cdots + \frac{1}{L_n} \tag{3.16}$$

where the equivalent inductances are calculated under assumption that there is no EM coupling among involved inductors.

1. By inspection of RC networks in Fig. 3.10, derive their time constants τ_0.

Fig. 3.10 Example 3.5-1

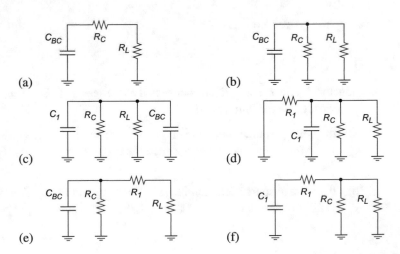

2. By inspection of RL networks in Fig. 3.11, derive their time constants τ_0.

Fig. 3.11 Example 3.5-2

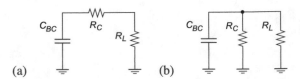

3. By inspection of RC networks in Fig. 3.12, derive their time constants τ_0.

Fig. 3.12 Example 3.5-3

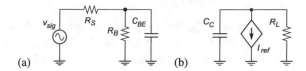

3.6 ***   **Non-linear Resistor**

1. Given a graph with two voltage/current (V/I) functions $\mathscr{A}$ and $\mathscr{B}$ that correspond to two unknown devices, Fig. 3.13, where the horizontal axis shows voltage and the vertical axis shows current,

   (a) $\mathscr{A}$: deduce its DC and AC resistances and type of device,

   (b) $\mathscr{B}$: deduce its DC and AC resistances and type of device.

Fig. 3.13 Example 3.6-1

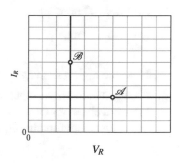

2. Given a family of voltage/current (V/I) functions $\mathscr{A}$ to $\mathscr{C}$, Fig. 3.14, where the horizontal axis shows voltage and the vertical axis shows current, deduce type of used devices, and in each case deduce:

   (a) $\mathscr{A}$: its R_{DC}, R_{AC}, and g_m,

   (b) $\mathscr{B}$: its R_{DC} and R_{AC}, and g_m,

   (c) $\mathscr{C}$: its R_{DC} and R_{AC}, and g_m.

Then, assuming a scale of 0.5 V/0.5 mA per square, calculate these resistances.

Fig. 3.14 Example 3.6-2

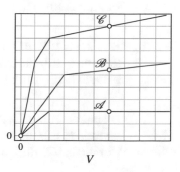

Solutions

Exercise 3.1, page 72

1. Resistance of a piece of physical material is determined by its geometry, i.e. cross-sectional area, length, and electrical resistivity of the material itself. In the special case of a cylindrical conductor, (3.5), its cross-section is circular, and therefore, we calculate

$$S = r^2\pi = \left(\frac{d}{2}\right)^2 \pi = 4.418 \, \text{nm}^2$$

which is substituted in (3.5) as

$$R_{DC} = \rho\frac{l}{S} \quad \therefore \quad l = \frac{R_{DC}S}{\rho}$$

and for three given materials, we calculate wire lengths as:

(a) 262.97 mm,   (b) 166.71 mm,

(c) 45.54 mm,   (d) 0.007 nm.

Note the high resistivity of silicon, the main material in IC manufacturing.

2. As a convention, the input and output of a function are shown in the horizontal and vertical axes, respectively. By Ohm's law, when a function is linear, either simple ratio or derivative of the voltage/current function equals to resistance as

$$R = \frac{\Delta V_{out}}{\Delta I_{in}} = \frac{V_{out}}{I_{in}} \quad \text{i.e. "slope"}$$

which is to say the function shown in the graph may represent a whole another universe; nevertheless, it may be used as a simple linear resistor. We note the slope of function $\mathscr{A}$ is $1\square/10\square$, of function $\mathscr{B}$ is $1\square/1\,\square$, and of function $\mathscr{C}$ is $10\square/1\,\square$. It is important to pay attention to which variable is "input" (i.e. denominator) and which is "output" (i.e. numerator), so that either resistance $R = V/I$ or its conductance $g_m = 1/R = I/V$ is correctly calculated. Therefore, we calculate

(a) $\mathscr{A}$: ratio is $1\square/10\square$; therefore, $R = 1\,\text{V}/(10 \times 1\,\text{mA}) = 100\,\Omega$.
     $\mathscr{B}$: ratio is $1\square/1\square$; therefore, $R = 1\,\text{V}/1\,\text{mA} = 1\,\text{k}\Omega$.
     $\mathscr{C}$: ratio is $10\square/1\square$; therefore, $R = (10 \times 1\,\text{V})/1\,\text{mA} = 10\,\text{k}\Omega$.

(b) $\mathscr{A}$: ratio is $1\square/10\square$; therefore, $R = 1\,\text{V}/(10 \times 10\,\mu\text{A}) = 10\,\text{k}\Omega$.
     $\mathscr{B}$: ratio is $1\square/10\square$; therefore, $R = 1\,\text{V}/10\,\mu\text{A} = 100\,\text{k}\Omega$.
     $\mathscr{C}$: ratio is $1\square/10\square$; therefore, $R = (10 \times 1\,\text{V})/10\,\mu\text{A} = 1\,\text{M}\Omega$.

(c) $\mathscr{A}$: ratio is $1\square/10\square$; therefore, $R = 1\,\text{V}/(10 \times 1\,\mu\text{A}) = 100\,\text{k}\Omega$.
     $\mathscr{B}$: ratio is $1\square/10\square$; therefore, $R = 1\,\text{V}/1\,\mu\text{A} = 1\,\text{M}\Omega$.
     $\mathscr{C}$: ratio is $1\square/10\square$; therefore, $R = (10 \times 1\,\text{V})/1\,\mu\text{A} = 10\,\text{M}\Omega$.

3. By definition, different forms of the equation for power are

$$P = VI = V\frac{V}{R} = \frac{V^2}{R}$$

therefore, given V and R, we calculate

(a) $P = 1\,\text{mW}$,   (b) $P = 100\,\text{mW}$,

(c) $P = 10\,\text{W}$,   (d) $P = 1\,\text{kW}$.

4. By definition, different forms of the equation for power are

$$P = VI = IRI = I^2 R$$

Therefore, given I and R, we calculate

(a) $P = 1\,\text{mW}$,   (b) $P = 10\,\text{mW}$,

(c) $P = 100\,\text{mW}$,   (d) $P = 1\,\text{W}$.

5. By definition, different forms of the equation for power are

$$P = VI = V\frac{V}{R} = \frac{V^2}{R} \quad \therefore \quad V = \sqrt{PR}$$

Therefore, given P and R, we calculate

(a) $V = 1\,\text{V}$,   (b) $V = 3.162\,\text{V}$,

(c) $V = 10\,\text{V}$,   (d) $V = 31.623\,\text{V}$.

6. By definition, different forms of the equation for power are

$$P = VI = IRI = I^2 R \quad \therefore \quad I = \sqrt{\frac{P}{R}}$$

Therefore, given P and R, we calculate

(a) $I = 1\,\text{mA}$,   (b) $I = 3.162\,\text{mA}$,

(c) $I = 10\,\text{mA}$,   (d) $I = 31.623\,\text{mA}$.

7. (a) Because in this example the expression for power is a sinusoidal function, the peak and the average powers are simply calculated as $P_m = V_m I_m = 2\,\text{V} \times 1.5\,\text{mA} = 3\,\text{mW}$, and $\langle P \rangle = (3\,\text{mW} + 0\,\text{mW})/2 = 1.5\,\text{mW}$.

Or, formally, by definition, we write (see Example 1.4-1(b))

$$\langle P \rangle = \frac{1}{T} \int_0^T I_m \sin \omega t \times V_m \sin \omega t = \frac{V_m I_m}{T} \int_0^T \sin \omega t \times \sin \omega t$$

$$= \frac{P_m}{T} \int_0^T \sin^2(\omega t)\,dt = \frac{P_m}{2T} \int_0^T [1 - \cos(2x)]\,dx$$

$$= \frac{P_m}{2T} x \Big|_0^T - \frac{3}{2T} \int_0^T \cos(2x)\,dx \overset{0}{\nearrow} = \frac{P_m}{2\cancel{T}} \cancel{T} = \frac{(1.5 \times 2)\,\text{mW}}{2}$$

$$= \frac{3}{2}\,\text{mW}$$

(b) At any moment, the resistor value is

$$R = \frac{v(t)}{i(t)} = \frac{2\,\text{V}}{{}^{3}/_{2}\,\text{mA}} = \frac{4}{3}\,\text{k}\Omega \approx 1.333\,\text{k}\Omega$$

(c) The equivalent DC voltage V that is needed to generate a power level equal to the average power, i.e. V_{rms}, is

$$V_{\text{rms}} = \sqrt{\langle P \rangle\, R} = \sqrt{\frac{3}{2}\,\text{mW} \times \frac{4}{3}\,\text{k}\Omega} = \sqrt{2}\,\text{V} \approx 1.414\,\text{V}$$

Exercise 3.2, page 73

1. By definition of capacitive impedance $(C \neq 0)$

$$Z_C = \frac{1}{j\omega C} = -j\frac{1}{\omega C} = -j\frac{1}{2\pi f C}$$

the two extreme limits are calculated simply as

(a) $DC \Rightarrow f = 0 \quad \therefore \quad \lim_{f \to 0} Z_C = \infty$

(b) $f = \infty \quad \therefore \quad \lim_{f \to \infty} Z_C = 0$

2. By definition of capacitive impedance $(C \neq 0)$ is

$$Z_C = \frac{1}{j\omega C} = -j\frac{1}{\omega C} = -j\frac{1}{2\pi f C}$$

which is strictly complex, that is to say $\Re(Z_C) = 0$; therefore, its phase is fixed by phase of "$-j$", i.e. $-90°$. For given $C = 159$ nF, we calculate the module $|Z_C|$ as

(a) $f = 1$ GHz $\therefore Z_C = 1$ mΩ,

(b) 100 MHz $\therefore Z_C = 10$ mΩ,

(c) 1 MHz $\therefore Z_C = 1\,\Omega$,

(d) 1 kHz $\therefore Z_C = 1$ kΩ,

which illustrated how the same capacitor presents very different Z_C to different components of signal's frequency spectrum.

3. By definition of plate capacitor, (3.5)

$$C = \epsilon_0 \frac{S}{d}$$

First, we calculate the area of semicircular plate as

$$S = \frac{r^2 \pi}{2}$$

Then, the equivalent electrical network consists of "n" capacitors in parallel, so that the total capacitance is

$$C_{tot} = n\,C \times \beta = n\epsilon_0 \frac{S}{d} \times \beta$$

$$= n\epsilon_0 \frac{r^2 \pi}{2d} \times \beta$$

where β is the multiplication constant to model the total overlapping area of this rotating capacitor.

(a) In the initial position, the rotating plates are fully overlapping, and thus, we set $\beta = 1$ and, given the data, calculate

$$C_{tot} = n\epsilon_0 \frac{r^2 \pi}{2d} \times \beta$$

$$= 5 \times 8.854 \times 10^{-12}\,\text{F/m}$$

$$\times \frac{(10\,\text{mm})^2 \pi}{2 \times 1\,\text{mm}} \times 1$$

$$= 6.954\,\text{pF}$$

(b) When rotating plates are set to $\alpha = \pi/4$ position, that is to say the overlapping area is reduced by $1/4$, which sets the capacitor size to $\beta = 3/4$ of the initial value, i.e.

$$C_{tot} = 6.954\,\text{pF} \times 0.75$$

$$= 5.216\,\text{pF}$$

(c) When rotating plates are set to $\alpha = \pi/2$ position, that is to say the overlapping area is reduced by $1/2$, which sets the capacitor size to $\beta = 1/2$ of the initial value, i.e.

$$C_{tot} = 6.954\,\text{pF} \times 0.5$$
$$= 3.477\,\text{pF}$$

4. By definition of capacitive impedance $(C \neq 0)$

$$Z_C = \frac{1}{j\omega C} \quad \therefore \quad C = \left| \frac{1}{j\, 2\pi f Z_C} \right|$$

given the data, we calculate

(a) $Z_C = 1\,\Omega \quad \therefore \quad C = 15.9\,\text{nF}$,

(b) $Z_C = 100\,\Omega \quad \therefore \quad C = 159.15\,\text{pF}$,

(c) $Z_C = 1\,\text{k}\Omega \quad \therefore \quad C = 15.9\,\text{pF}$.

5. By inspection of the given graph, we make two estimates as follows:

(a) $f < 20\,\text{MHz}$: In this region, the impedance slope is negative, which means it is capacitive impedance. Thus, for example at $f = 1\,\text{MHz}$, we see that $Z_C \approx 1.5\,\text{k}\Omega$. Given the frequency and impedance, we estimate $C \approx 100\,\text{pF}$. In other words, its impedance must be

$$Z_C = -j\,\frac{1}{2\pi \times 100\,\text{pF} \times f}$$

and its phase is obviously $-90°$.

(b) $f > 200\,\text{MHz}$: In this region, the impedance slope is positive, which means it is inductive impedance. Thus, for example at $f = 1\,\text{GHz}$, we see that $Z_L \approx 150\,\Omega$. Given the frequency and impedance, we estimate $L \approx 25\,\text{nF}$. In other words, its impedance must be

$$Z_L = +j\,2\pi \times 25\,\text{nH} \times f$$

and its phase is obviously $+90°$.

Every real physical component possess its "self-resonant" frequency where its electrical behaviour drastically changes to the point to invert its basic function from, like here, capacitive to inductive.

6. Relation between voltage and current at capacitor terminals is given by

$$i_C(t) = C\,\frac{dv(t)}{dt} \tag{3.17}$$

which is to say, current flowing through a capacitor is proportional to the first derivative of its voltage function. By inspection of $v(t)$ function, we see that it is, in effect, a piecewise linear function. As a consequence, its first derivative is easily deduced. The constant intervals correspond to zero value of the first derivative. Rapid step function corresponds to first derivative in the form of an impulse, i.e. Dirac function. In other words, each "edge" of voltage function generates very short current pulse. In practice, this relationship is exploited to design an "edge detector" circuit.

The time derivative term in (3.17) is estimated as follows: obviously, the voltage change is $dv(t) = +1\,\text{V}$ at each positive edge (from $0\,\text{V}$ to $1\,\text{V}$), and $dv(t) = -1\,\text{V}$ at each negative edge (from $1\,\text{V}$ to $0\,\text{V}$). It now questions of how fast this change happens. It is not possible to see that resolution from the given graph. Let us see what happens if, for example, that $1\,\text{V}$ change happens during $dt = 1\,\mu\text{s}$, which is easily achieved by modern circuits. If that is the case, the current pulse amplitude is

$$i_C(t) = C\,\frac{dv(t)}{dt} = 3\,\mu\text{F}\,\frac{1\,\text{V}}{1\,\mu\text{s}} = 3\,\text{A}$$

This example illustrates the importance of accounting for the "transient" currents generated across capacitors by rapid voltage changes. Simulation of this example confirms the above analysis, see Fig. 3.15.

Fig. 3.15 Example 3.2-6

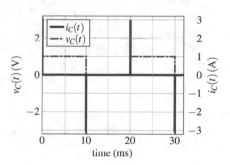

Exercise 3.3, page 75

1. By definition, the inductive impedance is

$$Z_L = j\omega L$$

and the two extreme limits are calculated simply as

(a) $DC \Rightarrow f = 0 \;\; \therefore \;\; \lim_{f \to 0} Z_L = 0,$
    task $f = \infty \;\; \therefore \;\; \lim_{f \to \infty} Z_L = \infty.$

We note the symmetry between the capacitive and inductive impedances.

2. Direct implementation of (3.9) results in $|Z_L| = 2\pi f L$; therefore, we calculated inductive impedances as:

  $f = 1\,\text{kHz} \;\; \therefore$

(a) $Z_L = 2\pi\,1\text{kHz} \times (1/2\pi)\text{mH}$

    $= 1\,\Omega$

$$f = 1\,\text{MHz} \quad \therefore$$

(b) $Z_L = 2\pi\,1\,\text{MHz} \times (1/2\pi)\,\text{mH}$

   $= 1\,\text{k}\Omega$

   $f = 100\,\text{MHz} \quad \therefore$

(c) $Z_L = 2\pi\,100\,\text{MHz} \times (1/2\pi)\,\text{mH}$

   $= 100\,\text{k}\Omega$

   $f = 1\,\text{GHz} \quad \therefore$

(d) $Z_L = 2\pi\,1\,\text{GHz} \times (1/2\pi)\,\text{mH}$

   $= 1\,\text{M}\Omega.$

3. Design of coil inductors is more difficult relative to the other two components. The reason is that containing magnetic field inside a given volume is never perfect. As a consequence, the final value of inductance is very sensitive to geometry of the coil, pitch p between rings coil, length l versus number of turns N ratio, cross-sectional area A of the coil core, etc. (see Fig. 3.16).

   In the first approximation, we make a few assumptions: we assume a single layer, ideal cylindrical inductor with air core, perfectly wound up (that is to say, there are no gaps between the turns, i.e. $p/d = 1$). That being the case, the cross-sectional area of the air core is $A = r^2\pi$ and $\mu = \mu_0\,\mu_r = \mu_0 = 4\pi \times 10^{-7}\,\text{H/m}$ (the air permeability is $\mu_r = 1$). Assuming perfect no-gaps winding, the coil length is simply $l = N \times d = N \times 2a$.

   We can substitute that ideal expression for l into (3.5) to calculate the number of turns as

   $$L = \frac{\pi r^2 \mu_0 N^2}{l} = \frac{\pi r^2 \mu_0 N^{\cancel{2}}}{2\cancel{N}a} \quad \therefore$$

   $$N = \frac{2aL}{\pi r^2 \mu_0} = 26.7 \approx 27\ \text{turns}$$

   which results in the coil length of $l = N \times 2a = 5.4\,\text{mm}$. Of course, this is very rough first estimate, and in practice, we use much more elaborated calculations to design manufacturable inductors.

Fig. 3.16 Example 3.3-3

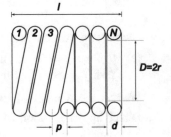

4. By definition of inductive impedance

$$Z_L = j\omega L$$

given the data, we calculate

(a) $Z_L = 1\,\Omega$ $\therefore$ $C = 15.9\,\text{nH}$,

(b) $Z_L = 100\,\Omega$ $\therefore$ $C = 1.59\,\mu\text{H}$,

(c) $Z_L = 1\,\text{k}\Omega$ $\therefore$ $C = 15.9\,\mu\text{H}$,

which illustrated how the same inductor presents very different Z_L to different components of signal's frequency spectrum.

5. By inspection of the given graph, we make two estimates as follows:

(a) $f < 100\,\text{MHz}$: In this region, the impedance slope is positive, which means it is inductive impedance. Thus, for example at $f = 10\,\text{MHz}$, we see that $Z_L \approx 200\,\Omega$. Given the frequency and impedance, we estimate $L \approx 3.2\,\mu\text{H}$. In other words, its impedance must be

$$Z_L = +j\,2\pi \times 3.2\,\mu\text{H} \times f$$

and its phase is obviously $+90°$.

(b) $f > 1\,\text{GHz}$: In this region, the impedance slope is positive, which means it is capacitive impedance. Thus, for example at $f = 10\,\text{GHz}$, we see that $Z_C \approx 150\,\Omega$. Given the frequency and impedance, we estimate $L \approx 0.1\,\text{pF}$. In other words, its impedance must be

$$Z_C = -j\,\frac{1}{2\pi \times 0.1\,\text{pF} \times f}$$

and its phase is obviously $-90°$.

Every real physical component possess its "self-resonant" frequency where its electrical behaviour drastically changes to the point to invert its basic function from, like here, capacitive to inductive.

6. Relation between voltage and current at inductor terminals is given by

$$v_C(t) = L\,\frac{di(t)}{dt} \tag{3.18}$$

which is to say, voltage across an inductor is proportional to the first derivative of its current function. By inspection of $i(t)$ function, we see that it is, in effect, a piecewise linear function. As a consequence, its first derivative is easily deduced. The constant intervals correspond to zero value of the first derivative. Rapid step function corresponds to first derivative in the form of an impulse, i.e. Dirac function. In other words, each "edge" of current function generates very short voltage pulse. In practice, this relationship is exploited to design an "edge detector" circuit.
The time derivative term in (3.18) is estimated as follows: obviously, the current change is $di(t) = +1\,\text{mA}$ at each positive edge (from $0\,\text{A}$ to $1\,\text{mA}$), and $di(t) = -1\,\text{mA}$ at each negative edge (from $1\,\text{mA}$ to $0\,\text{A}$). It now questions of how fast this change happens. It is not possible to see that resolution from the given graph. Let us see what happens if, for example, that $1\,\text{mA}$ change happens during $dt = 1\,\mu\text{s}$, which is easily achieved by modern circuits. If that is the case, the voltage pulse amplitude is

$$v_C(t) = L\,\frac{di(t)}{dt}$$

$$= 3\,\mu\text{H}\,\frac{1\,\text{mA}}{1\,\mu\text{s}} = 3\,\text{mV}$$

This example illustrates the importance of accounting for the "transient" voltages generated across inductors by rapid current changes. Simulation of this example confirms the above analysis, see Fig. 3.17.

Fig. 3.17 Example 3.3-6

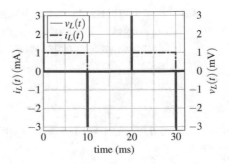

Exercise 3.4, page 76

1. The equivalent impedance of multiple impedances connected in series is equal to the sum of involved impedances, i.e. $Z_{eq} = Z_1 + Z_2 + \cdots + Z_n$. Useful observation is that the equivalent impedance is always greater than the greatest among of the involved impedances.

   (a) According to definitions of impedances for these two elements (3.7) and (3.9), the equivalent impedance is calculated as the module

$$Z_{RL}(\omega) = Z_R + Z_L = R + j\omega L = 1\,\text{k}\Omega + j\omega \times (1/20\pi)\,\text{mH}$$

Thus, by using the Pythagorean theorem (i.e. complex algebra), it follows:

$$|Z_{RL}(\omega)|^2 = R^2 + (\omega L)^2 \Rightarrow |Z_{RL}(\omega)| = \sqrt{R^2 + (\omega L)^2} \tag{3.19}$$

$$\therefore$$

$$|Z_{RL}(\omega_1)| = \sqrt{(1\,\text{k}\Omega)^2 + (2\pi \times 1\,\text{MHz} \times (1/20\pi)\,\text{mH})^2}$$

$$= \sqrt{(1\,\text{k}\Omega)^2 + (100\,\text{kHz} \times 1\,\text{mH})^2} = 1\,\text{k}\Omega$$

$$|Z_{RL}(\omega_2)| = \sqrt{(1\,\text{k}\Omega)^2 + (1\,\text{MHz} \times 1\,\text{mH})^2}$$

$$= \sqrt{(1000\,\Omega)^2 + (1000\,\Omega)^2} = \sqrt{2}\,\text{k}\Omega$$

$$|Z_{RL}(\omega_3)| = \sqrt{(1\,\text{k}\Omega)^2 + (10\text{MHz} \times 1\,\text{mH})^2} = 10\,\text{k}\Omega$$

We note that at low relatively LF $f = 1\,\text{MHz}$ the equivalent impedance is dominated by the resistor ($Z_L \approx 0$), while the inductor impedance becomes more dominant as the frequency increases.

Phase φ of a complex number is calculated by definition of $\tan(\varphi)$ function. Both real and imaginary parts of Z_{RL} are positive; therefore, the associated phase φ is in the first quadrant, i.e.

$$\tan \varphi = \frac{\Im(Z_{RL})}{\Re(Z_{RL})} \tag{3.20}$$

$$\therefore$$

$$\varphi_1(\omega_1) = \arctan\left[\frac{\omega_1 L}{R}\right] = \arctan\left[\frac{100\,\text{kHz} \times 1\,\text{mH}}{1\,\text{k}\Omega}\right] = 6°$$

$$\varphi_2(\omega_2) = \arctan\left[\frac{\omega_2 L}{R}\right] = \arctan\left[\frac{1\,\text{MHz} \times 1\,\text{mH}}{1\,\text{k}\Omega}\right] = 45°$$

$$\varphi_3(\omega_3) = \arctan\left[\frac{\omega_3 L}{R}\right] = \arctan\left[\frac{10\,\text{MHz} \times 1\,\text{mH}}{1\,\text{k}\Omega}\right] = 84°$$

which is to say that phase of this RL network at 100 MHz is dominated by the inductor, while at 1 MHz it is close to zero (because of resistor and negligible inductance).

(b) When $\omega \to 0$, from (3.19), we find $|Z_{RL}(0)| = 1\,\text{k}\Omega$ (i.e. the network is resistive), and when $\omega \to \infty$, we find $|Z_{RL}(\infty)| = \infty$ (i.e. the network is inductive).

When $\omega \to 0$ from (3.20), we find $\varphi(0) = 0$ (i.e. the network is resistive, i.e. real), and when $\omega \to \infty$, we find $|\varphi(\infty)| = +(\pi/2)$ (i.e. the network is inductive, i.e. complex).

In summary, at HF a serial RL network is close to inductor and at LF it is close to resistor.

2. The equivalent conductance (which is by definition the inverse of impedance) of multiple impedances connected in parallel is equal to the sum of involved conductances, i.e. $1/Z_{eq} = 1/Z_1 + 1/Z_2 + \cdots + 1/Z_n$. Useful observation is that the equivalent impedance is always lower than the lowest among the involved impedances.

(a) According to definitions of impedances for these two elements (3.7) and (3.9), their parallel impedance $Z_{R||Z_L}$ is found as

$$\frac{1}{Z_{R||Z_L}} = \frac{1}{Z_R} + \frac{1}{Z_L} \Rightarrow Z_{R||Z_L}(\omega) = \frac{1}{\frac{1}{Z_R} + \frac{1}{Z_L}} = \frac{Z_R Z_L}{Z_R + Z_L}$$

$$\therefore \quad \text{(resistor is real and inductor is complex part)}$$

$$|Z_{R||Z_L}(\omega)| = \frac{1}{\sqrt{\frac{1}{Z_R^2} + \frac{1}{Z_L^2}}} = \frac{1}{\sqrt{\frac{1}{R^2} + \frac{1}{(\omega L)^2}}}$$

$$\therefore$$

$$Z_{R||Z_L}(1\,\text{MHz}) = \frac{1}{\sqrt{\frac{1}{(1\,\text{k}\Omega)^2} + \frac{1}{(2\pi \times 1\,\text{MHz} \times (1/20\pi)\,\text{mH})^2}}}$$

$$= \frac{1}{\sqrt{\frac{1}{(1\,\text{k}\Omega)^2} + \frac{1}{(100\,\Omega)^2}}} = 99.5\,\Omega$$

$$Z_{R||Z_L}(10\,\text{MHz}) = \frac{1}{\sqrt{\frac{1}{(1\,\text{k}\Omega)^2} + \frac{1}{(1\,\text{k}\Omega)^2}}} = \frac{\sqrt{2}}{2}\,\text{k}\Omega$$

$$Z_{R||Z_L}(100\,\text{MHz}) = \frac{1}{\sqrt{\frac{1}{(1\,\text{k}\Omega)^2} + \frac{1}{(10\,\text{k}\Omega)^2}}} = 995\,\Omega$$

which shows that in parallel connection at LF the equivalent impedance is dominated by inductor's low resistance that is bypassing the $1\,\text{k}\Omega$ resistor.

Phase φ is calculated as

$$\varphi(1\text{MHz}) = \arctan\left[\frac{R}{\omega_1 L}\right] = \arctan\left[\frac{1\,\text{k}\Omega}{100\,\text{kHz} \times 1\text{mH}}\right] = 84°$$

$$\varphi(10\,\text{MHz}) = \arctan\left[\frac{R}{\omega_2 L}\right] = \arctan\left[\frac{1\,\text{k}\Omega}{1\,\text{MHz} \times 1\,\text{mH}}\right] = 45°$$

$$\varphi(100\,\text{MHz}) = \arctan\left[\frac{R}{\omega_3 L}\right] = \arctan\left[\frac{1\,\text{k}\Omega}{10\,\text{MHz} \times 1\,\text{mH}}\right] = 6°$$

which is to say that phase of this RL network at $1\,\text{MHz}$ is dominated by the inductor, while at $100\,\text{MHz}$ it is close to zero (that is to say, resistive).

(b) When $\omega \to 0$, $|Z_{R||Z_L}(0)| = 0$ because it is the inductor's impedance that tends to zero, and when $\omega \to \infty$, then $|Z_{R||Z_L}(\infty)| = R$ because the inductor's impedance is tending to infinity. At the same time, when $\omega \to 0$, $\varphi(0) = +\pi/2$ because it is the inductor's impedance that dominates the equivalent impedance, and when $\omega \to \infty$, then $\varphi(\infty) = 0$ because it is now the resistor that is dominant.

In summary, the parallel RL network is dominated by inductor at LF and by resistor at HF (when reactance of the inductor is very high).

3. The equivalent impedance of multiple impedances connected in series is equal to the sum of involved impedances, i.e. $Z_{eq} = Z_1 + Z_2 + \cdots + Z_n$. Useful observation is that the equivalent impedance is always greater than the greatest among of the involved impedances.

(a) According to definitions of impedances for these two elements (3.7) and (3.8), the equivalent impedance is calculated as the module

$$Z_{RC}(\omega) = Z_R + Z_C = R + \frac{1}{j\omega C} = 100\,\Omega - j\frac{1}{2\pi\,f \times (1/2\pi)\mu\text{F}}$$

$$= 100\,\Omega - j\frac{1}{f \times 1\,\mu\text{F}}$$

Thus, by using the Pythagorean theorem (i.e. complex algebra), it follows:

$$|Z_{RC}(\omega)|^2 = R^2 + \left[\frac{1}{2\pi\,fC}\right]^2 \Rightarrow |Z_{RC}(\omega)| = \sqrt{R^2 + \left[\frac{1}{2\pi\,fC}\right]^2}$$

$\therefore$ (resistor is real and capacitor is imaginary part)

$$|Z_{RC}(\omega_1)| = \sqrt{(100\,\Omega)^2 + \left[\frac{1}{1\,\text{kHz} \times 1\,\mu\text{F}}\right]^2} = 1\,\text{k}\Omega$$

$$|Z_{RC}(\omega_2)| = \sqrt{(100\,\Omega)^2 + \left[\frac{1}{10\,\text{kHz} \times 1\,\mu\text{F}}\right]^2} = \sqrt{100^2 + 100^2} = 10\sqrt{2}\,\Omega$$

$$|Z_{RC}(\omega_3)| = \sqrt{(100\,\Omega)^2 + \left[\frac{1}{100\,\text{kHz} \times 1\,\mu\text{F}}\right]^2} = 100.5\,\Omega$$

We note that at low relatively LF $f = 1\,\text{kHz}$ the equivalent impedance is dominated by the capacitor ($Z_C \to \infty$), while the capacitor impedance becomes close to zero as the frequency increases.

Phase φ of a complex number is calculated by definition of $\tan(\varphi)$ function. Real part of Z_{RC} is positive, while the imaginary part is negative; therefore, the associated phase φ is in the fourth quadrant, i.e. negative

$$\tan\varphi = \frac{\Im(Z_{RC})}{\Re(Z_{RC})} \tag{3.21}$$

$$\therefore$$

$$\varphi_1(\omega_1) = \arctan\left[\frac{-1/\omega_1 C}{R}\right] = \arctan\left[\frac{-1}{1\,\text{kHz} \times 1\,\mu\text{F} \times 100\,\Omega}\right] = -84°$$

$$\varphi_2(\omega_2) = \arctan\left[\frac{-1/\omega_2 C}{R}\right] = \arctan\left[\frac{-1}{10\,\text{kHz} \times 1\,\mu\text{F} \times 100\,\Omega}\right] = -45°$$

$$\varphi_2(\omega_3) = \arctan\left[\frac{-1/\omega_3 C}{R}\right] = \arctan\left[\frac{-1}{100\,\text{kHz} \times 1\,\mu\text{F} \times 100\,\Omega}\right] = -6°$$

which is to say that phase of this RC network at 1 kHz is dominated by the capacitor, while at 100 kHz it is close to zero (thus dominated by resistor).

(b) When $\omega \to 0$, from (3.21), we find $|Z_{RC}(0)| \to \infty$ (i.e. the network is capacitive), and when $\omega \to \infty$, we find $|Z_{RC}(\infty)| = 100\,\Omega$ (i.e. the network is resistive).

When $\omega \to 0$ from (3.21), we find $\varphi(0) = -\pi/2$ (i.e. the network is capacitive), and when $\omega \to \infty$, we find $\varphi(\infty)| = 0$ (i.e. the network is real, i.e. resistive).

In summary, at HF a serial RC network is close to resistor and at LF it is close to capacitor.

4. The equivalent conductance (which is by definition the inverse of impedance) of multiple impedances connected in parallel is equal to the sum of involved conductances, i.e. $1/Z_{eq} = 1/Z_1 + 1/Z_2 + \cdots + 1/Z_n$. Useful observation is that the equivalent impedance is always lower than the lowest among the involved impedances.

(a) According to definitions of impedances for these two elements (3.7) and (3.8), their parallel impedance $Z_{R||Z_C}$ is found as

$$\frac{1}{Z_{R||Z_C}} = \frac{1}{Z_R} + \frac{1}{Z_C} \Rightarrow Z_{R||Z_C}(\omega) = \frac{1}{\frac{1}{Z_R} + \frac{1}{Z_C}} = \frac{Z_R Z_C}{Z_R + Z_C}$$

$\therefore$ (resistor is real and capacitor is the complex part)

$$|Z_{R||Z_C}(\omega)| = \frac{1}{\sqrt{\frac{1}{Z_R^2} + \frac{1}{Z_C^2}}} = \frac{1}{\sqrt{\frac{1}{R^2} + (\omega C)^2}} = \frac{1}{\sqrt{\frac{1}{R^2} + \left(2\pi f \, \frac{1\,\mu F}{2\pi}\right)^2}}$$

$$= \frac{1}{\sqrt{\frac{1}{R^2} + (f \times 1\,\mu F)^2}}$$

$\therefore$

$$Z_{R||Z_C}(1\,\text{kHz}) = \frac{1}{\sqrt{\frac{1}{100\,\Omega^2} + (1\,\text{kHz} \times 1\,\mu F)^2}} = 99.5\,\Omega$$

$$Z_{R||Z_C}(10\,\text{kHz}) = \frac{1}{\sqrt{\frac{1}{100\,\Omega^2} + (10\,\text{kHz} \times 1\,\mu F)^2}} = 100\frac{\sqrt{2}}{2}\,\Omega$$

$$Z_{R||Z_C}(100\,\text{kHz}) = \frac{1}{\sqrt{\frac{1}{100\,\Omega^2} + (100\,\text{kHz} \times 1\,\mu F)^2}} = 9.95\,\Omega$$

which shows that in parallel connection at HF the equivalent impedance is dominated by capacitor's low resistance that is bypassing the $100\,\Omega$ resistor.

Phase φ is calculated as

$$\varphi(1\,\text{kHz}) = \arctan\left[\frac{1/j\omega C}{R}\right] = \arctan\left[\frac{-1}{100\,\Omega \times 1\,\text{kHz} \times 1\,\mu F}\right] = -6°$$

$$\varphi(10\,\text{kHz}) = \arctan\left[\frac{-1}{100\,\Omega \times 10\,\text{kHz} \times 1\,\mu F}\right] = -45°$$

$$\varphi(100\,\text{kHz}) = \arctan\left[\frac{-1}{100\,\Omega \times 100\,\text{kHz} \times 1\,\mu F}\right] = -84°$$

which is to say that phase of this RC network at $100\,\text{kHz}$ is dominated by the capacitor, while at $1\,\text{kHz}$ it is close to zero (that is to say, resistive).

(b) When $\omega \to 0$, $|Z_{R||Z_C}(0)| = R$ because it is the capacitor's impedance that tends to infinity, and when $\omega \to \infty$, then $|Z_{R||Z_C}(\infty)| = 0$ because the capacitor's impedance is tending to zero and bypassing the resistor.

At the same time, when $\omega \to \infty$, $\varphi(0) = +\pi/2$ because it is capacitor's impedance that dominates the equivalent impedance, and when $\omega \to 0$, then $\varphi(\infty) = 0$ because it is now the resistor that is dominant; therefore, the equivalent impedance is very close to real, and thus, the phase is zero.

In summary, the parallel RC network is dominated by capacitor at HF and by resistor at LF (when reactance of the parallel capacitor is very high).

5. Shape of a pulse signal $v(t)$ is created by two constant voltage levels (i.e. the "low" and the "high"). Each level by itself is in effect the DC voltage while it lasts. Another way to say it is that a pulse shape could have been created with a DC voltage source and an ON/OFF switch. For that reason, given the initial condition in this example, for $(t < 0)$, we write $v(t < 0) = 0$ =const, and for $(t \geq 0)$, we write $v(t \geq 0) = V_0$ =const. In both cases, the derivative of a constant equals zero. At the same time, the initial condition for capacitor is $v_C(t = 0) = 0\,\mathrm{V}$.

Given the initial conditions, note that $i_C(t) = i_R(t)$, the KVL equation at $t = 0$ gives

$$v(t) = v_C(t) + v_R(t) = v_C(t) + i_C\,R \quad \therefore \quad V_0 = 0\,\mathrm{V} + i_C\,R$$

$$\therefore$$

$$i_C(t = 0) = \frac{V_0}{R}$$

which is the initial current at $t = 0$ flowing through RC branch.

Hence, after recalling equation (see Exercise 3.2-6, and Eq. (3.17))

$$i_C(t) = C\,\frac{dv_C(t)}{dt} \quad \therefore \quad v_C(t) = \frac{1}{C}\int i_C(t)\,dt = \frac{q(t)}{C} \tag{3.22}$$

For $(t \geq 0)$, given $v(t = 0) = V_0$, we write the KVL equation and its derivative as

$$V_0 = v_R(t) + v_C(t) = i(t)\,R + \frac{q(t)}{C} \tag{3.23}$$

therefore, the derivative of both sides (3.23) is

$$0 = R\,\frac{di(t)}{dt} + \frac{1}{C}\,\frac{dq(t)}{dt} \quad \therefore \quad 0 = \frac{di(t)}{dt} + \frac{1}{RC}\,i(t) \tag{3.24}$$

Solution of first order differential equation (3.24) is derived as

$$\frac{di(t)}{dt} = -\frac{1}{RC}\,i(t) \quad \therefore \quad \frac{di(t)}{i(t)} = -\frac{1}{RC}\,dt \quad \therefore \quad \int \frac{di(t)}{i(t)} = -\frac{1}{RC}\int dt$$

We conveniently choose the integration constant so that we can take advantage of logarithm identities, and write the equation solution as

$$\ln i(t) = -\frac{t}{RC} + \ln C \quad \therefore \quad \ln i(t) - \ln C = -\frac{t}{RC} \quad \therefore \quad \ln \frac{i(t)}{C} = -\frac{t}{RC}$$

$$\therefore$$

$$i(t) = C\,e^{-\frac{t}{RC}} = C\,e^{-\frac{t}{\tau_0}} \tag{3.25}$$

where

$$\tau_0 = RC \tag{3.26}$$

is the *time constant* of this RC network. This parameter is among the most important and first ones to determine in any RC circuit analysis.

Taking into account the initial condition $i(t_0) = V_0/R$, we set $t = 0$ and derive the final solution as

$$i(t_0) = \frac{V_0}{R} = C\, e^{-0/\tau_0}{}^{1} \quad \therefore \quad C = \frac{V_0}{R} \quad \therefore \quad i_C(t) = \frac{V_0}{R} e^{-t/\tau_0} \tag{3.27}$$

After substituting (3.27) back into (3.23), we derive expression for voltage across this capacitor as

$$V_0 = i(t)\, R + v_C(t) = \cancel{R}\, \frac{V_0}{\cancel{R}} e^{-t/\tau_0} + v_C(t)$$

$$\therefore$$

$$v_C(t) = V_0 \left(1 - e^{-t/\tau_0}\right) \tag{3.28}$$

Equations (3.27) and (3.28) describe how voltage and current across a capacitor follow abrupt changes in the DC voltage level across the capacitor terminals. Term used to denote this type of changes is *transient*, and it is, obviously, a very non-linear process.

To quickly evaluate this function, let us take a look at a few specific points:

- $t = 0$: initially, $v(0) = V_0 (1 - \exp(0)) = V_0$,
- $t = \tau_0$: $v(\tau_0) = V_0 (1 - \exp(-\tau_0/\tau_0)) = V_0 (1 - 1/e) \approx 0.632 V_0$,
- $t = 5\tau_0$: $v(5\tau_0) = V_0 (1 - \exp(-5\tau_0/\tau_0)) = V_0 (1 - \exp(-5)) \approx 0.993 V_0$,
- $t \to \infty$: $\lim_{t \to \infty} v(t) = V_0 (1 - \lim_{t \to \infty} \exp(-t/\tau_0)) = V_0 (1 - 0) = V_0$.

Points to note are as follows:

1. a capacitor is very good in passing fast, abrupt voltage changes, while, at the same time, it presents open circuit for DC currents;
2. theoretically, a capacitor never reaches V_0 voltage level, and it only keeps tending to it forever. Because of that, practical decision when a capacitor is "fully charged" is usually made at about $t = 5\tau_0$, because at that moment the capacitor voltage v_C is at over 99% of V_0.
3. time constant is identical for both serial and parallel RC networks.

Voltage discharging function of this RC branch is found by repeating the above analysis, however starting with the initial condition that $v_C(0) = V_0$, $v(t < 0) = V_0$, and $v(t \geq 0) = 0$. That being the case, we find simple decaying function

$$v_C(t) = V_0\, e^{-t/\tau_0} \tag{3.29}$$

To conclude, charging/discharging time constant is simply $\tau = RC$. That product is under our control, and we design circuits in respect to its time constant.

For example, given $R = 1\,\text{k}\Omega$ and $C = 1\,\mu\text{F}$, we find that $\tau = RC = 1\,\text{ms}$, and consequently, it takes approximately 5 ms for the capacitor to achieve 99% of its "full" voltage level, which is "close enough" (theoretically, 100% is *never* achieved), see Fig. 3.18.

One of the principal consequences of time constants is that if we are to clearly detect "high" and "low" states of capacitor's voltage we must wait at least $5\tau_0$ before switching the states. This is the fundamental physical limitation of how fast our circuits can operate.

Fig. 3.18 Example 3.4-5

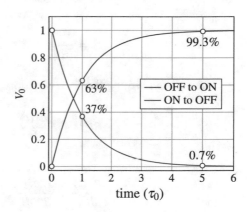

6. At any given moment in time, the source voltage $v(t)$ is split between voltages across the resistor $v_R(t)$ and the inductor $v_L(t)$, i.e. $v(t) = v_R(t) + v_L(t)$.

The step function shape of a pulse signal $v(t)$ is created by two constant voltage levels (i.e. the "low" and the "high"). Each level by itself is in effect the DC voltage while it lasts. Another way to say it is that a pulse shape could have been created with a DC voltage source and an ON/OFF switch. For that reason, given the initial condition in this example, for $(t < 0)$, we write $v(t < 0) = 0$ =const, and for $(t \geq 0)$, we write $v(t \geq 0) = V_0$ =const. In both cases, the derivative of a constant equals zero.

Given the initial conditions, the KVL equation at $t \geq 0$, after $v(t)$ switches from "low" to "high" level and stays there, gives

$$v(t) = v_L(t) + v_R(t) = v_L(t) + i(t)\,R \quad \therefore \quad V_0 = L\frac{di(t)}{dt} + i(t)\,R \qquad (3.30)$$

However, at $(t = 0)$, there is a large voltage generated across the inductor, and this is because there is a large rate of change in the current, i.e. $v_L(t) = L\,di(t)/dt$, which means that $i_R(0) = 0$ (because $v_R(0) = 0$). As a consequence at $t = 0$, we write

$$v(0) = V_0 = v_L(0) + \underbrace{v_R(0)}_{0} \quad \therefore \quad V_0 = L\frac{di(t)}{dt}\bigg|_{t=0}$$

On the other side, at $t \to \infty$ when the voltage/current transitions are over, i.e. when $di(t)/dt = 0$ (we keep in mind that $v(t \geq 0) = V_0 = $ const), we write

$$V_0 = \lim_{t \to \infty} (v_L(t) + v_R(t)) = \lim_{t \to \infty} \left(L \frac{di(t)}{dt}^{\ 0} + i(t)\, R \right)$$

$$\therefore$$

$$i(t) = \left. \frac{V_0}{R} \right|_{t \to \infty}$$

Hence, after writing the KVL equations, we solve (3.30) as

$$Ri(t) + L \frac{di(t)}{dL} = V_0 \quad \therefore \quad R \frac{di(t)}{dt} + L \frac{d^2 i(t)}{dt^2} = 0$$

$$\therefore$$

$$\frac{d^2 i(t)}{dt^2} + \frac{L}{R} \frac{di(t)}{dt} = 0 \tag{3.31}$$

This second order differential equation with constant coefficients has characteristic equation and its solutions as

$$y^2 + \frac{L}{R} y = 0 \quad \therefore \quad y \left(y + \frac{L}{R} \right) = 0 \quad \therefore \quad y_1 = 0, \text{ and } y_2 = -\frac{L}{R} = -\tau_0$$

where

$$\tau_0 = \frac{L}{R} \tag{3.32}$$

is the *time constant* of LR circuit, and the current change follows the natural growth law. Now we write the solution of (3.31) as

$$i(t) = C_1 + C_2\, e^{-t/\tau_0} \tag{3.33}$$

where the two integration constants are calculated from the initial conditions as

$$t = 0; \quad 0 = C_1 + C_2\, e^{0}{}^{\ 1} \quad \therefore \quad C_1 = -C_2$$

$$t \to \infty; \quad \frac{V_0}{R} = C_1 + C_2\, e^{-\infty}{}^{\ 0} \quad \therefore \quad C_1 = \frac{V_0}{R}$$

Now, we write (3.31) as

$$i(t) = \frac{V_0}{R} - \frac{V_0}{R}\, e^{-t/\tau_0} = \frac{V_0}{R} \left(1 - e^{-t/\tau_0} \right) \tag{3.34}$$

Equation (3.34) describes how current flowing through inductor follows step (i.e. ON/OFF) changes in the DC voltage level across the inductor terminals. Points to note are as follows:

1. An inductor is very good in passing fast, abrupt voltage changes, while, at the same time, it presents short circuit for DC currents.
2. Theoretically, an inductor never reaches the V_0/R current level, and it only keeps tending to it forever. Because of that, in practice, we accept that at about $t = 5\tau_0$ after there was an ON/OFF voltage change the inductor's current reaches its maximum/minimum level. At that moment, the inductor current $i_L(t)$ is at over 99% of the maximum level set by V_0/R (or, dropped below 1% in the other case).
3. Time constant is identical for both the serial and parallel RL networks.

Voltage discharging function of this RL branch is found by repeating the above analysis, however, this time starting with the initial condition that $i(0) = V_0/R$, $v(t < 0) = V_0$, and $v(t \geq 0) = 0$, i.e. $ON \rightarrow OFF$ voltage change. That being the case, we find simple decaying function

$$i_L(t) = \frac{V_0}{R} e^{-t/\tau_0} \tag{3.35}$$

To conclude, charging/discharging time constant is simply $\tau_0 = L/R$. That product is under our control, and we design circuits in respect to its time constant.

For example, given $R = 100\,\Omega$ and $L = 100\,\mu H$, we find that $\tau_0 = L/R = 1\,\mu s$, and consequently, it takes approximately $5\,\mu s$ for the inductor's current to achieve 99% of its "full" level, which is "close enough" (theoretically, 100% is *never* achieved), see Fig. 3.19.

One of the principal consequences of time constants is that if we are to clearly detect "high" and "low" states of capacitor's voltage, we must wait at least $5\tau_0$ before switching the states. This is fundamental physical limitation of how fast our circuits can operate.

Fig. 3.19 Example 3.4-6

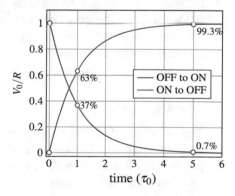

7. It takes at least $5\tau_0$ to charge/discharge RC (or RL network). That is to say, if period of the square wave entering, for example, RC network is equal to $T = 10\tau_0$, see graph Fig. 3.20 (top), there will be just enough time for the output voltage to reach 99% level. Then, the discharging would start immediately. Even so, the square wave is still recognizable.

In case of $T < 10\tau_0$, see graph Fig. 3.20 (bottom), the charging process will not be finished before discharging starts. Equally, discharging process will not have time to finish before the beginning of new charging phase. As a consequence, the output square signal is distorted too much; it is losing its amplitude and morphing into a triangular waveform whose average is $V_0/2$.

Fig. 3.20 Example 3.4-7

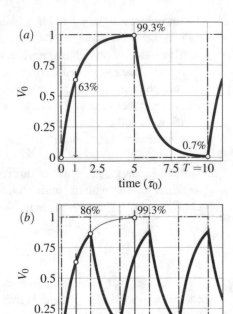

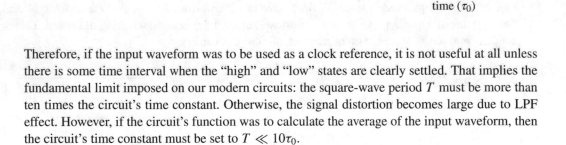

Therefore, if the input waveform was to be used as a clock reference, it is not useful at all unless there is some time interval when the "high" and "low" states are clearly settled. That implies the fundamental limit imposed on our modern circuits: the square-wave period T must be more than ten times the circuit's time constant. Otherwise, the signal distortion becomes large due to LPF effect. However, if the circuit's function was to calculate the average of the input waveform, then the circuit's time constant must be set to $T \ll 10\tau_0$.

Exercise 3.5, page 78

1. The main idea of how to derive time constant τ_0 is to use simple serial/parallel circuit reduction techniques until the original network is reduced to a single RC pair of components, so that it is possible to calculate $\tau_0 = RC$. Thus, the "R" and "C" in that equation should be interpreted as "the total equivalent resistance connected to the total equivalent capacitance".

   In case where several capacitors are found in the circuit but they are not directly connected to each other, we remove all but one and calculate its time constant while including all resistors in the network. Then, we repeat this calculation for all capacitors one by one.

   (a) Serial connection of resistors is replaced with its equivalent sum, and by inspection of the equivalent circuit, we write

   $$\tau_0 = R_{eq} C_{BC} = (R_C + R_L)C_{BC}$$

   where parallel connection of two resistors is shown in Fig. 3.21.

Fig. 3.21
Example 3.5-1(a)

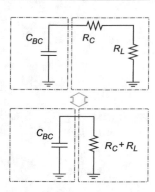

(b) Parallel connection of resistors is replaced with its equivalent resistance, see Fig. 3.22, and by inspection of the equivalent circuit, we write

$$\tau_0 = R_{eq}\, C_{BC} = (R_C||R_L)C_{BC}$$

where the equivalent parallel resistance of two resistors is

$$\frac{1}{R_{eq}} = \frac{1}{R_C} + \frac{1}{R_L} = \frac{R_C + R_L}{R_C R_L}$$

$$\therefore$$

$$R_{eq} = (R_C||R_L) = \frac{R_C R_L}{R_C + R_L}$$

Fig. 3.22
Example 3.5-1(b)

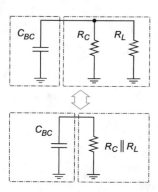

(c) Parallel connections of several resistors and capacitors are replaced by their equivalent networks, see Fig. 3.23, and by inspection, we write

$$\tau_0 = R_{eq}\, C_{eq} = (R_C||R_L)(C_1 + C_{BC})$$

where the equivalent parallel resistance of two resistors is

$$\frac{1}{R_{eq}} = \frac{1}{R_C} + \frac{1}{R_L} = \frac{R_C + R_L}{R_C R_L}$$

$$\therefore$$

$$R_{eq} = (R_C \| R_L) = \frac{R_C R_L}{R_C + R_L}$$

Fig. 3.23
Example 3.5-1(c)

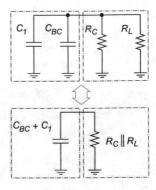

(d) Parallel connections of several resistors are replaced by their equivalent resistance, Fig. 3.24, and by inspection, we write

$$\tau_0 = R_{eq} C_1$$

$$= (R_C \| R_L \| R_1)(C_1 + C_1)$$

where the equivalent parallel resistance of three resistors is

$$\frac{1}{R_{eq}} = \frac{1}{R_C} + \frac{1}{R_L} + \frac{1}{R_1}$$

$$= \frac{R_L R_1 + R_C R_1 + R_C R_l}{R_C R_L R_1}$$

$$\therefore$$

$$R_{eq} = \frac{R_C R_L R_1}{R_L R_1 + R_C R_1 + R_C R_l}$$

Fig. 3.24
Example 3.5-1(d)

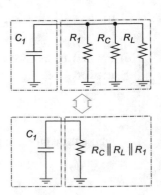

(e) Serial/parallel resistive networks are replaced by their equivalent resistance, Fig. 3.25. In this case, first we combine serial connection of two resistors to $R_S = R_1 + R_L$. Then, obviously, we see that $R_C \| R_S$, and thus by inspection of the equivalent circuit, we write

$$\tau_0 = R_{eq} \, C_{BC}$$
$$= R_C \| (R_L + R_1) C_{BC}$$

where the equivalent parallel/serial resistance of three resistors is

$$\frac{1}{R_{eq}} = \frac{1}{R_C} + \frac{1}{R_S} = \frac{1}{R_C} + \frac{1}{R_1 + R_L}$$
$$= \frac{R_1 + R_L + R_C}{R_C(R_1 + R_L)}$$
$$\therefore$$
$$R_{eq} = \frac{R_C(R_1 + R_L)}{R_1 + R_L + R_C}$$

Fig. 3.25
Example 3.5-1(e)

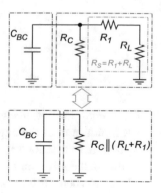

(f) Serial/parallel resistive networks are replaced by their equivalent resistance, Fig. 3.26. In this case, first we combine parallel connection of two resistors to $R_P = R_C + R_L$. Then, obviously, we see serial connection of $R_{eq} = R_1 + R_P$, and thus by inspection of the equivalent circuit, we write

$$\tau_0 = R_{eq} \, C_1$$
$$= [R_1 + (R_C \| R_L)] C_1$$

where the equivalent parallel resistance of two resistors is

$$\frac{1}{R_P} = \frac{1}{R_C} + \frac{1}{R_L} = \frac{R_C + R_L}{R_C R_L}$$
$$\therefore$$

$$R_P = (R_C || R_L) = \frac{R_C R_L}{R_C + R_L}$$

Now we can calculate the total equivalent serial resistance as

$$R_{eq} = R_1 + R_P = R_1 + \frac{R_C R_L}{R_C + R_L}$$

$$= \frac{R_1(R_C + R_L) + R_C R_L}{R_C + R_L}$$

$$= \frac{R_1 R_C + R_1 R_L + R_C R_L}{R_C + R_L}$$

Fig. 3.26
Example 3.5-1(f)

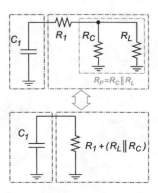

2. The main idea of how to derive time constant τ_0 is to use simple serial/parallel circuit reduction techniques until the original network is reduced to a single RL pair of components, so that it is possible to calculate $\tau_0 = L/R$. Thus, the "R" and "L" in that equation should be interpreted as "the total equivalent resistance connected to the total equivalent inductance".

   In case where several inductors are found in the circuit but they are not directly connected to each other, we remove all but one and calculate its time constant while including all resistors in the network. Then, we repeat this calculation for all capacitors one by one.

   We keep in mind that all components connected to GND symbol are indeed physically connected among themselves through the ground point. That is to say, current does flow through loops that include the ground node, as indicated in schematics of this exercise. Following the current loop passing through the given inductor helps us figure out serial/parallel connections of the equivalent resistance that are connected to its terminals.

   (a) By following the current loop passing through L, Fig. 3.27, we conclude that two resistors are connected in series. By inspection of the equivalent network, we write

$$\tau_0 = \frac{L}{R_{eq}} = \frac{L}{R_1 + R_2}$$

Fig. 3.27
Example 3.5-2(a)

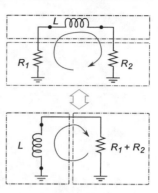

(b) By following the current loop passing through L, Fig. 3.28, we conclude that two resistors are connected in parallel (i.e. because both of them are connected to L). By inspection of the equivalent network, we write

$$\tau_0 = \frac{L}{R_{eq}} = \frac{L}{R_C \| R_L}$$

where

$$R_C \| R_L = \frac{R_C R_L}{R_C + R_L}$$

Fig. 3.28
Example 3.5-2(b)

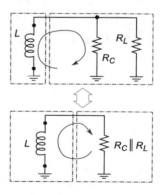

3. If it exists, a voltage source is replaced by its internal resistance $R = 0$, that is to say with a short connection. Similarly, a current source is replaced with its internal resistance $R = \infty$, that is to say an open connection.

(a) The existing voltage source is replaced by short connection. By doing so, we find that $R_S \| R_B$, and by inspection of the equivalent network, Fig. 3.29, we write

$$\tau_0 = R_{eq} C_{BE} = (R_S \| R_B) C_{BE}$$

where

$$R_S || R_B = \frac{R_S R_B}{R_S + R_B}$$

Fig. 3.29
Example 3.5-3(a)

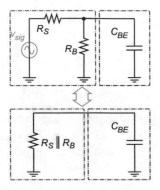

(b) The existing current source is replaced by open connection. By doing so, it is effectively removed from the network, and by inspection of the remaining equivalent network, Fig. 3.30, we write

$$\tau_0 = R_L C_C$$

Fig. 3.30
Example 3.5-3(b)

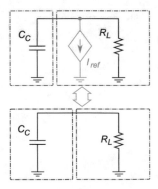

Exercise 3.6, page 79

1. The main idea for interpreting values of resistances and its inverse (i.e. conductances, a.k.a g_m) comes from understanding that by Ohm's law, resistance (or conductance) is fundamentally a *ratio of two numbers*. When the numerator in a ratio is in units of volts and denominator in units of ampere, then this ratio is written as

$$\frac{\text{number of volts}}{\text{number of amperes}} \overset{\text{def}}{=} \text{number of ohms} \overset{\text{def}}{=} R \tag{3.36}$$

Equivalently, when the numerator in a ratio is in units of ampere and denominator in units of volts, then this ratio is written as

$$\frac{\text{number of amperes}}{\text{number of volts}} = \frac{1}{\text{ohm}} \stackrel{\text{def}}{=} \text{number of siemens} \stackrel{\text{def}}{=} g_m \qquad (3.37)$$

The only question left to ask is: where do we find those two numbers? They are maybe calculated from data given in a circuit schematic, or they may be given explicitly in a form of table, or they are given explicitly in a form of graph whose one axis is declared in units of ampere and the other in units of volts. As long as there are two such axes in the plot, thus two numbers, we can calculate their ratios (3.36) and (3.37). It is not important at all which axis is in what units. Great advantage of the graph form is that we can deduce not only the V/I ratio for one volt/ampere pair of numbers, but for many. What is more, we can easily see how sensitive this ratio is by observing the graph's *slope* at any given point.

(a) $\mathscr{A}$: By inspection of this V/I characteristic, first we establish that the horizontal axis is graded to measure voltage, and the vertical axis is graded to measure current. We can try to write the ratio (3.36) assuming that each square represents the smallest unit, both for voltage and current. Thus, we only count number of squares in each direction, without specifically writing the units (e.g. mV, μA, etc.). We choose several pairs of voltage and current numbers as

$$R_{DC} = \frac{V_R}{I_R} = \frac{0}{3} = \frac{1}{3} = \frac{2}{3} = \cdots$$

We must conclude that this V/I characteristics is mathematically ambiguous because it does not have a unique interpretation or value for R. That is to say, there are infinitely many values of voltages that are paired with only one value of current. This method of calculating R, i.e. by the simple ratio of two fixed numbers, is said to be "static" or "DC". In the case of *constant current*, it produces ambiguous values for R. Consequently, its inverse ratio (3.37) is also ambiguous.

We can ask question: then, what is the physical interpretation of characteristics $\mathscr{A}$, i.e. is there a physical element with such characteristics?

Before answering that question, we can calculate R not only as "static" but also as "dynamic" or "AC". That is to say, we can calculate resistance to small *changes* of voltages and currents. In calculus, this ratio is known as the "derivative". Thus, after finding that $I_R =$const, we write

$$R_{AC} = \frac{\Delta V_R}{\Delta I_R} = \frac{\Delta V_R}{\Delta \text{const}^{\nearrow 0}} = \infty$$

This ratio is not ambiguous. It simply states that this characteristics belongs to a device that is infinitely capable of resisting *change* of its current, which holds its value regardless of voltage difference at its terminals.

Therefore, characteristics $\mathscr{A}$ belongs to *ideal current source*. Even though being ideal mathematical model, ideal current source is one of our main devices in circuit analysis. Physical realization of an ideal current source is, of course, not possible because it would have to be capable of infinite power (recall that $P = V \times I$). Nevertheless, it sets our theoretical limit for design of current source circuits: higher dynamic resistance means closer to the ideal model, i.e. better current source circuit.

In summary, there are two types of resistances: DC ("static") and AC ("dynamic"). Horizontal characteristics in I/V graph (i.e. showing current versus voltage relationship) unmistakably belongs to an ideal current source.

(b) $\mathscr{B}$: By inspection of this V/I characteristic, this time we find that it is perfectly vertical. We choose several pairs of voltages and current numbers as

$$R_{DC} = \frac{V_R}{I_R} = \frac{3}{0} = \frac{3}{1} = \frac{3}{2} = \cdots$$

We must conclude that this V/I characteristics is also mathematically ambiguous because it does not have a unique interpretation or value for R_{DC}. There are infinitely many values of currents that are paired with only one value of voltage. Consequently, its inverse ratio (3.37) is also ambiguous.

We calculate again AC resistance, after finding that V_R =const, as

$$R_{AC} = \frac{\Delta V_R}{\Delta I_R} = \frac{\overbrace{\Delta \mathrm{const}}^{0}}{\Delta I_R} = 0$$

This ratio is not ambiguous. It simply states that this characteristics belongs to a device that is infinitely capable of resisting *change* of its voltage, which holds its value regardless of current flowing through its terminals.

Therefore, characteristics $\mathscr{B}$ belongs to *ideal voltage source*. Even though being ideal mathematical model, ideal voltage source is one of our main devices in circuit analysis. Physical realization of an ideal current source is, of course, not possible because it would have to be capable of infinite power. Nevertheless, it sets our theoretical limit for design of voltage source circuits: lower dynamic resistance means closer to the ideal model, i.e. better voltage source circuit.

Horizontal characteristics in I/V graph (i.e. showing current versus voltage relationship) unmistakably belongs to an ideal voltage source whose AC resistance equals to zero.

2. These functions are obviously piecewise linear, and because they show (V/I) relationship they can be interpreted either as resistance or its equivalent conductance. As opposed to the linear resistors studied so far, each of these functions is *non-linear*, which is very important distinction. Piecewise linear approximations are easier to study because each linear section can be analysed separately. In addition, being linear, we can use finite $\Delta x / \Delta y$ ratio, instead of the exact derivative dx/dy ratio that must be used in case of continuous smooth functions. Being the ratios, we can always evaluate relative resistance or conductance by slope of the linear section that forms hypotenuse of a right-angled triangle, see Fig. 3.31.

It is important not to mix units of the vertical and horizontal cathetus in the given triangle. For example, in the graph given in this example, the vertical axis is in units of current and the horizontal axis is in the units of voltage. In case of function $\mathscr{B}$, there are two linear regions, each forming its own triangle whose slopes are equal to *conductances*. This is because mathematical definition of slope is equivalent to *tangent* of the formed angle. And tangent is, by definition, the ratio of the facing (here, vertical) and forming (here, horizontal) cathetus.

Fig. 3.31 Example 3.6-2

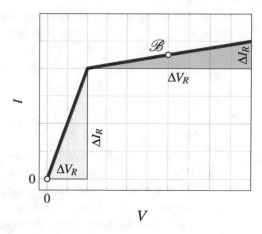

(a) $\mathscr{A}$: There are two linear sections: one from zero to two unit voltages, and one from two to ten unit voltages. Therefore, we do calculations as follows:

a. $V \in (0, 2)$: In this interval, the function is indistinguishable from already seen functions of regular linear resistors—it is linear and it crosses $(V, I) = (0, 0)$ origin. That is to say, voltage across a resistor is directly proportional to its current: zero current produces zero voltage and each subsequent unity increase in current results in unity increase in voltage. For the sake of argument, we calculate DC resistance at a few points, for example,

$$R_{DC} = \frac{V}{I} = \frac{1}{1} = \frac{2}{2} = 1 \quad \text{therefore} \quad g_m(DC) = \frac{1}{R_{DC}} = \frac{I_R}{V} = 1$$

AC calculations result in

$$R_{AC} = \frac{\Delta V}{\Delta I_R} = \frac{2}{2} = 1 \quad \text{therefore,} \quad g_m(AC) = \frac{1}{R_{AC}} = \frac{\Delta I_R}{\Delta V} = 1$$

As expected for a linear resistor, both R and g_m exist and do not change from one point to another. In addition, both DC and AC calculations produce the same results. Given a scale of 0.5 V/0.5 mA per square, we calculate $R = 1\,\text{k}\Omega$ and $g_m = 1\,\text{mS}$.

b. $V \in (2, 10)$: In this interval, we see that $I = 2 = $ constant, which is to say that both R_{DC} and $g_m(DC)$ are ambiguous (see Examples 3.6–1). On the other hand, AC calculations result in

$$R_{AC} = \frac{\Delta V}{\Delta I_R^{\nearrow 0}} = \frac{\Delta V}{0} = \infty \quad \text{therefore} \quad g_m = \frac{1}{R_{AC}} = \frac{\Delta I_R^{\nearrow 0}}{\Delta V} = 0$$

which is to say that this device generates its own current and holds it constant $I = 2$ units by having infinite resistance to its change.

We must conclude therefore that in "low" input voltage interval this non-linear device behaves as a regular $R = 1\,\text{k}\Omega$ resistor; however, for "high" input voltages, it behaves as an ideal current source $I = 2$. Given a scale of 0.5 V/0.5 mA per square, we calculate $I = 1\,\text{mA}$ and $g_m = 0\,\text{S}$.

(b) $\mathscr{B}$: Following this line of reasoning, we do calculations as follows.

a. $V \in (0, 3)$: In this interval, this is a regular linear resistor, so we calculate its resistances as

$$R_{DC} = R_{AC} = \frac{V}{I} = \frac{3}{5} \approx \frac{2}{3.3} = 0.6$$

$$\therefore$$

$$g_m(DC) = g_m(AC) = \frac{1}{R_{DC}} = \frac{I_R}{V} = \frac{5}{3}$$

Given a scale of 0.5 V/0.5 mA per square, we calculate $R = 600\,\Omega$ and $g_m = (5/3)\,\mathrm{mS}$.

b. $V \in (3, 10)$: In this interval, this is a linear resistance; however, it is *not* a regular linear resistor. The reason is that this linear function would not cross $(V, I) = (0, 0)$ origin if the voltage is reduced to zero. That is to say, there would be a non-zero current even when the voltage equals zero and that is possible only if this device is a current source. We calculate DC resistance at a few points, for example,

$$R_{DC} = \frac{V}{I} = \frac{3}{5} \neq \frac{6}{5.5} \neq \frac{10}{6}$$

therefore, it is ambiguous. Consequently, value of its $g_m(DC)$ is also not unique. On the other hand, AC calculations result in

$$R_{AC} = \frac{\Delta V}{\Delta I_R} = \frac{7}{1} = 7 \text{ therefore, } g_m(AC) = \frac{1}{R_{AC}} = \frac{\Delta I_R}{\Delta V} = \frac{1}{7}$$

both being well defined. Given a scale of 0.5 V/0.5 mA per square, we calculate $R_{AC} = 7\,\mathrm{k}\Omega$, which is equivalent to $g_m(AC) = 143\,\mu\mathrm{S}$.

Being the case, we must conclude that this device is a *non-ideal* current source, i.e. it allows a small change in its current. As opposed to the ideal model, resistance is high but not infinite. Consequently, the generated current is not perfectly constant, instead it is a weak function of its voltage—it starts as $I_{DC} = 2.5\,\mathrm{mA}$ at $V = 1.5\,\mathrm{V}$ and finishes as $I_{DC} = 3.0\,\mathrm{mA}$ at $V = 10\,\mathrm{V}$.

We must conclude therefore that inside "low" input voltage interval this non-linear device behaves as a regular $R = 600\,\Omega$ resistor; however, for "high" input voltages, it behaves as a "real" current source with $I = 5$ to 6. Given a scale of 0.5 V/0.5 mA per square, we approximate current source as $I \approx 3\,\mathrm{mA}$ and $R = 7\,\mathrm{k}\Omega$, i.e. $g_m = 143\,\mu\mathrm{S}$.

(c) $\mathscr{C}$: In this case, there are three linear intervals, and thus, by repeating the previous arguments, we conclude as follows:

a. $V \in (0, 1)$: This is a linear resistor, $R = 1/6 = 167\,\Omega$, $g_m = 6\,\mathrm{mS}$.
b. $V \in (1, 2)$: This is a real current source, $I_{DC} = 6$ to $8 \approx 3.5\,\mathrm{mA}$, $R_{AC} = 1/2 = 500\,\Omega$, $g_m = 2\,\mathrm{mS}$.
c. $V \in (2, 10)$: This is a real current source, $I_{DC} = 8$ to $10 \approx 4.5\,\mathrm{mA}$, $R_{AC} = 8/2 = 4\,\mathrm{k}\Omega$, $g_m = 250\,\mu\mathrm{S}$.

Here, we note that these piecewise linear functions may be used to approximate non-linear smooth characteristics of BJT/MOS transistors.

Multistage Interface

4

Over the time, we have developed what is known as "top-to-bottom then bottom-to-top design flow", that is to say that a complicated system is designed hierarchically; during its development stage systems are often split into more than ten levels of hierarchy. Once the hierarchy chain is established and each of the stages is replaced by its equivalent Thévenin or Norton model, each of the blocks is considered to be a "black box" described by its input and output impedances and its transfer function.

4.1 Important to Know

1. The total gain of multistage system

$$A = A_1 \times A_2 \times A_3 \times \cdots \times A_n \tag{4.1}$$

$$A_{\mathrm{dB}} = A_{1_{\mathrm{dB}}} + A_{1_{\mathrm{dB}}} + \cdots \times A_{n_{\mathrm{dB}}} \tag{4.2}$$

2. Voltage divider

$$i_R = \frac{v_{in}}{Z_1 + Z_2} \quad \text{and} \quad v_{out} = i_R Z_2 \quad \therefore \quad v_{out} = \frac{v_{in}}{Z_1 + Z_2} Z_2$$

$$\therefore$$

$$A_V \stackrel{\text{def}}{=} \frac{v_{out}}{v_{in}} = \frac{Z_2}{Z_1 + Z_2} = \frac{1}{1 + \dfrac{Z_1}{Z_2}} \tag{4.3}$$

3. RC interface

$$A_V = \left| \frac{1}{1 + j\omega\, RC} \right| = \frac{1}{\sqrt{1 + (\omega RC)^2}} \quad \text{and,} \quad \omega_0 = \frac{1}{\tau} = \frac{1}{RC} \tag{4.4}$$

© Springer Nature Switzerland AG 2021
R. Sobot, *Wireless Communication Electronics by Example*,
https://doi.org/10.1007/978-3-030-59498-5_4

4. RL interface

$$A_V = \left| \frac{1}{1 - j\dfrac{R}{\omega L}} \right| = \frac{1}{\sqrt{1 + \left(\dfrac{R}{\omega L}\right)^2}} \quad \text{and,} \quad \omega_0 = \frac{1}{\tau} = \frac{R}{L} \tag{4.5}$$

5. Current divider

$$i_{in} = i_1 + i_2 \quad \text{and} \quad v_{Z_1} = v_{Z_2} = v_1 = i_{in}\,(Z_1 \| Z_2)$$

$$\therefore$$

$$i_2 = \frac{v_{Z_2}}{Z_2} = \frac{i_{in}\,(Z_1 \| Z_2)}{Z_2} = i_{in}\,\frac{Z_1\,\cancel{Z_2}}{Z_1 + Z_2}\,\frac{1}{\cancel{Z_2}}$$

$$\therefore$$

$$A_i \overset{\text{def}}{=} \frac{i_2}{i_{in}} = \frac{1}{1 + \dfrac{Z_2}{Z_1}} \tag{4.6}$$

6. RC current divider

$$A_i = \frac{1}{1 + \dfrac{Z_C}{R}} = \frac{1}{1 + \dfrac{1}{j\omega RC}} = \frac{j\omega RC}{1 + j\omega RC} \quad \text{and,} \quad \omega_0 = \frac{1}{\tau} = \frac{1}{RC} \tag{4.7}$$

7. RL current divider

$$A_i = \frac{1}{1 + \dfrac{Z_L}{R}} = \frac{1}{1 + j\dfrac{\omega L}{R}} = \frac{1}{1 + j\dfrac{\omega}{\omega_0}} \quad \text{and,} \quad \omega_0 = \frac{1}{\tau} = \frac{R}{L} \tag{4.8}$$

8. Maximum power transfer condition

$$Z_S = Z_L^* \tag{4.9}$$

9. Power loss due to mismatch: reflection coefficient and return loss

$$\Gamma = \frac{Z_2 - Z_1}{Z_2 + Z_1} \tag{4.10}$$

$$RL = 10 \log(|\Gamma|^2) = 20 \log |\Gamma| \tag{4.11}$$

4.2   Exercises

4.1 *   Resistive Voltage Divider

1. Derive an expression for the maximum possible power P_{max} that can be delivered by a realistic voltage generator, i.e. with non-zero internal resistance, to a purely resistive load, see Fig. 4.1. A realistic voltage signal source is modelled with v_{in} and its internal resistance R_1, while resistance R_2 serves as the load.

Fig. 4.1 Example 4.1-1

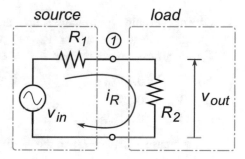

2. Using schematic in Fig. 4.1, design voltage reference $V_1 = V_{out}$ generated from given V_{in} voltage source. Additional constrain is that the equivalent resistance looking into node ① node is R_{out}. Calculate R_1 and R_2 resistors of this voltage divider. **Data:** $V_{in} = 5\,\text{V}$, $V_1 = V_{out} = 2\,\text{V}$, $R_{out} = 6\,\text{k}\Omega$.

3. Using schematic in Fig. 4.1, design voltage reference at node ① so that:

   (a) $v_{out} = 0.1\,v_{in}$,

   (b) $v_{out} = 0.5\,v_{in}$,

   (c) $v_{out} = 0.9\,v_{in}$.

   In each case, derive the equivalent resistance at the output node, then rank (V_{in}, R_1) sources using the criteria of "which one is the best voltage source"?

4.2 *   RC and RL Voltage Dividers

1. Given Thévenin type voltage source and capacitive load C, the goal is to design a divider for a specific voltage gain $A_V = V_{out}/V_{in}$, Fig. 4.2. Assume a sine signal voltage source $v_{in}(t) = V_{in}\sin\omega t$, the source resistance $R = 1\,\text{k}\Omega$, and:

   (a) $V_{out} = 0.1\,V_{in}$,

   (b) $V_{out} = 0.5\,V_{in}$,

   (c) $V_{out} = 0.9\,V_{in}$.

   In each case, calculate C and the $-3\,\text{dB}$ frequency at the following frequencies: $10\,\text{kHz}$, $1\,\text{MHz}$, and $1\,\text{GHz}$.

Fig. 4.2 Example 4.2-1

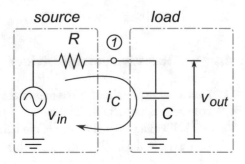

2. Given Thévenin type voltage source and inductive load L, the goal is to design a divider for a specific voltage gain $A_V = V_{out}/V_{in}$, Fig. 4.3. Assume a sine signal voltage source $v_{in}(t) = V_{in} \sin \omega t$, the source resistance $R = 1 \, k\Omega$, and:

(a) $V_{out} = 0.1 \, V_{in}$,
(b) $V_{out} = 0.5 \, V_{in}$,
(c) $V_{out} = 0.9 \, V_{in}$.

In each case, calculate L and the $-3 \, dB$ frequency at the following frequencies: 10 kHz, 1 MHz, and 1 GHz.

Fig. 4.3 Example 4.2-2

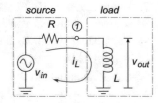

4.3 *   Current Dividers

1. Given Norton type current source and resistive load R_2, the goal is to design a current divider for a specific gain $A_i = i_2/i_{in}$, Fig. 4.4. Assume a sine signal current source $i_{in}(t) = I_{in} \sin \omega t$. Derive expression for the load resistance R_2, so that:

(a) $i_2 = 0.1 \, i_{in}$,
(b) $i_2 = 0.5 \, i_{in}$,
(c) $i_2 = 0.9 \, i_{in}$.

Given the data, in each case, calculate R_2. Comment on the calculated results.
Data: $R_1 = 100 \, k\Omega$.

Fig. 4.4 Example 4.3-1

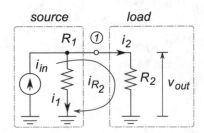

2. Given Norton type current source and capacitive load C, the goal is to design a current divider for a specific gain $A_i = i_2/i_{in}$, Fig. 4.5. Assume a sine signal current source $i_{in}(t) = I_{in} \sin \omega t$. Derive expression for the load capacitance C, so that:

(a) $i_2 = 0.1 \, i_{in}$,

(b) $i_2 = 0.5 \, i_{in}$,

(c) $i_2 = 0.9 \, i_{in}$.

In each case, calculate C and the -3 dB frequency at the following frequencies: 10 kHz, 1 MHz, and 1 GHz.
Data: $R_1 = 1 \, k\Omega$.

Fig. 4.5 Example 4.3-2

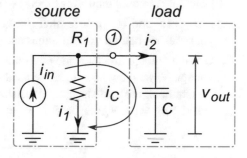

3. Given Norton type current source and inductive load L, the goal is to design a current divider for a specific gain $A_i = i_2/i_{in}$, Fig. 4.6. Assume a sine signal current source $i_{in}(t) = I_{in} \sin \omega t$. Derive expression for the load inductance L, so that:

(a) $i_2 = 0.1 \, i_{in}$,

(b) $i_2 = 0.5 \, i_{in}$,

(c) $i_2 = 0.9 \, i_{in}$.

In each case, calculate L and the -3 dB frequency at the following frequencies: 10 kHz, 1 MHz, and 1 GHz.
Data: $R_1 = 100 \, k\Omega$.

Fig. 4.6 Example 4.3-3

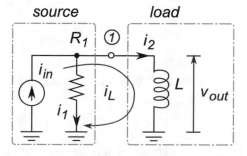

4.4 *   Interfacing

1. An antenna, modelled as $Z = 50\,\Omega$, is connected to the input node of RF amplifier whose input resistance is: (a) $R_{in} = 50\,\Omega$, (b) $R_{in} = 100\,\Omega$, and (c) $R_{in} = 1\,k\Omega$.
   In each case, calculate the reflection coefficient, the return loss, and the mismatch loss.

2. A three stage communication system is driven by an ideal Thévenin source, and it consists of:

   1. *RF amplifier*: $R_{in} = 5\,k\Omega$, $A_V = 10$, and $R_{out} = 25\,\Omega$,
   2. *Mixer*: $R_{in} = 1\,k\Omega$, $A_V = 1$, and $R_{out} = 100\,\Omega$,
   3. *IF amplifier*: $R_{in} = 10\,k\Omega$, $A_V = 6\,dB$, and $R_{out} = 50\,\Omega$.

   At the output of IF amplifier, there is a load resistor $R_L = 1\,k\Omega$
   Calculate the total gain A_V of this system and express it in V/V and dB units.

3. Design LP and HP passive filters, both with the 3 dB frequency at $f_0 = 1\,kHz$. Briefly explain your reasons behind your solutions and propose possible alternatives.

4. Given a voltage divider made of R_1 and R_2 that generates DC reference V_{ref} from power supply voltage V_{DD}, show how it can be used to drive a $R_L = 100\,\Omega$ resistive load. This load may be, for example, the next stage input resistance.
   Data: $V_{DD} = 10\,V$, $V_{ref} = 5\,V$, $R_1 = 10\,k\Omega$, $R_L = 100\,\Omega$.

Solutions

Exercise 4.1, page 111

1. By definition of power and with the application of voltage divider transfer function (4.3), we write

$$P \overset{\text{def}}{=} i_R\, v_{out} = i_R^2\, R_2 = \left[\frac{v_{in}}{R_1 + R_2} \right]^2 R_2 = \frac{v_{in}^2\, R_2}{(R_1 + R_2)^2}$$

$$= \frac{v_{in}^2}{R_1} \frac{\dfrac{R_2}{R_1}}{\left(1 + \dfrac{R_2}{R_1}\right)^2} = \frac{v_{in}^2}{R_1} \frac{x}{(1 + x)^2} \qquad (4.12)$$

after substitution of $R_2/R_1 = x$. We find that the function $f(x)$

$$f(x) = \frac{x}{(1 + x)^2} \qquad \therefore \qquad f'(x) = \frac{1 - x}{(1 + x)^3} \qquad (4.13)$$

where $x \geq 0$, has a maximum for $x = 1$ (found when derivative $f'(x) = 0$), leading into $\max[(f(x))] = 1/4$, hence

$$P_{max} = \frac{v_{in}^2}{4\,R_1} \qquad (4.14)$$

The conclusion is that the maximum power (4.14) that can be generated by a voltage generator v_{in} whose internal resistance is R_1 is achieved for $x = 1$, i.e. $R_1 = R_2$. Generalized complex form of this conclusion represents one of the key design guidelines in RF circuit design.

2. Voltage divider transfer function (4.3) by itself gives us only the R_1/R_2 ratio not the two resistors separately. That is to say, unless given additional constrains, there are two unknowns (R_1, R_2) and only equation. Therefore, one of the two resistances must be chosen arbitrary, then the second one is also determined by the given ratio. This is because, mathematically, it takes two design conditions to uniquely calculate both resistors, not only one I/O voltage ratio.

That being said, from (4.3), assuming a purely resistive divider, we write

$$\frac{v_{out}}{v_{in}} = \frac{1}{1 + \dfrac{R_1}{R_2}} \quad \therefore \quad \frac{R_1}{R_2} = \frac{v_{in}}{v_{out}} - 1 \tag{4.15}$$

Looking into node ①, after shorting the voltage source (its internal resistance is zero), the equivalent resistance is found by inspection to be the parallel resistance $R_{out} = R_1 \| R_2$. Now, we have system of two equations and two unknown variables. There are many ways to solve this system, for example,

$$\left. \begin{array}{l} \dfrac{R_1}{R_2} = \dfrac{v_{in}}{v_{out}} - 1 \\[3mm] R_{out} = R_1 \| R_2 = \dfrac{R_1 R_2}{R_1 + R_2} = \dfrac{R_1}{\frac{R_1}{R_2} + 1} \end{array} \right\} \quad \therefore \quad R_{out} = \frac{R_1}{\left(\dfrac{v_{in}}{v_{out}} - 1 \right) + 1} = R_1 \frac{v_{out}}{v_{in}}$$

Finally, given the data, we calculate

$$R_1 = R_{out} \frac{v_{in}}{v_{out}} = 6\,\text{k}\Omega \frac{5\,\text{V}}{2\,\text{V}} = 15\,\text{k}\Omega$$

and

$$\frac{R_1}{R_2} = \frac{v_{in}}{v_{out}} - 1 \quad \therefore \quad R_2 = \frac{R_1}{\dfrac{v_{in}}{v_{out}} - 1} = \frac{15\,\text{k}\Omega}{\dfrac{5\,\text{V}}{2\,\text{V}} - 1} = 15\,\text{k}\Omega \frac{2}{3} = 10\,\text{k}\Omega$$

3. From (4.3), assuming a purely resistive divider, we see that the voltage ratio V_{out}/V_{in} is controlled strictly by the resistor ratio $k = R_1/R_2$. That is to say, unless given an additional constrain, only the ratio k is needed to set the voltage ratio, not specific values of R_1 and R_2 separately. Therefore,

$$\frac{R_1}{R_2} = \frac{v_{in}}{v_{out}} - 1$$

(a) Given $v_{out} = 0.1\, v_{in}$, we write

$$\frac{R_1}{R_2} = \frac{v_{in}}{v_{out}} - 1 = \frac{\cancel{v_{in}}}{0.1\,\cancel{v_{in}}} - 1 = 9 \quad \therefore \quad R_1 = 9\,R_2$$

which may be interpreted as the source resistance being much greater than the load resistance. As a consequence, only 10% of the input source voltage is transferred to the output loading resistor.

Looking into node ①, the two resistors appear connected in parallel, and thus, the equivalent resistance is derived and approximated as

$$R_{eq} = R1 \| R_2 = \frac{1}{R_1} + \frac{1}{R_2} = \frac{R_1 R_2}{R_1 + R_2} = \frac{9 R_2 R_2}{9 R_2 + R_2} = \frac{9 R_2 \cancel{R_2}}{10 \cancel{R_2}}$$

$$= \frac{9}{10} R_2 \approx R_2$$

Therefore, we conclude that this is not a good voltage transfer interface. If anything, much greater source resistance relative to the load resistance is a property of a good current source.

(b) Given $v_{out} = 0.5 \, v_{in}$, we write

$$\frac{R_1}{R_2} = \frac{v_{in}}{v_{out}} - 1 = \frac{\cancel{v_{in}}}{0.5 \, \cancel{v_{in}}} - 1 = 1 \quad \therefore \quad R_1 = R_2$$

which may be interpreted as the voltage divider by factor two circuits. Looking into node①, the two resistors appear connected in parallel, and thus, the equivalent resistance is derived and approximated as

$$R_{eq} = R1 \| R_2 = \frac{1}{R_1} + \frac{1}{R_2} = \frac{R_1 R_2}{R_1 + R_2} = \frac{R_2 R_2}{R_1 + R_2} = \frac{R_2 \cancel{R_2}}{2 \, \cancel{R_2}}$$

$$= \frac{1}{2} R_2 \quad \text{or, equally} \quad = \frac{1}{2} R_1$$

If the intended application of this circuit was to serve as the voltage divider by factor two, then it is perfect. However, if this source is the incoming signal (e.g. from antenna), then 50% of voltage amplitude is lost. Not as inefficient as the previous case, but not perfect either.

(c) Given $v_{out} = 0.9 \, v_{in}$, we write

$$\frac{R_1}{R_2} = \frac{v_{in}}{v_{out}} - 1 = \frac{\cancel{v_{in}}}{0.9 \, \cancel{v_{in}}} - 1 = \frac{1}{9} \quad \therefore \quad R_1 = \frac{1}{9} R_2 \quad \text{or, equally} \quad R_2 = 9 R_1$$

which may be interpreted as the load resistance being much greater than the source resistance. Consequently, 90% of the input source voltage is transferred to the output loading resistor. Looking into node①, the two resistors appear connected in parallel, and thus, the equivalent resistance is derived and approximated as

$$R_{eq} = R1 \| R_2 = \frac{1}{R_1} + \frac{1}{R_2} = \frac{R_1 R_2}{R_1 + R_2} = \frac{9 R_1 R_1}{R_1 + 9 R_1} = \frac{9 R_1 \cancel{R_1}}{10 \cancel{R_1}}$$

$$= \frac{9}{10} R_1 \approx R_1$$

Therefore, we conclude that, although not ideal, this is a very good voltage transfer interface. What is more, much greater load resistance relative to the source resistance is a property of a good voltage source.

Exercise 4.2, page 111

1. Voltage divider transfer function (4.3) includes two impedances, pure resistance of source $Z_1 = R$, and capacitive impedance $Z_2 = Z_C = 1/j\omega C$. Therefore, we derive the RC voltage divider transfer function as

$$A_V \overset{\text{def}}{=} \frac{v_{out}}{v_{in}} = \frac{Z_2}{Z_1 + Z_2} = \frac{\frac{1}{j\omega C}}{R + \frac{1}{j\omega C}} = \frac{1}{1 + j\omega RC} \tag{4.16}$$

where we note that the RC impedance ratio is complex

$$\frac{Z_1}{Z_2} = \frac{R}{\frac{1}{j\omega C}} = j\omega RC$$

Module of transfer function (4.16) is

$$|A_V| = \left| \frac{1}{1 + j\omega RC} \right| = \frac{1}{\sqrt{1 + (\omega RC)^2}}$$

$$\therefore$$

$$C = \frac{\sqrt{(1/A_V)^2 - 1}}{2\pi f R} \tag{4.17}$$

In each case, the $-3\,$dB frequency is calculated as $\tau_0 = RC$; therefore, $\omega_0 = 1/RC$, that is, $f_0 = 1/2\pi RC$, and the direct implementations of (4.17) are summarized as

$\underline{R = 1\,\text{k}\Omega \text{ and } f = 10\,\text{kHz}:}$

(a) $A_V = 0.1$, $C = 158.357\text{nF}$, $f_0 = 1\,\text{kHz}$,

(b) $A_V = 0.5$, $C = 27.566\,\text{nF}$, $f_0 = 5.773\,\text{kHz}$,

(c) $A_V = 0.9$, $C = 7.708\,\text{nF}$, $f_0 = 20.65\,\text{kHz}$.

$\underline{R = 1\,\text{k}\Omega \text{ and } f = 1\text{M}\,\text{Hz}:}$

(a) $A_V = 0.1$, $C = 1.583\,\text{nF}$, $f_0 = 100\,\text{kHz}$,

(b) $A_V = 0.5$, $C = 275.664\,\text{pF}$, $f_0 = 577.3\,\text{kHz}$,

(c) $A_V = 0.9$, $C = 77.08\,\text{pF}$, $f_0 = 2.06\,\text{MHz}$.

$\underline{R = 1\,\text{k}\Omega \text{ and } f = 1\text{G}\,\text{Hz}:}$

(a) $A_V = 0.1$, $C = 1.583\,\text{pF}$, $f_0 = 100\,\text{MHz}$,

(b) $A_V = 0.5$, $C = 275.664\,\text{fF}$, $f_0 = 577.3\,\text{MHz}$,

(c) $A_V = 0.9$, $C = 77.08\,\text{fF}$, $f_0 = 2.06\,\text{GHz}$.

2. Voltage divider transfer function (4.3) includes two impedances, pure resistance of source $Z_1 = R$, and inductive impedance $Z_2 = Z_L = j\omega L$. Therefore, we derive the RL voltage divider transfer function as

$$A_V \overset{\text{def}}{=} \frac{v_{out}}{v_{in}} = \frac{Z_2}{Z_1 + Z_2} = \frac{j\omega L}{R + j\omega L} = \frac{1}{1 - j\frac{R}{\omega L}} \tag{4.18}$$

where we note that the RC impedance ratio is complex

$$\frac{Z_1}{Z_2} = \frac{R}{j\omega L} = -j\frac{R}{\omega L}$$

Module of transfer function (4.18) is

$$|A_V| = \left|\frac{1}{1 - j\dfrac{R}{\omega L}}\right| = \frac{1}{\sqrt{1 + \left(\dfrac{R}{\omega L}\right)^2}}$$

$$\therefore$$

$$L = \frac{R}{2\pi\, f\, \sqrt{(1/A_V)^2 - 1}} \tag{4.19}$$

In each case, the $-3\,\mathrm{dB}$ frequency is calculated as $\tau_0 = L/R$; therefore, $\omega_0 = R/L$, that is, $f_0 = R/2\pi L$, and the direct implementations of (4.19) are summarized as

$R = 1\,\mathrm{k\Omega}$ and $f = 10\,\mathrm{kHz}$:

(a) $A_V = 0.1$, $L = 1.599\,\mathrm{mH}$, $f_0 = 100\,\mathrm{kHz}$,

(b) $A_V = 0.5$, $L = 9.188\,\mathrm{mH}$, $f_0 = 17.5\,\mathrm{kHz}$,

(c) $A_V = 0.9$, $L = 32.86\,\mathrm{mH}$, $f_0 = 4.85\,\mathrm{kHz}$.

$R = 1\,\mathrm{k\Omega}$ and $f = 1\mathrm{M\,Hz}$:

(a) $A_V = 0.1$, $L = 15.99\,\mathrm{\mu H}$, $f_0 = 10\,\mathrm{MHz}$,

(b) $A_V = 0.5$, $L = 91.88\,\mathrm{\mu H}$, $f_0 = 1.75\,\mathrm{MHz}$,

(c) $A_V = 0.9$, $L = 328.6\,\mathrm{\mu H}$, $f_0 = 485\,\mathrm{kHz}$.

$R = 1\,\mathrm{k\Omega}$ and $f = 1\mathrm{G\,Hz}$:

(a) $A_V = 0.1$, $L = 15.99\,\mathrm{nH}$, $f_0 = 10\,\mathrm{GHz}$,

(b) $A_V = 0.5$, $L = 91.88\,\mathrm{nH}$, $f_0 = 1.75\,\mathrm{GHz}$,

(c) $A_V = 0.9$, $L = 328.6\,\mathrm{nH}$, $f_0 = 485\,\mathrm{MHz}$.

;

Exercise 4.3, page 112

1. Two resistors, the "internal" source resistance R_1 and the load resistance R_2, are seen by the ideal current source as the parallel resistance $R_{eq} = R_1 || R_2$. By direct application of Kirchhoff's laws, we derive the current gain A_i as

$$\left.\begin{array}{l} i_{in} = i_1 + i_2 \\[4pt] v_{R_1} = v_{R_2} = v_1 = i_{in}\,(R_1 || R_2) \end{array}\right\} \quad \therefore\ i_2 = \frac{v_{R_2}}{R_2} = \frac{i_{in}\,(R_1 || R_2)}{R_2} = i_{in}\,\frac{R_1\,\cancel{R_2}}{R_1 + R_2}\,\frac{1}{\cancel{R_2}}$$

Therefore, the current gain is

$$A_i = \frac{i_2}{i_{in}} = \frac{1}{1 + \frac{R_2}{R_1}} \quad \therefore \quad R_2 = \left(\frac{i_{in}}{i_2} - 1\right) R_1 \tag{4.20}$$

Given the data, direct application of (4.20) results in the following:

(a) $i_2 = 0.1\, i_{in}$, so we calculate

$$R_2 = \left(\frac{\cancel{i_{in}}}{0.1\, \cancel{i_{in}}} - 1\right) R_1 = 9R_1 = 900\,\text{k}\Omega$$

(b) $i_2 = 0.5\, i_{in}$, so we calculate

$$R_2 = \left(\frac{\cancel{i_{in}}}{0.5\, \cancel{i_{in}}} - 1\right) R_1 = R_1 = 100\,\text{k}\Omega$$

(c) $i_2 = 0.9\, i_{in}$, so we calculate

$$R_2 = \left(\frac{\cancel{i_{in}}}{0.9\, \cancel{i_{in}}} - 1\right) R_1 = \frac{1}{9}R_1 = 11.11\,\text{k}\Omega$$

In the best case scenario, the source current i_{in} is delivered unattenuated (i.e. 100'%) to the receiving load R_2, i.e. $i_2 = i_{in}$. This is possible only in two cases: either if $R_1 \to \infty$ and/or $R_2 = 0$. In both cases, $i_2 = i_{in}$, as we find by the following limits:

$$\lim_{R_1 \to \infty} A_i = \frac{1}{1 + \cancelto{0}{\frac{R_2}{\infty}}} = 1 \Rightarrow i_2 = i_{in}$$

$$\lim_{R_2 \to 0} A_i = \frac{1}{1 + \cancelto{0}{\frac{0}{R_1}}} = 1 \Rightarrow i_2 = i_{in}$$

The above two limits are summarized by the following approximation:

$$R_1 \gg R_2 \Rightarrow i_2 \approx i_{in} \tag{4.21}$$

where (4.21) is the "back of the envelope" guideline for design of current interfaces, or equally, the measure of "good current source".

It is important to notice that in this case of current divider the relationship (4.21) between the source and loading resistors is exactly opposite to the result in the case of voltage divider. This conclusion is essential to understanding fundamental difference between the voltage and current interface models. As a concluding remark, we note that being frequency independent circuit, resistive current divider indiscriminately scales both the DC and AC voltage signals.

2. The equivalent impedance seen by the current source $Z_{eq} = R||Z_C$ is

$$Z_{eq} = Z_R||Z_C \quad \therefore \quad \frac{1}{Z_{eq}} = \frac{1}{R} + j\omega C = \frac{1 + j\omega RC}{R}$$

$$\therefore$$

$$Z_{eq} = R\frac{1}{1 + j\omega RC} \quad \therefore \quad |Z_{eq}| = \frac{R}{\sqrt{1 + (\omega RC)^2}} = \frac{R}{\sqrt{1 + \left(\dfrac{\omega}{\omega_0}\right)^2}}, \qquad (4.22)$$

where $\omega_0 = 1/RC$ is the $-3\,\mathrm{dB}$ frequency. We note that the equivalent resistance is complex; in other words, there is the signal propagation delay between the input and output points measured by "phase" of this impedance.

Starting from the general expression for transfer function of current divider, see (4.6), substitution $R_2 = Z_C = 1/j\omega C$ enables us to calculate module of gain as

$$|A_i| = \frac{i_2}{i_{in}} = \left|\frac{1}{1 + \dfrac{Z_C}{R}}\right| = \left|\frac{1}{1 + \dfrac{1}{j\omega RC}}\right| = \frac{1}{\sqrt{1 + \left(\dfrac{1}{\omega RC}\right)^2}}$$

Therefore, we find

$$C = \frac{1}{2\pi f\, R\sqrt{\left(\dfrac{1}{A_i}\right)^2 - 1}} \qquad (4.23)$$

In each case, the $-3\,\mathrm{dB}$ frequency is calculated as $\tau_0 = RC$; therefore, $\omega_0 = 1/RC$, that is, $f_0 = 1/2\pi RC$, and the direct implementations of (4.23) are summarized as:

$\underline{R = 100\,\mathrm{k\Omega} \text{ and } f = 10\,\mathrm{kHz}}$:

(a) $A_i = 0.1$, $C = 15.995\,\mathrm{pF}$, $f_0 = 100\,\mathrm{kHz}$,

(b) $A_i = 0.5$, $C = 91.888\,\mathrm{pF}$, $f_0 = 17.32\,\mathrm{kHz}$,

(c) $A_i = 0.9$, $C = 328.6\,\mathrm{pF}$, $f_0 = 4.843\,\mathrm{kHz}$.

$\underline{R = 100\,\mathrm{k\Omega} \text{ and } f = 1\,\mathrm{MHz}}$:

(a) $A_i = 0.1$, $C = 159.95\,\mathrm{fF}$, $f_0 = 10\,\mathrm{MHz}$,

(b) $A_i = 0.5$, $C = 918.88\,\mathrm{fF}$, $f_0 = 1.732\,\mathrm{MHz}$,

(c) $A_i = 0.9$, $C = 3.286\,\mathrm{pF}$, $f_0 = 484.322\,\mathrm{kHz}$.

$\underline{R = 100\,\mathrm{k\Omega} \text{ and } f = 1\,\mathrm{GHz}}$:

(a) $A_i = 0.1$, $C = 0.159\,\mathrm{fF}$, $f_0 = 10\,\mathrm{GHz}$,

(b) $A_i = 0.5$, $C = 0.918\,\mathrm{fF}$, $f_0 = 1.732\,\mathrm{GHz}$,

(c) $A_i = 0.9$, $C = 3.286\,\mathrm{fF}$, $f_0 = 484.322\,\mathrm{MHz}$.

3. The equivalent impedance seen by the current source $Z_{eq} = R||Z_L$ is

$$Z_{eq} = Z_R||Z_L \quad \therefore \quad \frac{1}{Z_{eq}} = \frac{1}{R} + \frac{1}{j\omega L} = \frac{R + j\omega L}{j\omega RL}$$

$$\therefore$$

$$Z_{eq} = \frac{j\omega RL}{j\omega L + R} \quad \therefore \quad |Z_{R||L}| = \frac{|j\omega L|}{\left|1 + j\omega \dfrac{L}{R}\right|} = \frac{\omega L}{\sqrt{1 + \left(\dfrac{\omega}{R/L}\right)^2}}, \qquad (4.24)$$

where $\omega_0 = R/L$ is the $-3\,\mathrm{dB}$ frequency. We note that the equivalent resistance is complex; in other words, there is the signal propagation delay between the input and output points measured by "phase" of this impedance.

After substitution of $R_2 = Z_L = j\omega L$ into (4.6), we derive the transfer function of RL current divider as

$$|A_i| = \frac{i_2}{i_{in}} = \left|\frac{1}{1 + \dfrac{Z_L}{R}}\right| = \left|\frac{1}{1 + j\dfrac{\omega L}{R}}\right| = \frac{1}{\sqrt{1 + \left(\dfrac{2\pi f}{R/L}\right)^2}} \qquad (4.25)$$

Therefore, we find

$$L = \frac{R\sqrt{\left(\dfrac{i_{in}}{i_2}\right)^2 - 1}}{2\pi f} \qquad (4.26)$$

In each case, the $-3\,\mathrm{dB}$ frequency is calculated as $\tau_0 = L/R$; therefore, $\omega_0 = R/L$, that is, $f_0 = R/2\pi L$, and the direct implementations of (4.26) are summarized as:

$\underline{R = 100\,\mathrm{k\Omega} \text{ and } f = 10\,\mathrm{kHz}:}$

(a) $A_i = 0.1$, $L = 15.836\,\mathrm{H}$, $f_0 = 1\,\mathrm{kHz}$,

(b) $A_i = 0.5$, $L = 2.757\,\mathrm{H}$, $f_0 = 5.773\,\mathrm{kHz}$,

(c) $A_i = 0.9$, $L = 770.8\,\mathrm{mH}$, $f_0 = 20.65\,\mathrm{kHz}$.

$\underline{R = 100\,\mathrm{k\Omega} \text{ and } f = 1\,\mathrm{MHz}:}$

(a) $A_i = 0.1$, $L = 158.36\,\mathrm{mH}$, $f_0 = 100\,\mathrm{kHz}$,

(b) $A_i = 0.5$, $L = 27.567\,\mathrm{mH}$, $f_0 = 577.3\,\mathrm{kHz}$,

(c) $A_i = 0.9$, $L = 7.708\,\mathrm{mH}$, $f_0 = 2.065\,\mathrm{MHz}$.

$\underline{R = 100\,\mathrm{k\Omega} \text{ and } f = 1\,\mathrm{GHz}:}$

(a) $A_i = 0.1$, $L = 158.36\,\mathrm{\mu H}$, $f_0 = 100\,\mathrm{MHz}$,

(b) $A_i = 0.5$, $L = 27.567\,\mathrm{\mu H}$, $f_0 = 577.3\,\mathrm{MHz}$,

(c) $A_i = 0.9$, $L = 7.708\,\mathrm{\mu H}$, $f_0 = 2.065\,\mathrm{GHz}$.

Exercise 4.4, page 114

1. Given the data, by definition

$$\Gamma = \frac{Z_2 - Z_1}{Z_2 + Z_1} \quad \therefore \quad RL = 20\log|\Gamma| \ \text{ and } \ ML = \frac{1}{1 - |\Gamma|^2}$$

Therefore, matching of antenna $Z_1 = 50\,\Omega$ and RF amplifier whose input resistance is Z_2, we calculate:

(a) $R_{in} = 50\,\Omega$: $\Gamma = 0$, $RL = -\infty$, and $ML = 0$,

(b) $R_{in} = 100\,\Omega$: $\Gamma = 1/3$, $RL = -9.54\,\text{dB}$, and $ML = 1.125 = 1\,\text{dB}$,

(c) $R_{in} = 1\,\text{k}\Omega$: $\Gamma = 19/21$, $RL = -0.87\,\text{dB}$, and $ML = 5.51 = 14.8\,\text{dB}$.

2. A three stage communication system with the source and load is modelled as three stages that progressively create four voltage divider interfaces, as in Fig. 4.7.

Fig. 4.7 Example 4.4-2

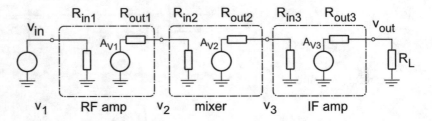

1. *RF amplifier*: $R_{in1} = 5\,\text{k}\Omega$, $A_{V1} = 10$, and $R_{out1} = 25\,\Omega$.
2. *Mixer*: $R_{in2} = 1\,\text{k}\Omega$, $A_{V2} = 1$, and $R_{out2} = 100\,\Omega$.
3. *IF amplifier*: $R_{in3} = 10\,\text{k}\Omega$, $A_{V3} = 6\,\text{dB}$, and $R_{out3} = 50\,\Omega$.
4. Load resistor $R_L = 1\,\text{k}\Omega$.

The total gain is calculated either as product of gains in V/V units, or as sum of gains in dB, as calculated at each interfacing points v_1, v_2, v_3, and v_{out}.

Following the signal path from the input to output terminals, by inspection of each voltage divider interface, we write:

1. $\underline{v_1}$: (ideal source, thus) $v_1 = v_{in} \quad \therefore \quad \underline{A_1 = 1 = 0\,\text{dB}}$,
2. $\underline{A_{V1}}$: (internal voltage source), $\underline{A_{V1} = 10 = 20\,\text{dB}}$,
3. $\underline{v_2}$: (voltage divider gain, R_{in2}, R_{out1}), $\underline{A_2 = 1000/1025 = -0.214\,\text{dB}}$,
4. $\underline{A_{V2}}$: (internal voltage source), $\underline{A_{V2} = 1 = 0\,\text{dB}}$,
5. $\underline{v_3}$: (voltage divider gain, R_{in3}, R_{out2}), $\underline{A_3 = 100/101 = -0.086\,\text{dB}}$,
6. $\underline{A_{V3}}$: (internal voltage source), $\underline{A_{V3} = 2 = 6\,\text{dB}}$,
7. $\underline{v_{out}}$: (voltage divider gain, R_L, R_{out3}), $\underline{A_3 = 100/105 = -0.424\,\text{dB}}$.

The total gain A_V as the product of all gains:

$$A_V = 1 \times 10 \times \frac{40}{41} \times 1 \times \frac{100}{101} \times 2 \times \frac{20}{21} = 18.399 = 25.3\,\text{dB}$$

while the sum of gains in dB results in the same result as

$$A = 0\,dB + 20\,dB - 0.21\,dB + 0\,dB - 0.086\,dB + 6\,dB - 0.424\,dB = 25.3\,dB$$

3. *LPF:* In order to achieve LPF, it is necessary to suppress HF tones, or to put it differently, to pull voltage at the output terminal to the ground by a short connection. Component that has zero resistance at HF is a capacitor. Therefore, placing a capacitor C between the output terminal and ground creates a short connection at HF. At the same time, the input signal voltage is applied by Thévenin type source whose resistance is R. Consequently, an RC voltage divider is created, see Exercise 4.4-1.

Time constant determines the 3 dB frequency as $\omega_0 = 1/\tau_0 = 1/RC$. At the same time, ω_0 is accepted as the upper boundary of LPF (lower boundary being DC). Since it is the product of RC that defines time constant, not the individual values of R and C components, to set $f_0 = 1\,kHz$ there are an infinite number of possibilities. In the absence of additional constrains, either R or C is chosen arbitrary and the other is calculated. For example, if $R = 1\,k\Omega$, then $C = 1/(2\pi f_0 R) = (1/2\pi)\,\mu F \approx 159\,nF$.

Alternative solution is to design a "current mode" LPF, that is to say to block the HF current at the output node by creating RC current divider where L is placed at the "load" position. Consequently, at HF, the Z_L tends to infinity, thus suppressing the output current, see Exercise 4.4–3.

Time constant of RL network is $\tau_0 = L/R$, and the rest of arguments are as same as in RC case.

HPF In order to achieve HPF, it is necessary to suppress LF tones, or to put it differently, to block LF voltage signal on the path between the source and load. Component that has infinite resistance at LF is a capacitor. Therefore, placing a capacitor C between the output terminal and ground creates an open connection at DC. Therefore, if the signal is delivered to a load R, then a CR voltage divider is crated whose transfer function is HPF.

After accounting for the complementarity between the C and L components, the rest of the discussion is as same as in LPF case.

4. In order to create $V_{ref} = 5\,V$ reference DC voltage from $V_{DD} = 10\,V$ power supply voltage, obviously the divider ratio is two, which means that $R_2 = R_1 = 10\,k\Omega$, see Fig. 4.8. Without the load connected, the DC reference voltage is set to exactly $V_{ref} = 5\,V$. However, after the load is connected to the output terminal, its resistance is added in parallel to R_2, and the voltage divider network is altered.

Fig. 4.8 Example 4.4-4

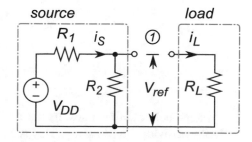

By inspection of the new circuit, we write

$$R_2||R_L = 10\,\text{k}\Omega||100\,\Omega = 99\,\Omega \approx 100\,\Omega$$

which is very close to the load resistance. As a consequence, we must recalculate the reference voltage as

$$V_{\text{ref}} = i_S\,(R_2||R_L) = \frac{v_S}{R_1 + (R_2||R_L)}\,(R_2||R_L) = \frac{v_S}{\dfrac{R_1}{R_2||R_L}+1} = \frac{10\,\text{V}}{\dfrac{10\,\text{k}\Omega}{99\,\Omega}+1}$$

$$= 98\,\text{mV} \approx 100\,\text{mV}$$

We find that connecting a low impedance load to high impedance voltage source has a disastrous consequence. The original $V_{\text{ref}} = 5\,\text{V}$ reference is created assuming that there is no significant current flowing out of node①. To rephrase, the assumption is that the loading stage has very large (ideally infinite) input resistance. Otherwise, instead of R_2, there is much lower effective resistance in its place. This case exposes the fundamental interfacing problem and why we must know driver/load resistances relationship.

Trivial solution to this problem would have been to recreate the voltage divider by using much smaller resistances so that $R_1 = R_2 \ll R_L$. Let us say, for the sake of argument, $R_1 = R_2 = 10\,\Omega$ (which is ten times lower than R_L), and again $V_{\text{ref}} = 5\,\text{V}$. But, this time, the DC current flowing through R_1, R_2 branch is

$$I_S = \frac{10\,\text{V}}{10\,\Omega + 10\,\Omega} = 0.5\,\text{A}$$

which, compared with the original $I_S = 0.5\,\text{mA}$, represents a thousand time increase in current consumption. The problem is that this high current is not necessary at all. The same reference voltage can be created with much smaller current (as the original circuit version shows). However, the circuit energy source (i.e. battery) is drained thousand times faster. Therefore, from practical perspective, this trivial solution is not acceptable.

Once the problem is understood, the solution is to use "voltage buffer", i.e. an amplifier whose input gain is very high (ideally, infinite), the output resistance is very low (ideally, zero), and gain $A_V = 1$, see Fig. 4.9. We note that practical realization of such buffer can be made using, for example, an operational amplifier. This buffer is put in between the voltage divider and the load.

Fig. 4.9 Example 4.4-4

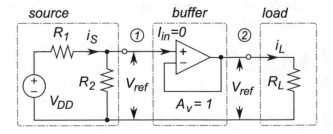

Voltage buffer simply transfers the full $V_{\text{ref}} = 5\,\text{V}$ voltage to the load without disturbing the voltage divider R_1, R_2 setup. This is because $I_{in} = 0$, which is equivalent to infinite resistance connected in parallel to R_2.

In addition, for the two solutions, it is instructional to calculate and compare powers delivered to the load. In the case of direct connection, the total power delivered to R_L is

$$P_{R_L} = v\,i = \frac{v_1^2}{R_L} = \frac{(98\,\text{mV})^2}{100\,\Omega} = 96\,\mu\text{W} \approx 100\,\mu\text{W}$$

This result is to be compared with ideal case when the full $v_1 = 5\,\text{V}$ is available at node②,

$$P_{R_L} = v\,i = \frac{v_1^2}{R_L} = \frac{(5\,\text{V})^2}{100\,\Omega} = 250\,\text{mW}$$

that is to say, the ratio of delivered powers is

$$P = \frac{250\,\text{mW}}{99\,\mu\text{W}} \approx 2500 = 68\,\text{dB}$$

This example demonstrates the importance of "buffering" the reference voltage level and maximizes the power delivered to the load.

Basic Semiconductor Devices

5

Three terminal active devices are capable of controlling large current flow at its output terminal if a small input signal is applied to its input terminal. That is to say, large waveform at the output terminal is faithful replica of the small input side waveform, thus the amplification. Key point, however, is that the device merely *controls* the output current flow. The signal energy used both at the input and output sides is provided by the *external* energy sources, battery, for example. In a mechanical analogy, function of three terminal devices is similar to a water tap that controls water flow (it does not create it).

5.1   Important to Know

1. The thermal voltage

$$V_T = \frac{kT}{q} \approx 25\,\text{mV} \tag{5.1}$$

2. Diode voltage/current relationship

$$I_D(V_D) = I_S \left[\exp\left(\frac{V_D}{n\,V_T}\right) - 1 \right] = I_S \left[\exp\left(\frac{q\,V_D}{n\,k\,T}\right) - 1 \right] \tag{5.2}$$

3. Small signal g_m gain

$$g_m \stackrel{\text{def}}{=} \left.\frac{dI_D}{dV_D}\right|_{(I_0, V_0)} = \frac{d}{dV_D}\left[I_S \exp\left(\frac{V_D}{V_T}\right) - I_S \right]_{(I_0, V_0)} = \frac{1}{V_T}\underbrace{\left[I_S \exp\left(\frac{V_0}{V_T}\right) \right]}_{I_0}$$

$$\therefore$$

$$g_m = \frac{I_0}{V_T}\ \ [\text{S}] \tag{5.3}$$

© Springer Nature Switzerland AG 2021
R. Sobot, *Wireless Communication Electronics by Example*,
https://doi.org/10.1007/978-3-030-59498-5_5

4. Small signal resistance

$$r_d \overset{\text{def}}{=} \frac{1}{g_m} = \frac{V_T}{I_0} \ [\Omega] \tag{5.4}$$

5. Varicap diode capacitance (an example model)

$$C_D = \frac{C_{D0}}{(1 - 2V_D)^{\frac{5}{4}}} \tag{5.5}$$

6. BJT model

$$I_E = I_C + I_B \tag{5.6}$$

$$I_C = \beta I_B \tag{5.7}$$

$$I_C = I_S \left[\exp\left(\frac{V_{BE}}{V_T} \right) - 1 \right] \tag{5.8}$$

7. Relationships among r_e, r_π, and g_m

$$\frac{1}{g_m} = r_e = \frac{r_\pi}{\beta + 1} \approx \frac{r_\pi}{\beta} \quad \text{and,} \quad r_\pi = \frac{\beta}{g_m} = \frac{V_T}{I_B} \tag{5.9}$$

8. Collector resistance

$$\frac{1}{r_o} \overset{\text{def}}{=} \frac{\partial I_C}{V_{CE}}\bigg|_{V_{BE}=\text{const.}} \quad \therefore \quad r_o = \frac{V_A}{I_{C0}} \tag{5.10}$$

9. MOSFET model

$$I_D = \frac{1}{2} \mu_n C_{ox} \frac{W}{L} (V_{GS} - V_t)^2 \ (1 + \lambda V_{DS}) = \frac{1}{2} \mu_n C_{ox} \frac{W}{L} V_{OV}^2 \ (1 + \lambda V_{DS}) \tag{5.11}$$

$$I_G = 0 \tag{5.12}$$

$$I_D = I_S \tag{5.13}$$

10. MOSFET g_m gain

$$g_m \overset{\text{def}}{=} \frac{dI_D}{dV_{GS}}\bigg|_{(I_{D0}, V_{GS0})} = \frac{2\,I_D}{V_{OV}}\bigg|_{I_D=I_{D0}} = \frac{2\,I_D}{V_{GS} - V_t} \tag{5.14}$$

11. JFET model

$$I_D = I_{DSS} \left[1 - \frac{V_{GS}}{V_P} \right]^2 \tag{5.15}$$

12. JFET g_m gain

$$g_m \stackrel{\text{def}}{=} \frac{dI_D}{dV_{GS}}\bigg|_{V_{DS}=\text{const.}} = \frac{2\,I_{DSS}}{|V_P|}\left(1 - \frac{V_{GS}}{V_P}\right) \tag{5.16}$$

5.2 Exercises

5.1 * Diode

1. Given a diode and resistor network, calculate the required R to set up the diode's biasing point to (I_{D0}, V_{D0}), see Fig. 5.1.
   Data: $(I_{D0}, V_{D0}) : (10\,\text{mA}, 0.695\,\text{V})$, $V_{CC} = 5\,\text{V}$.

Fig. 5.1 Example 5.1–1

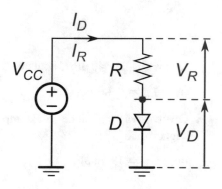

2. Assuming the same diode/resistor network as in Example 5.1–1, given the data and assuming constant I_S, calculate diode's biasing point (I_{D0}, V_{D0}).
   Data: $V_{CC} = 5\,\text{V}$, $R = 430.5\,\Omega$, $V_T = 25\,\text{mV}$, $I_S = 8.445 \times 10^{-15}\,\text{A}$

3. Given a diode and its voltage/current characteristics, see Fig. 5.2, estimate and compare DC biasing (I_0, V_0) at two points $\mathscr{A}$ and $\mathscr{B}$. In addition, determine diode's DC resistance at these two points.

Fig. 5.2 Example 5.1-3

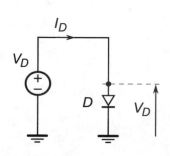

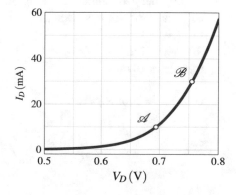

4. Further to Example 5.1-3, estimate diode's small signal parameters: $g_m(\mathscr{A})$, $g_m(\mathscr{B})$, $r_d(\mathscr{A})$, and $r_d(\mathscr{B})$.

5. Following to Examples 5.1-3, and 4, assuming a small signal variation of the diode voltage $v_D(t) = 10\,\text{mV}\,\sin(2\pi \times 1\,\text{kHz} \times t)$, calculate the total small signal diode's current $i_D(t)$ at the biasing points $\mathscr{A}$ and $\mathscr{B}$.

6. Given the exact V/I diode transfer function and its approximated form in (5.2), calculate the error in percents between the exact and approximated diode current equations, assuming the room temperature of $T = 17°C$, if:

   (a) $V_D = 120\,\text{mV}$,   (b) $V_D = 60\,\text{mV}$.

5.2 *   Varicap Diode

1. Given a varicap diode whose voltage/capacitance transfer function is given by

$$C_D = \frac{C_{D0}}{(1 - 2V_D)^{\frac{5}{4}}},\tag{5.17}$$

   where $C_{D0} = 140\,\text{pF}$. Calculate its capacitance if the applied voltage is

   (a) $V_D = -4.6\,\text{V}$,   (b) $V_D = -1.9\,\text{V}$.

2. Given a varicap diode in Example 5.2-1, calculate the diode voltage required to create the capacitance

   (a) $C_D = 19.56\,\text{pF}$,   (b) $C_D = 7.68\,\text{pF}$.

5.3 *   BJT

1. Estimate ΔV_{BE} if the collector current is increased ten times. Assume the room temperature.

2. Given a BJT transistor, calculate its g_m gain, emitter resistance r_e, and base resistance r_π, assuming $\beta = 200$, relatively large v_{BE}, and the biasing current,

   (a) $I_{C0} = 1\,\text{mA}$,   (b) $I_{C0} = 2\,\text{mA}$,   (c) $I_{C0} = 10\,\text{mA}$.

3. Given a BJT transistor and resistor R_E connected between its emitter and the ground (i.e. the case of "degenerated emitter"), see Fig. 5.3, estimate the input impedance perceived at the base node. BJT biasing current is given as
   (a) $I_{C0} = 1\,\text{mA}$,
   (b) $I_{C0} = 2\,\text{mA}$,
   (c) $I_{C0} = 10\,\text{mA}$.

   Data: $\beta = 200$, $R_E = 500\,\Omega$.

Fig. 5.3 Example 5.3-3

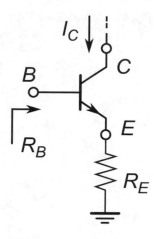

5.4 * Temperature Dependence

1. By simulation of a typical n-type BJT transistor, create a plot that shows β dependence on the collector current. In addition, use temperature as the parameter and show plots at three typical temperatures: $-55°C$, $25°C$, and $125°C$, similar to the graph in Fig. 5.4.

   If the collector current is held constant $I_C = 1\,\text{mA}$, estimate how many times the base current changes over the full temperature range?

Fig. 5.4 Example 5.4-1

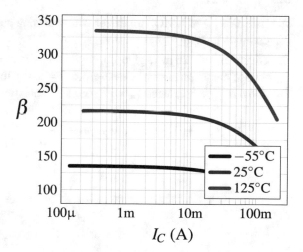

2. By simulation of a typical diode, create a plot that shows the diode current dependence on its voltage. In addition, use temperature as the parameter and show plots at three typical temperatures: $-55°C$, $25°C$, and $125°C$, similar to the graph in Fig. 5.5. Given, for example, $V_D = 755\,\text{mV}$, estimate how many times the diode current changes over the full temperature range.

Fig. 5.5 Example 5.4-2

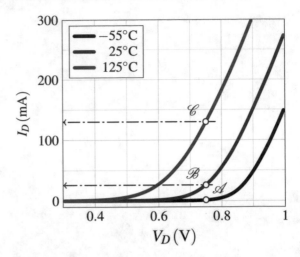

3. By simulation of a typical NPN BJT, show its g_m dependence relative:

   (a) to the collector current, at three typical temperatures $100°C$, $25°C$, and $-40°C$;
   (b) to the temperature, at three typical collector currents $100\,mA$, $10\,mA$, and $1\,mA$.

5.5 *   FET

1. Estimate g_m of NMOS transistor whose biasing point is set as $I_D = 1\,mA$ and $V_{OV} = 1\,V$. Then, estimate its r_o if $V_A = -100\,V$.

2. Given $I_D = 25\,mA$, give approximate estimates of V_{GS}, V_P, I_{DSS}, g_m, and r_o for a JFET whose characteristics are in Fig. 5.6. By using the estimated values, evaluate I_D as per (5.15) and compare estimated g_m with the exact analytical definition.

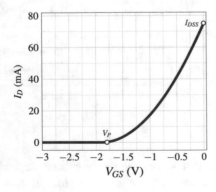
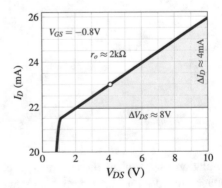

Fig. 5.6 Example 5.5-2

Solutions

Exercise 5.1, page 129

1. In the given configuration, the resistor and diode currents are the same (the two components are connected in series). Therefore, by inspection of the schematic diagram, we write

$$\left.\begin{array}{l} V_{CC} = V_D + V_R \\ I_D = I_R \end{array}\right\} \quad \therefore \quad I_R = \frac{V_R}{R} = \frac{V_{CC} - V_D}{R}$$

Therefore,

$$R = \frac{V_{CC} - V_D}{I_D} = \frac{5\,\mathrm{V} - 0.695\,\mathrm{V}}{10\,\mathrm{mA}} = 430.5\,\Omega$$

2. By inspection of the given circuit network and data, we can only write the following system of equations:

$$\left.\begin{array}{l} I_D = I_S \left[\exp\left(\frac{V_D}{V_T}\right) - 1\right] \\ I_D = I_R = \dfrac{V_{CC} - V_D}{R} \end{array}\right\} \quad \therefore$$

$$\frac{V_{CC} - V_D}{R} = I_S \left[\exp\left(\frac{V_D}{V_T}\right) - 1\right] \tag{5.18}$$

This case illustrates a classic mathematical problem. Given (5.18), it is not possible to write analytical expression for V_D because on the left side there is a linear equation, and on the right side there is a non-linear equation.

Graphical method of solving this type of equations shows that the condition $I_D = I_R$ is found at the intersect of the diode's current function (exponential, red) and the resistor's current function (linear, blue). While keeping in mind that $V_{CC} = V_D + V_R = V_E + I_R R$, this linear function is deduced from two points: when $V_D = V_{CC}$, then $V_R = 0, \therefore I_R = 0$; and when $V_D = 0$, then $I_R = V_{CC}/R$. A high resolution graph enables us to estimate the coordinates of the crossing point, i.e. the equation solution, with good accuracy, see Fig. 5.7.

Fig. 5.7 Example 5.1-2

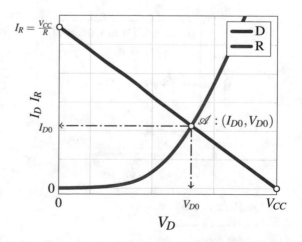

In order to use the numerical (i.e. iterative) method, we rearrange (5.18) as

$$V_{D_{(n+1)}} = V_T \ln\left[\frac{V_{CC} - V_{D_{(n)}}}{R\,I_S} + 1\right], \tag{5.19}$$

where $(n+1)$th index refers to iteration after (n)th iteration is solved. Specifically, we start by choosing an arbitrary initial value $V_{D_{(0)}} = 1\,\mathrm{V}$, and we calculate $V_{D_{(1)}}$, i.e. the next iterated solution of (5.19), as

$$V_{D_{(1)}} = 25\,\mathrm{mV}\ln\left[\frac{5\,\mathrm{V} - 1\,\mathrm{V}}{430.5\,\Omega \times 8.445 \times 10^{-15}\mathrm{A}} + 1\right] = 0.693\,\mathrm{mV} \tag{5.20}$$

Therefore, the result of the first iteration is $V_D = 0.693\,\mathrm{mV}$. We now use this value in the second iteration by substituting $V_{D_{(n)}} = 0.693\,\mathrm{mV}$ into (5.19), which produces

$$V_{D_{(2)}} = 25\,\mathrm{mV}\ln\left[\frac{5\,\mathrm{V} - 0.693\,\mathrm{mV}}{430.5\,\Omega \times 8.445 \times 10^{-15}\mathrm{A}} + 1\right] = 0.695\,\mathrm{mV} \tag{5.21}$$

We repeat again and substitute $V_{D_{(2)}} = 0.695\,\mathrm{mV}$ into (5.19), which results in

$$V_{D_{(3)}} = 25\,\mathrm{mV}\ln\left[\frac{5\,\mathrm{V} - 0.695\,\mathrm{mV}}{430.5\,\Omega \times 8.445 \times 10^{-15}\mathrm{A}} + 1\right] = 0.695\,\mathrm{mV} \tag{5.22}$$

Since $V_{D_{(3)}} = V_{D_{(2)}}$, we accept this $1\,\mathrm{mV}$ resolution result and stop the iterative process. As an exercise, we can repeat this example with some other initial values and confirm that the iterative process is indeed rapid. For example, even choosing $V_{D_0} = 4.999\,\mathrm{V}$[1], it converges to the same result.

3. Given the voltage/current characteristics of a device, we determine its electrical properties by inspection, see Fig. 5.8.
   Biasing points $\mathscr{A}$ and $\mathscr{B}$ are estimated by reading their coordinates, i.e.
   $\mathscr{A} : (I_0, V_0) = (10\,\mathrm{mA}, 695\,\mathrm{mV})$, and $\mathscr{B} : (I_0, V_0) = (30\,\mathrm{mA}, 755\,\mathrm{mV})$.
   Therefore, each point in the characteristics is uniquely described by its coordinates that are interpreted as one unique pair of mutually dependent DC current and voltage, referred to as the "biasing point".

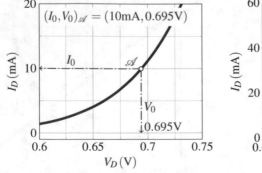
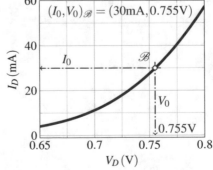

Fig. 5.8 Example 5.1-3

---

[1]Logarithmic function is defined only for the positive arguments.

DC resistance: By Ohm's law, the DC resistance is the ratio of the associated voltage and current, and thus,

$$R_{DC}(\mathscr{A}) \stackrel{\text{def}}{=} \frac{V_0}{I_0}\bigg|_{\mathscr{A}} = \frac{695\,\text{mV}}{10\,\text{mA}} = 69.5\,\Omega$$

$$R_{DC}(\mathscr{B}) \stackrel{\text{def}}{=} \frac{V_0}{I_0}\bigg|_{\mathscr{B}} = \frac{755\,\text{mV}}{30\,\text{mA}} = 25.1\,\Omega$$

We note that increase in the voltage of only 8.6% caused 200% increase in the diode current, which illustrates the non-linear voltage/current relationship. Consequently, the overall DC resistance is *reduced* to 64%. Diodes, and pn-junctions in general, are typical examples of a non-linear resistance.

4. "Small signal" device parameters are associated with their behaviour due to the presence of a non-static (a.k.a AC) signal. "Small" in this context simply means that, for example, the voltage amplitude of the applied sine signal is much smaller than the value of the biasing voltage V_{D0}. In the engineering jargon, being smaller ten or more times qualifies for being "small". We keep in mind that after DC biasing point is chosen, the small input voltage signal is actually added, which creates voltage fluctuations whose *average* is the DC voltage. Mathematically, the total input signal is created as the sum of a number (i.e. DC) and a zero-average sine waveform (i.e. AC).

Zooming-in close to either $\mathscr{A}$ or $\mathscr{B}$, biasing point helps us to graphically estimate AC parameters of a non-linear device, such as a diode. This practical and quick estimation method is based on right-angled triangle geometrical interpretation of the first derivative, i.e. linear approximation at the given point, while precise results are obtained by numerical simulations. Due to similarity between small and large triangles where both hypotenuses are aligned, we use the large triangle to make better estimate of its tangent, as shown here.

(a) g_m *gain:* In the graph, the output variable is I_D and the input variable is V_D, and thus by definition of gain as the output/input ratio, we calculate the inverse of resistance. In addition, we find the ratio of small change of current ΔI_D and small change of voltage ΔV_D, from two catheti of a large triangle (see Fig. 5.9) as

$\mathscr{A}$ (10 mA, 695 mV): The length of vertical cathetus is proportional to variation of the output current, and the length of horizontal cathetus is proportional to variation of the input voltage, and thus,

$$\left.\begin{array}{l} \Delta I_D \approx 20\,\text{mA} - 0 = 20\,\text{mA} \\[2mm] \Delta V_D \approx 750\,\text{mV} - 639\,\text{mV} = 111\,\text{mV} \end{array}\right\} g_m(\mathscr{A}) \approx \frac{\Delta I_D}{\Delta V_D}\bigg|_{\mathscr{A}} = 180\,\text{mS}$$

$\mathscr{B}$ (30 mA, 755 mV): Similarly,

$$\left.\begin{array}{l} \Delta I_D \approx 52\,\text{mA} - 0 = 52\,\text{mA} \\[2mm] \Delta V_D \approx 800\,\text{mV} - 693\,\text{mV} = 107\,\text{mV} \end{array}\right\} g_m(\mathscr{B}) \approx \frac{\Delta I_D}{\Delta V_D}\bigg|_{\mathscr{B}} = 486\,\text{mS}$$

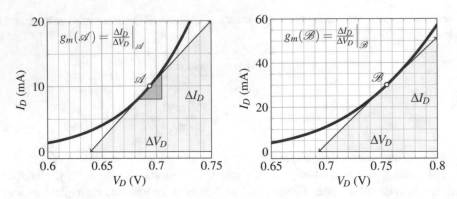

Fig. 5.9 Example 5.1-4

(b) _AC resistance:_ _Small signal_ resistance r_d (a.k.a. AC resistance) is the resistance to _change_ in voltage ΔV_D due to change in ΔI_D at the chosen biasing point. Thus, by definition, it is inverse to g_m, and we write

$$r_d(\mathscr{A}) \overset{\text{def}}{=} \frac{1}{g_m}\bigg|_{\mathscr{A}} \approx \frac{\Delta V_D}{\Delta I_D}\bigg|_{\mathscr{A}} = \frac{111\,\text{mV}}{20\,\text{mA}} = 5.55\,\Omega \;;$$

$$r_d(\mathscr{B}) \overset{\text{def}}{=} \frac{1}{g_m}\bigg|_{\mathscr{B}} \approx \frac{\Delta V_D}{\Delta I_D}\bigg|_{\mathscr{B}} = \frac{107\,\text{mV}}{52\,\text{mA}} = 2\,\Omega$$

It is useful to note that, in any case, the resistance of a conducting diode is relatively low, thus often approximated to zero.

5. At biasing point $\mathscr{A}$: $(10\,\text{mA}, 695\,\text{mV})$, we calculate the AC current as

$$i_D(t) = \frac{v_D(t)}{r_d(\mathscr{A})} \overset{\text{def}}{=} g_m(\mathscr{A}) \times v_D(t) = 180\,\text{mS} \times 10\,\text{mV}\,\sin(2\pi \times 1\,\text{kHz} \times t)$$

$$= 1.8\,\text{mA}\,\sin(2\pi \times 1\,\text{kHz} \times t)$$

$$\therefore$$

$$i_D(t) = I_0 + i_D(t) = 10\,\text{mA} + 1.8\,\text{mA}\,\sin(2\pi \times 1\,\text{kHz} \times t)$$

At biasing point $\mathscr{B}$: $(30\,\text{mA}, 755\,\text{mV})$, we calculate the AC current as

$$i_D(t) = \frac{v_D(t)}{r_d(\mathscr{B})} \overset{\text{def}}{=} g_m(\mathscr{A}) \times v_D(t) = 500\,\text{mS} \times 10\,\text{mV}\,\sin(2\pi \times 1\,\text{kHz} \times t)$$

$$= 5\,\text{mA}\,\sin(2\pi \times 1\,\text{kHz} \times t)$$

$$\therefore$$

$$i_D(t) = I_0 + i_D(t) = 30\,\text{mA} + 5\,\text{mA}\,\sin(2\pi \times 1\,\text{kHz} \times t)$$

Fig. 5.10 Example 5.1-5

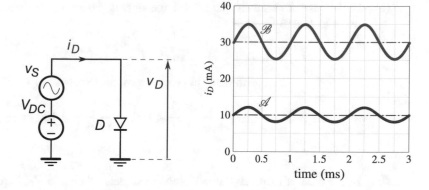

SPICE simulation confirms these hand calculated results, Fig. 5.10. Observe how AC and DC components are added in the simulation schema and how output waveforms are measured. As an exercise, repeat this example by assuming some other amplitude of $v_D(t)$ and/or some other biasing points.

6. The goal of this exercise is to study under what conditions it is valid to use the approximated version of (5.2). First, we calculate

$$V_T \overset{\text{def}}{=} \frac{kT}{q} = \frac{k(273.15\text{K} + T\,[^\circ\text{C}])}{q} = \frac{1.380 \times 10^{-23}\text{J/K}\,(273.15\text{K} + 17^\circ\text{C})}{1.602 \times 10^{-19}}$$
$$= 25\,\text{mV}$$

and note that at 17°C, we calculate $V_T = 25\,\text{mV}$, which is to say that in hand calculation we use this convenient voltage as "at the room temperature" value, even though that in reality the room temperature is a little bit higher (in fact, a pn-junction is a bit hotter than 17°C). For the sake of argument, at 30°C, we calculate $V_T \approx 26\,\text{mV}$, and thus the voltage approximation is still valid. Next, it is the argument of exponential function that is important, so that we calculate two ratios as

(a) $x_1 = \dfrac{120\,\text{mV}}{25\,\text{mV}} = 4.8$ $\therefore$ $e^{x_1} = 121$ and $e^{x_1} - 1 = 120$

(b) $x_2 = \dfrac{60\,\text{mV}}{25\,\text{mV}} = 2.4$ $\therefore$ $e^{x_2} = 11$ and $e^{x_2} - 1 = 10$

We use the above numbers as reminder that it is the *ratio* of pn-junction voltage relative to approximately 25 mV that, due to the exponential function, very quickly generates a large number. Consequently, we calculate the approximation errors as follows:

(a) In the case of relatively "large" voltage signals relative to V_T, here only 4.8 times or more, we calculate the error as

$$I_{D1} = I_S \left(e^{x_1} - 1 \right) \quad \text{and} \quad I_{D2} = I_S \left(e^{x_1} - 1 \right)$$

$$\therefore \quad \text{(difference in percentages, the exact vs. approximated)}$$

$$\text{err } [\%] = \frac{\cancel{I_S} \left(e^{x_1} - e^{x_1} + 1 \right)}{\cancel{I_S} \left(e^{x_1} - 1 \right)} \times 100 = \frac{1}{e^{x_1} - 1} \times 100 = \frac{1}{120} \times 100 = 0.83\%$$

(b) In the case of relatively "small" voltage signals relative to V_T, here even 2.8 times or less, we calculate the error as

$$I_{D1} = I_S \left(e^{x_2} - 1 \right) \quad \text{and} \quad I_{D2} = I_S \left(e^{x_2} - 1 \right)$$

$$\therefore \quad \text{(difference in percentages, the exact vs. approximated)}$$

$$\text{err } [\%] = \frac{\cancel{I_S} \left(e^{x_2} - e^{x_2} + 1 \right)}{\cancel{I_S} \left(e^{x_2} - 1 \right)} \times 100 = \frac{1}{e^{x_2} - 1} \times 100 = \frac{1}{10} \times 100 = 10\%$$

We conclude that for "large" diode voltage using the approximated equation for diode current is justified and the error is less than 1%. However, for "small" voltage signals, 60 mV or less, the error becomes 10% or larger, and consequently, the exact form of (5.2) should be used.

Exercise 5.2, page 130

1. From transfer characteristics (5.17), we simply calculate

(a) $V_D = -4.6$ V: $C_D = \dfrac{C_{D0}}{(1 - 2V_D)^{\frac{5}{4}}} = \dfrac{140 \, \text{pF}}{(1 - 2(-4.6))^{\frac{5}{4}}} \approx 8.6 \, \text{pF}$

(b) $V_D = -1.9$ V: $C_D = \dfrac{C_{D0}}{(1 - 2V_D)^{\frac{5}{4}}} = \dfrac{140 \, \text{pF}}{(1 - 2(-1.9))^{\frac{5}{4}}} \approx 21.3 \, \text{pF}$

We note that varicap diodes are reverse biased. This is because this diode is not used to turn ON or OFF its current. Instead, its pn-junction capacitance is a function of the reverse bias voltage, which controls the pn-junction width and, therefore, its capacitance.[2] In addition, due to exponent "5/4", the numerical calculations are very sensitive, as well as the capacitance of a real varicap diode. In practice, the varicap diode voltage is well controlled so that we can precisely "tune" its capacitance.

2. From the transfer function (5.17), we derive

$$V_D = \frac{1}{2} \left[1 - \left(\frac{C_{D0}}{C_D} \right)^{\frac{4}{5}} \right]$$

and we calculate

---

[2] See lecture on the capacitance of a plate capacitor.

(a) $C_D = 19.56\,\text{pF}$: $V_D = \dfrac{1}{2}\left[1 - \left(\dfrac{140\,\text{pF}}{19.56\,\text{pF}}\right)^{\frac{4}{5}}\right] = -1.9\,\text{V}$

(b) $C_D = 7.678\,\text{pF}$: $V_D = \dfrac{1}{2}\left[1 - \left(\dfrac{140\,\text{pF}}{7.678\,\text{pF}}\right)^{\frac{4}{5}}\right] = -4.6\,\text{V}$

Thus, to set $C_D = 7.678\,\text{pF}$, the diode must be biased with $V_D = -4.6\,\text{V}$, and to set $C_D = 19.559\,\text{pF}$, the diode must be biased with $V_D = -1.9\,\text{V}$.

Exercise 5.3, page 130

1. In accordance to a diode's V/I transfer function, assuming $I_S = \text{const.}$ and relatively large V_{BE}, for two different emitter diode currents (therefore, two different base–emitter voltages), we write

$$I_C = I_S \exp\left(\frac{V_{BE1}}{V_T}\right) \quad \text{and} \quad 10 \times I_C = I_S \exp\left(\frac{V_{BE2}}{V_T}\right)$$

$$\therefore \quad \text{(ratio of two currents)}$$

$$\frac{10 \times \cancel{I_C}}{\cancel{I_C}} = \frac{\cancel{I_S} \exp\left(\frac{V_{BE2}}{V_T}\right)}{\cancel{I_S} \exp\left(\frac{V_{BE1}}{V_T}\right)} = \exp\left(\frac{V_{BE2} - V_{BE1}}{V_T}\right) = \exp\left(\frac{\Delta V_{BE}}{V_T}\right)$$

$$\therefore$$

$$\Delta V_{BE} = V_T \ln 10 = 25\,\text{mV} \times 2.3026 = 57.567\,\text{mV} \approx 60\,\text{mV}$$

That is to say, if the base–emitter voltage is increased by about $60\,\text{mV}$, the collector current is increased ten times, which illustrates their strong exponential relationship.

2. With the assumption of "relatively large" v_{BE}, we use the approximated equation form for collector current

$$I_C = I_S \exp\left(\frac{V_{BE}}{V_T}\right)$$

At a given biasing point (I_{C0}, V_{BE0}), the transconductance gain g_m of a BJT is the ratio of the output current *change*, i.e. ΔI_C, and the input voltage *change*, i.e. ΔV_{BE}, that is causing the output current. More accurately, it is the ratio of the infinitely small changes, and in other words, it is the derivative of the collector current relative to the base–emitter voltage[3]

$$g_m \overset{\text{def}}{=} \frac{dI_C}{dV_{BE}}\bigg|_{(I_{C0}, V_{BE0})} = \frac{d}{dV_{BE}}\left[I_S \exp\left(\frac{V_{BE}}{V_T}\right)\right] = \frac{I_S \exp\left(\dfrac{V_{BE}}{V_T}\right)}{V_T} = \frac{I_C}{V_T}$$

$$\therefore$$

$$g_m\big|_{(I_{C0}, V_{BE0})} = \frac{I_{C0}}{V_T} \tag{5.23}$$

Given the data, we calculate the following:

---

[3] See textbook for more detailed discussion on small signal parameters.

(a) Convenience of (5.23) is that by knowing the BJT biasing current we easily calculate the corresponding g_m gain and, therefore, the emitter resistance r_e as

$$g_m = \frac{I_C}{V_T} = \frac{1\,\text{mA}}{25\,\text{mV}} = 40\,\text{mS} \quad \therefore \quad r_e = \frac{1}{g_m} = \frac{1}{40\,\text{mS}} = 25\,\Omega$$

Given β and r_e, we "project" the emitter resistance to the base node and calculate $r_\pi = (\beta + 1)r_e \approx 200 \times 25\,\Omega = 5\,\text{k}\Omega$.

(b) BJT biasing current directly determines its g_m gain and emitter resistance r_e, and indirectly (through β and r_e) its base resistance,

$$g_m = \frac{I_C}{V_T} = \frac{2\,\text{mA}}{25\,\text{mV}} = 80\,\text{mS} \quad \therefore \quad r_e = \frac{1}{g_m} = \frac{1}{80\,\text{mS}} = 12.5\,\Omega$$

Given β and r_e, we "project" the emitter resistance to the base node and calculate $r_\pi = (\beta + 1)r_e \approx 200 \times 12.5\,\Omega = 2.5\,\text{k}\Omega$.

(c) BJT biasing current directly determines its g_m gain and emitter resistance r_e, and indirectly (through β and r_e) its base resistance,

$$g_m = \frac{I_C}{V_T} = \frac{10\,\text{mA}}{25\,\text{mV}} = 400\,\text{mS} \quad \therefore \quad r_e = \frac{1}{g_m} = \frac{1}{400\,\text{mS}} = 2.5\,\Omega$$

Given β and r_e, we "project" the emitter resistance to the base node and calculate $r_\pi = (\beta + 1)r_e \approx 200 \times 2.5\,\Omega = 500\,\Omega$.

In summary, the BJT biasing current is the first design parameter to be decided upon, then following (5.23), we find that g_m, r_e, and r_π are not independent and are already predetermined through their reciprocal relationships.

3. Assuming "relatively large" v_{BE}, we use the approximated equation form for collector current

$$I_C = I_S \exp\left(\frac{V_{BE}}{V_T}\right)$$

and, therefore, (5.23) to calculate g_m gain and subsequently r_e. However, in BJT emitter branch, there are two resistors in series, i.e. the internal r_e and the external R_E, see Fig. 5.11. The consequence of this connection is that the *total* resistance of emitter branch is $R'_E = r_e + R_E$.

Fig. 5.11 Example 5.3-3

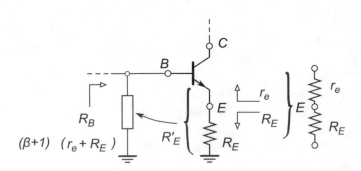

Given the data, we calculate

(a) $I_C = 1\,\text{mA}$, $R_E = 500\,\Omega$, $\beta = 200$:

$$g_m = \frac{I_C}{V_T} = \frac{1\,\text{mA}}{25\,\text{mV}} = 40\,\text{mS} \quad \therefore \quad r_e = \frac{1}{g_m} = \frac{1}{40\,\text{mS}} = 25\,\Omega$$

Given β and r_e, we "project" the total emitter resistance R'_E to the base node and calculate
$R_B = (\beta + 1)R'_E \approx 200 \times 525\,\Omega = 105\,\text{k}\Omega \approx 100\,\text{k}\Omega$.

(b) $I_C = 2\,\text{mA}$, $R_E = 500\,\Omega$, $\beta = 200$:

$$g_m = \frac{I_C}{V_T} = \frac{2\,\text{mA}}{25\,\text{mV}} = 80\,\text{mS} \quad \therefore \quad r_e = \frac{1}{g_m} = \frac{1}{80\,\text{mS}} = 12.5\,\Omega$$

Given β and r_e, we "project" the total emitter resistance R'_E to the base node and calculate
$R_B = (\beta + 1)R'_E \approx 200 \times 512.5\,\Omega = 102.5\,\text{k}\Omega \approx 100\,\text{k}\Omega$.

(c) $I_C = 10\,\text{mA}$, $R_E = 500\,\Omega$, $\beta = 200$:

$$g_m = \frac{I_C}{V_T} = \frac{10\,\text{mA}}{25\,\text{mV}} = 400\,\text{mS} \quad \therefore \quad r_e = \frac{1}{g_m} = \frac{1}{400\,\text{mS}} = 2.5\,\Omega$$

Given β and r_e, we "project" the total emitter resistance R'_E to the base node and calculate
$R_B = (\beta + 1)R'_E \approx 200 \times 502.5\,\Omega = 100.5\,\text{k}\Omega \approx 100\,\text{k}\Omega$.

In summary, as opposed to $r_\pi = \beta r_e$, the addition of the external $R_E \gg r_e$ increases the base resistance to approximately $R_B = \beta R_E$. This large input resistance is much better suited for interfacing with voltage source connected to the base. What is more, the dependence of the base resistance on the collector current is, for all practical purposes, removed. For these reasons, degenerated emitter BJT is extensively used in amplifier circuits.

Exercise 5.4, page 131

1. By inspection of temperature dependent curves, we find that $\beta(-55°\text{C}, 1\,\text{mA}) = 135$ and $\beta(125°\text{C}, 1\,\text{mA}) = 335$. Direct implementation of (5.7) leads into conclusion that

$$I_B(-55°\text{C}, 1\,\text{mA}) = \frac{1\,\text{mA}}{\beta(-55°\text{C}, 1\,\text{mA})} = \frac{1\,\text{mA}}{135} = 7.41\,\mu\text{A}$$

and

$$I_B(125°\text{C}, 1\,\text{mA}) = \frac{1\,\text{mA}}{\beta(-55°\text{C}, 1\,\text{mA})} = \frac{1\,\text{mA}}{335} = 2.98\,\mu\text{A}$$

that is, if the collector current is to be held constant, the base current must change

$$n = \frac{I_B(-55°\text{C}, 1\,\text{mA})}{I_B(125°\text{C}, 1\,\text{mA})} = \frac{7.41\,\mu\text{A}}{2.98\,\mu\text{A}} = 2.48 \text{ times } = 248\%$$

which is the significant change that illustrates the need for using the temperature compensation techniques in high precision applications (e.g. A/D converters), or in circuits intended for harsh environments (e.g. space, military). However, if the environment temperature is stable (e.g. biomedical implants), then β factor is inherently stable over a relatively wide range of biasing currents.

2. By inspection of temperature dependent curves, we find that at $V_D = 755\,\text{mV}$, the diode is practically turned off if the environment is cooled to $-55°\text{C}$, point $\mathscr{A}$, and thus, its current is orders of magnitude lower than the estimated $25\,\text{mA}$ at $25°\text{C}$ (point $\mathscr{B}$) and $130\,\text{mA}$ at $125°\text{C}$, point $\mathscr{C}$.

   This remarkable temperature dependence that is associated with any pn-junction (including the base–emitter diode of a BJT, for example) illustrates the need for using the temperature compensation techniques in high precision circuits for applications like A/D converters, or in circuits intended for harsh environments (e.g. space, military).

3. Equation (5.23) shows that g_m gain is directly proportional to I_C and inversely proportional to V_T, both of which, by themselves, are temperature dependent. SPICE simulation illustrates this temperature sensitivity, see Fig. 5.12.

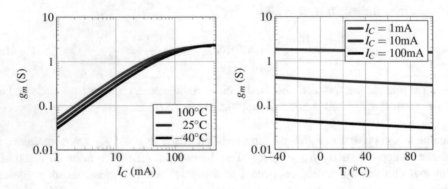

Fig. 5.12 Example 5.4-3

The simulation shows that g_m dependence is not exactly as it would be implied by (5.23). This is because there is also strong $I_S(T)$ function involved. Nevertheless, with a careful design, it is possible to achieve a very good temperature compensation of $g_m(T)$ function.

Exercise 5.5, page 132

1. Direct implementation of (5.14) gives

$$g_m = \frac{2\,I_D}{V_{OV}}\bigg|_{I_D=I_{D0}} = \frac{2 \times 1\,\text{mA}}{1\,\text{V}} = 2\,\text{mS}$$

   By definition of Early voltage, we find

$$r_o = \frac{V_A}{I_{D0}} = \frac{|-100\,\text{V}|}{1\,\text{mA}} = 100\,\text{k}\Omega$$

2. By inspection of graph in Fig. 5.6(left), we estimate: $I_D = 25\,\text{mA} \rightarrow V_{GS} = 800\,\text{mV}$, as well as $V_P \approx -1.8\,\text{V}$, and $I_{DSS} \approx 75\,\text{mA}$. In order to estimate g_m, we draw a tangent at (I_D, V_{GS}) : $(25\,\text{mA}, 800\,\text{mV})$ biasing point, see Fig. 5.13(left), which enables us to estimate

$$g_m \approx \frac{\Delta I_D}{\Delta V_{GS}} \approx \frac{62\,\text{mA}}{1.35\,\text{V}} \approx 46\,\text{mS}$$

By magnifying graph I_D vs. V_{DS}, see Fig. 5.13(right), we estimate the output resistance as

$$r_o \approx \frac{\Delta V_{DS}}{\Delta I_D} \approx \frac{8\,\text{V}}{4\,\text{mA}} = 2\,\text{k}\Omega$$

Calculation of I_D as per (5.15) gives

$$I_D = I_{DSS}\left[1 - \frac{V_{GS}}{V_P}\right]^2 = 75\,\text{mA}\left[1 - \frac{-0.8\,\text{V}}{-1.8\,\text{V}}\right]^2 \approx 23\,\text{mA}$$

which is also shown in the output characteristics, along with the finite output resistance r_0. Similarly, when g_m gain is calculated analytically, we find

$$g_m \overset{\text{def}}{=} \frac{dI_D}{dV_{GS}}\bigg|_{V_{DS}=\text{const.}} = \frac{2\,I_{DSS}}{|V_P|}\left(1 - \frac{V_{GS}}{V_P}\right) = \frac{2 \times 75\,\text{mA}}{1.8\,\text{V}}\left(1 - \frac{0.8\,\text{V}}{1.8\,\text{V}}\right)$$

$$= 46.3\,\text{mS}$$

which is very close to the estimated value, thus illustrating the applicability of the graph method.

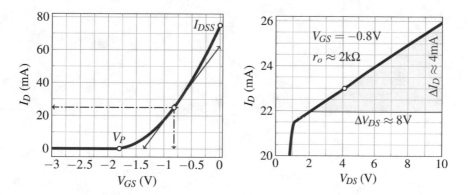

Fig. 5.13 Example 5.5-2. In order to keep the same scale as the other graphs in this section, please replace with the updated pdf files

Transistor Biasing

Active devices pose a design challenge due to their non-linear voltage–current characteristics. Inherently, there are two very different resistances that are found at each point of the V–I transfer characteristics, one static and the other one dynamic. When fixed voltage/current stimulus is applied at the terminals of an active device, then by Ohm's law, we find static (i.e. DC) resistance at the given point simply by calculating V/I ratio. However, when this stimulus signal varies around its static value, resistance to that change is then calculated by using the first derivative mathematical operation. Depending on the specific shape of a non-linear function, this dynamic (i.e. AC) resistance is very much dependent on the function's curvature at the given static point, that is to say on its derivative.

6.1   Important to Know

1. BJT biasing conditions for constant current mode of operations

$$V_{BE} \geq V_t \text{ and, } V_{CE} \geq V_{CE}(min) \tag{6.1}$$

2. Voltage divider

$$V_B = \frac{V_{CC}}{R_1 + R_2} R_2 \tag{6.2}$$

3. Sensitivity

$$S_x^y = \frac{\dfrac{dy}{y}}{\dfrac{dx}{x}} = \frac{x}{y}\frac{dy}{dx} \tag{6.3}$$

4. Maximum collector resistance (CE amplifier)

$$R_C(max) = \frac{V_{CC} - V_{CE}(min)}{I_{C0}} \tag{6.4}$$

© Springer Nature Switzerland AG 2021
R. Sobot, *Wireless Communication Electronics by Example*,
https://doi.org/10.1007/978-3-030-59498-5_6

6.2   Exercises

6.1 *   BJT Biasing: NPN

1. For a single NPN BJT transistor, draw the schematic symbol and indicate potentials at the three terminals, i.e. the V_C, V_B, and V_E, and their relationship assuming the transistor is turned on, i.e. it is operating in the forward active region. Repeat the exercise using a PNP BJT transistor.

2. If V_{BE} voltage of a BJT transistor changes by 18 mV, how large is the change of I_C as expressed in dB? What if V_{BE} changes by 60 mV? (note: $V_T = kT/q = 25$ mV)

3. Given BJT whose biasing point is set by the power supply voltage and resistive divider, see Fig. 6.1, calculate its biasing voltage (V_{BE0}).
   Data: $V_{CE} = 10$ V, $R_1 = 143.85$ kΩ, $R_2 = 10$ kΩ.

Fig. 6.1 Example 6.1-3

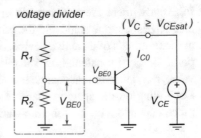

4. Further to Example 6.1-3, first, design a biasing voltage reference so that BJT base–emitter voltage is set to V_{BE0}. Then, calculate the variation in V_{BE0} if discrete resistors that have standard industrial values and ±10% tolerance are used.
   Data: $V_{CE} = 10$ V, $V_{BE0} = 0.650$ V.

5. Given the BJT transistor, calculate R_{BB} so that the biasing point is set to (V_{BE}, I_C), Fig. 6.2. Then, replace the initial ideal value of R_{BB} with the closest 10% standard value and evaluate again the actually implemented biasing point.
   Data: $\beta = 100$, $V_{CE} = 10$ V, (V_{BE}, I_C) = (650 mV, 1 mA).

Fig. 6.2 Example 6.1-5

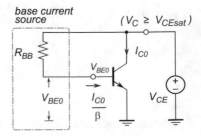

6.2 *   Sensitivity

1. Given the BJT biased by a voltage divider reference, calculate the sensitivity of base–emitter voltage due to collector–emitter variations $S_{V_{BE0}}^{V_{CE}}$.

2. Assuming the biasing technique used in Example 6.1-1, calculate the sensitivity $S_{I_{C0}}^{\beta}$ of collector current I_{C0} due to variations of β.

3. Assuming the voltage divider biasing technique used in Example 6.1-5, calculate the sensitivity $S_{I_{C0}}^{\beta}$ of collector current I_{C0} due to variations of β.

4. Given the BJT biased by resistive voltage divider and the external R_E in Fig. 6.3,

   (a)   derive the condition to minimize I_C sensitivity due to β variations,

   (b)   design the biasing network to set the biasing point to (V_{BE0}, I_{C0}).

   Data: $(V_{BE0}, I_{C0}) = (650\,\text{mV}, 1\,\text{mA})$ $V_{CC} = 10\,\text{V}$, $\beta = 100$, $R_E = 1\,\text{k}\Omega$, $V_{CEsat} = 0.1\,\text{V}$.

Fig. 6.3 Example 6.2-4

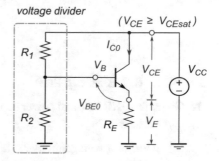

5. We use a typical BJT transistor biased with voltage divider, see Fig. 6.1, as well as its "degenerated emitter" version as in Fig. 6.3. Design the two biasing networks so that both amplifiers are biased with the same biasing current I_{C0_1}. Using SPICE simulation, compare the temperature dependence of collector's current in these two cases. For the simulations, use the standard industrial temperature range interval, i.e. $(-40\,°\text{C}, +80\,°\text{C})$. Repeat the exercise at I_{C0_2} and compare the two outcomes. **Data:** $I_{C0_1} = 1\,\text{mA}$, $I_{C0_2} = 5\,\text{mA}$.

6.3 *   Collector Resistor

1. Assuming that the biasing current of BJT transistor is set to I_{C0} in Fig. 6.4, calculate the range of values of resistor R_C.
   Data: $V_{CC} = 10\,\text{V}$, $I_{C0} = 1\,\text{mA}$, $V_{CE}(min) = 0.1\,\text{V}$.

Fig. 6.4 Example 6.3-1

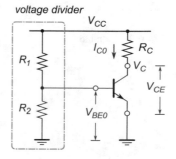

2. Assuming that the biasing point of NPN BJT transistor is as same as in Example 6.1-1, with the addition of R_E, calculate the maximum value of resistor R_C, see Fig. 6.5.
   Data: $R_E = 1\,k\Omega$.

Fig. 6.5 Example 6.3-2

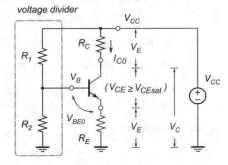

6.4 *   BJT Biasing: PNP

1. Given the PNP BJT and power supply V_{EC}, design two types of biasing networks, the voltage and current based.

   (a) Design a voltage divider reference to set $V_{EB0} = 650\,mV$, see Fig. 6.6. Note the symmetry of base–emitter voltage relative to NPN BJT setup.

   (b) Design the base-current reference using R_{BB} to set up the biasing point to (V_{EB0}, I_{C0}) given β of this PNP transistor, see Fig. 6.7.

   Data: $V_{EB0} = 650\,mV$, $I_{C0} = 1\,mA$, $V_{EC} = 10\,V$, $\beta = 100$.

Fig. 6.6 Example 6.4-1

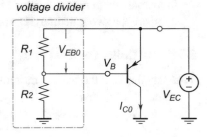

Fig. 6.7 Example 6.4-1

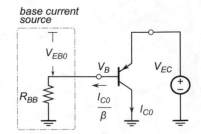

Solutions

Exercise 6.1, page 146

1. Relative potentials at NPN and PNP BJT transistor that are required to set the forward active mode are shown in Fig. 6.8. In addition, the equivalent "back to back" diode models are used to specifically illustrate polarizations of BE and BC diodes.

$$\frac{I_{C2}}{I_{C1}} = \exp\left(\frac{18\,\text{mV}}{25\,\text{mV}}\right) = 2 = 6\,\text{dB}$$

2. Given data, where $V_{BE2} = V_{BE2} + 8mV$, the ratio of two collector currents is

3. By inspection of the schematic diagram, we write

$$V_{BE0} = I_{R_2} R_2 = \frac{V_{CE}}{R_1 + R_2} R_2 = V_{CE} \frac{1}{1 + \dfrac{R_1}{R_2}} \tag{6.5}$$

Direct implementation of (6.5), after neglecting the base current, results in

$$V_{BE0} = V_{CE} \frac{1}{1 + \dfrac{R_1}{R_2}} = 10\,\text{V} \frac{1}{1 + \dfrac{143.85\,\text{k}\Omega}{10\,\text{k}\Omega}} = 10\,\text{V} \times 65 \times 10^{-3} = 650\,\text{mV}$$

Also, we always verify that $V_{CE} > V_{CE}(min)$, which in this case is true because $10\,\text{V} \gg 100\,\text{mV}$, and thus we conclude that BJT is in the constant current mode.

4. We calculate the voltage divider ratio as

$$V_{BE0} = I_{R_2} R_2 = \frac{V_{CE}}{R_1 + R_2} R_2 = V_{CE} \frac{1}{1 + \dfrac{R_1}{R_2}},$$

and therefore

$$1 + \frac{R_1}{R_2} = \frac{V_{CE}}{V_{BE0}} \Rightarrow \frac{R_1}{R_2} = \frac{V_{CE}}{V_{BE0}} - 1 \quad \therefore \quad \frac{R_1}{R_2} = \frac{10\,\text{V}}{0.650\,\text{V}} - 1 = 14.385$$

Without additional constraints, this ratio is set by choosing, for example, $R_2 = 1\,\text{k}\Omega$, which results in $R_1 = 14.385\,\text{k}\Omega$.

The standard discrete resistor values with $\pm 10\%$ tolerance are based on these numbers (10, 12, 15, 18, 22, 27, 33, 39, 47, 56, 68, 82)[1] followed by the exponent. Therefore, $1 \times 10^3\,\Omega \pm 10\%$ is a standard resistor value, while the number that is closest to $14.385\,\text{k}\Omega$ is $R = 15 \times 10^3\,\Omega \pm 10\%$. With this standard value of R_1, the biasing voltage becomes

$$V_{BE0} = 10\,\text{V} \frac{1}{1 + \dfrac{15\,\text{k}\Omega}{1\,\text{k}\Omega}} = 625\,\text{mV} \tag{6.6}$$

---

[1] Search the external resources for this standard.

Thus, if made of standard resistors, this biasing circuit reduces V_{BE0} by 25 mV relative to the initial value of 650 mV. Relative to the intended V_{BE1}, this new V_{BE2} causes the collector current to change by a factor x, so we calculate

$$I_C = I_S \exp\left(\frac{V_{BE1}}{V_T}\right) \quad \text{and} \quad x \times I_C = I_S \exp\left(\frac{V_{BE2}}{V_T}\right)$$

$$\therefore$$

$$\frac{x \times \cancel{I_C}}{\cancel{I_C}} = \frac{\cancel{I_S} \exp\left(\frac{V_{BE2}}{V_T}\right)}{\cancel{I_S} \exp\left(\frac{V_{BE1}}{V_T}\right)} = \exp\left(\frac{V_{BE2} - V_{BE1}}{V_T}\right) = \exp\left(\frac{\Delta V_{BE}}{V_T}\right)$$

$$\therefore$$

$$x = \exp\left(\frac{25\,\text{mV}}{25\,\text{mV}}\right) = e = 2.718$$

Consequently, after substituting the standard resistor values, the actually realized biasing current is increased 2.7 times relative to the ideal case. In addition, after accounting for the $\pm 10\%$ resistor variations, we find the extreme variations by calculating minimum and maximum[2] of (6.6) as

$$V_{BE0}(min) = 10\,\text{V}\, \frac{1}{1 + \dfrac{15\,\text{k}\Omega + 10\%}{1\,\text{k}\Omega - 10\%}} = 517\,\text{mV}$$

$$V_{BE0}(max) = 10\,\text{V}\, \frac{1}{1 + \dfrac{15\,\text{k}\Omega - 10\%}{1\,\text{k}\Omega + 10\%}} = 753\,\text{mV}$$

That is, after accounting for worst case scenario with standard $\pm 10\%$ resistor tolerances, the actual biasing voltage variance is $\Delta V_{BE0} = 753\,\text{mV} - 517\,\text{mV} = 236\,\text{mV}$. Thus, the actually achieved biasing voltage is $V_{BE0} = 625\,\text{mV} \pm 118\,\text{mV}$. Furthermore, accounting for this variation, the collector biasing current may change by the factor of $x \approx 12 \times 10^3$. This large variation factor is found even before accounting for the power supply, temperature and β factor variations and the component ageing.

In conclusion, this example illustrates that the design of biasing networks must be undertaken with appreciation for all approximations and assumptions made in the initial "ideal" design. Nevertheless, in practice, the voltage divider reference is used very often for electronic circuits that are not intended for high precision applications. Otherwise, more expensive low-tolerance components must be used, including the custom "trimming" technique used in IC designs.

5. DC voltage equation is written by inspection

$$V_{CE} = V_{R_{BB}} + V_{BE} = R_{BB}\, I_B + V_{BE}$$

$$\therefore$$

$$R_{BB} = \frac{V_{CE} - V_{BE}}{I_B}, \tag{6.7}$$

---

[2]A rational function increases when its denominator decreases, and vice versa.

where

$$I_C = \beta I_B \qquad \therefore \qquad I_B = \frac{I_C}{\beta} \tag{6.8}$$

Direct implementation of (6.8) and (6.7) results in

$$I_B = \frac{I_C}{\beta} = \frac{1\,\mathrm{mA}}{100} = 10\,\mathrm{\mu A} \Rightarrow R_{BB} = \frac{V_{CE} - V_{BE}}{I_B} = \frac{10\,\mathrm{V} - 0.65\,\mathrm{V}}{10\,\mathrm{\mu A}} = 935\,\mathrm{k\Omega}$$

Although there are many possible ways to implement 935 kΩ resistance, for the sake of argument, let us use serial connection of two 470 kΩ to create $R_{BB} = 940\,\mathrm{k\Omega} \pm 10\%$ resistor. Thus, in order to achieve the required $I_C = 1\,\mathrm{mA}$, according to (6.7), we calculate the equivalent V_{BE} voltage as

$$V_{BE} = V_{CE} - I_B R_{BB} = 10\,\mathrm{V} - 10\,\mathrm{\mu A} \times 940\,\mathrm{k\Omega} = 600\,\mathrm{mV}$$

Similar to the voltage divider biasing technique in Exercise 6.1-4, the biasing technique based on the control of BJT base current by R_{BB} also shows similar sensitivity to the resistor values.

Exercise 6.2, page 146

1. By inspection of a BJT biased by voltage divider circuit, we write

$$V_{BE0} = I_{R_2} R_2 = \frac{V_{CE}}{R_1 + R_2} R_2 = V_{CE} \frac{1}{1 + \dfrac{R_1}{R_2}} \tag{6.9}$$

By definition, the sensitivity of a variable y due to variations in variable x is calculated as

$$S_x^y = \frac{\dfrac{dy}{y}}{\dfrac{dx}{x}} = \frac{x}{y} \frac{dy}{dx} \tag{6.10}$$

Therefore, using result (6.9), for a given ratio R_1/R_2, we write

$$\frac{dV_{BE0}}{dV_{CE}} = \frac{d}{dV_{CE}} \left(V_{CE} \frac{R_2}{R_1 + R_2} \right) = \frac{R_2}{R_1 + R_2} \stackrel{(6.5)}{=} \frac{V_{BE0}}{V_{CE}}$$

$$\therefore$$

$$S_{V_{CE}}^{V_{BE0}} = \frac{\dfrac{dV_{BE0}}{V_{BE0}}}{\dfrac{dV_{CE}}{V_{CE}}} = \frac{V_{CE}}{V_{BE0}} \frac{dV_{BE0}}{dV_{CE}} = \frac{\cancel{V_{CE}}}{\cancel{V_{BE0}}} \frac{\cancel{V_{BE0}}}{\cancel{V_{CE}}} = 1$$

which is to say that any variation in V_{CE} directly propagates to the variation of B_{BE0}. This result is important because it shows that voltage divider generates voltage reference V_{BE0} whose variation directly follows the variation of voltage source that is used. Consequently, this ΔV_{BE0} variation then translates to the collector current ΔI_{C0} variation. We keep in mind that standard commercial power supply voltages vary $V_{CC} \pm 10\%$, which directly results in $V_{BE0} \pm 10\%$, which may not be acceptable for more demanding applications.

2. By inspection of a BJT biased by R_{BB} circuit schematic, we write

$$V_{CE} = V_{R_{BB}} + V_{BE0} = R_{BB} I_{B0} + V_{BE0}$$

$$\therefore$$

$$R_{BB} = \frac{V_{CE} - V_{BE0}}{I_{B0}}, \qquad (6.11)$$

where

$$I_{C0} = \beta I_{B0} \qquad \therefore \qquad I_{B0} = \frac{I_{C0}}{\beta} \qquad (6.12)$$

After substituting (6.11) into (6.12), we write

$$I_{C0} = \beta \frac{V_{CE} - V_{BE0}}{R_{BB}}$$

which leads into

$$\frac{dI_{C0}}{d\beta} = \frac{d}{d\beta} \left(\beta \frac{V_{CE} - V_{BE0}}{R_{BB}} \right) = \frac{V_{CE} - V_{BE0}}{R_{BB}} \overset{(6.11)}{=} \frac{I_{C0}}{\beta}$$

$$\therefore$$

$$S_{\beta}^{I_{C0}} = \frac{\dfrac{dI_{C0}}{I_{C0}}}{\dfrac{d\beta}{\beta}} = \frac{\beta}{I_{C0}} \frac{dI_{C0}}{d\beta} = \frac{\beta}{I_{C0}} \frac{I_{C0}}{\beta} = 1$$

This result is important because it shows that in this technique, the variation of I_{C0} directly follows the variation of transistor's β factor. We keep in mind that commercial BJT transistors have β that is very strong function of process (i.e. large spread among components) and the temperature.

3. In order to find expression for collector current, first, the resistive voltage divider circuit is transformed into its equivalent version using Thévenin generator as in Fig. 6.8,

Fig. 6.8 Example 6.2-3

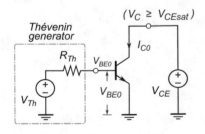

where $R_{Th} = R_1 || R_2$ and $V_{Th} = V_{CE} R_2/(R_1 + R_2)$. After this transformation, by inspection, we write

$$V_{Th} = V_{R_{Th}} + V_{BE0} = I_{B0} R_{Th} + V_{BE0} = \frac{I_{C0}}{\beta} R_{Th} + V_{BE0}$$

$$\therefore$$

$$I_{C0} = \beta \, \frac{V_{Th} - V_{BE0}}{R_{Th}}$$

which leads into

$$\frac{dI_{C0}}{d\beta} = \frac{d}{d\beta} \left(\beta \, \frac{V_{Th} - V_{BE0}}{R_{Th}} \right) = \frac{V_{Th} - V_{BE0}}{R_{Th}} = \frac{I_{C0}}{\beta}$$

$$\therefore$$

$$S_\beta^{I_{C0}} = \frac{\dfrac{dI_{C0}}{I_{C0}}}{\dfrac{d\beta}{\beta}} = \frac{\beta}{I_{C0}} \frac{dI_{C0}}{d\beta} = \frac{\beta}{\cancel{I_{C0}}} \frac{\cancel{I_{C0}}}{\beta} = 1$$

This result shows that the sensitivity $S_\beta^{I_{C0}}$ of voltage divider biasing technique is as same as the sensitivity for base-current technique, as shown in Example 6.1-2.

In conclusion, a voltage reference derived from voltage source through means of passive resistors directly follows variations of the voltage source itself. In addition, variations of the components themselves, i.e. resistance, and β factor, as well as process and temperature variations present significant challenge to the designer who must apply some compensation techniques to stabilize the reference voltage and currents.

4. The equivalent circuit, where the resistive voltage divider reference is replaced by its equivalent Thévenin generator, is shown in Fig. 6.9.

Fig. 6.9 Example 6.2-4

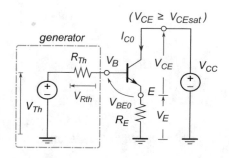

Condition to minimize I_C sensitivity: by inspection, we write KVL equation at the base node as

$$V_{Th} = V_{R_{Th}} + V_{BE0} + V_E = I_{B0} R_{Th} + V_{BE0} + I_{E0} R_E$$

$$= \frac{I_{C0}}{\beta} R_{Th} + V_{BE0} + \frac{\beta + 1}{\beta} I_{C0} R_E$$

$$\therefore$$

$$I_{C0} = \frac{V_{Th} - V_{BE0}}{\dfrac{R_{Th}}{\beta} + \dfrac{\beta + 1}{\beta} R_E} = (V_{Th} - V_{BE0}) \frac{\beta}{R_{Th} + (\beta + 1) R_E}$$

which leads into

$$\frac{dI_{C0}}{d\beta} = (V_{Th} - V_{BE0}) \frac{d}{d\beta} \left(\frac{\beta}{R_{Th} + (\beta + 1) R_E} \right)$$

$$= (V_{Th} - V_{BE0}) \frac{R_{Th} + R_E}{(R_{Th} + (\beta + 1) R_E)^2}$$

$$= \frac{I_{C0}}{\beta} \frac{R_{Th} + R_E}{R_{Th} + (\beta + 1) R_E}$$

$$\therefore$$

$$S_{\beta}^{I_{C0}} = \frac{\dfrac{dI_{C0}}{I_{C0}}}{\dfrac{d\beta}{\beta}} = \frac{\beta}{I_{C0}} \frac{dI_{C0}}{d\beta} = \frac{\beta}{\cancel{I_{C0}}} \frac{\cancel{I_{C0}}}{\beta} \frac{R_{Th} + R_E}{R_{Th} + (\beta + 1) R_E} < 1 \qquad (6.13)$$

because (R_{Th}, R_E, β) are positive numbers and $(\beta \gg 1)$, and thus the denominator in (6.13) is greater than its numerator. The conclusion is that relative to the previous biasing techniques where $S_{\beta}^{I_{C0}} = 1$, the degenerated emitter technique is always less sensitive and thus preferred option. What is more, from (6.13), we can say that

$$S_{\beta}^{I_{C0}} = \frac{R_{Th} + R_E}{R_{Th} + (\beta + 1) R_E} \frac{\beta}{\beta} = \frac{\dfrac{R_{Th}}{\beta} + \dfrac{R_E}{\beta}}{\dfrac{R_{Th}}{\beta} + \dfrac{(\beta + 1)}{\beta} R_E} \qquad (6.14)$$

Assuming $\beta \gg 1$, it follows that

$$\frac{(\beta + 1)}{\beta} \approx 1 \Rightarrow \frac{R_{Th}}{\beta} + \frac{(\beta + 1)}{\beta} R_E \approx \frac{R_{Th}}{\beta} + R_E,$$

and then if

$$\frac{R_{Th}}{\beta} \ll R_E \Rightarrow \frac{R_{Th}}{\beta} + R_E \approx R_E$$

which, for $\beta \gg 1$, leads into

$$S_{\beta}^{I_{C0}} \approx \frac{\cancelto{0}{\dfrac{R_{Th}}{\beta}} + \cancelto{0}{\dfrac{R_E}{\beta}}}{R_E} \approx 0 \qquad (6.15)$$

which gives us important guideline for the design of biasing network that is independent of β variations. Under condition

$$\frac{R_{Th}}{\beta} \ll R_E \tag{6.16}$$

we conclude that, given condition (6.16), the degenerated emitter technique and modern transistors are used and that the sensitivity $S_\beta^{I_{C0}}$ can be practically eliminated. A reminder is that, in practice, "much smaller" means ten times or more.

Design of the biasing network: by inspection of the given schematic diagram, we calculate DC voltage V_B by writing KVL for BJT base as

$$V_B = V_{R_E} + V_{BE} = R_E I_E + V_{BE} = R_E \frac{\beta+1}{\beta} I_C + V_{BE}$$

$$\therefore$$

$$V_B = 1\,k\Omega \frac{100+1}{100} \times 1\,mA + 650\,mV = 1.660\,V$$

Therefore, the voltage divider reference needed to generate V_B is calculated as

$$V_{BE0} = I_{R_2} R_2 = \frac{V_{CE}}{R_1 + R_2} R_2 = V_{CE} \frac{1}{1 + \dfrac{R_1}{R_2}}$$

$$\therefore$$

$$\frac{R_1}{R_2} = \frac{V_{CE}}{V_{BE0}} - 1 = \frac{10\,V}{1.660\,V} - 1 = 5.024 \approx 5$$

To minimize the influence of β variations, by following (6.16) and $R_{th} = R_1 \| R_2$, we write

$$\frac{R_{Th}}{\beta} \ll R_E \quad \therefore \quad \frac{R_1 \| R_2}{\beta} \ll R_E \quad \therefore \quad \frac{R_1 R_2}{R_1 + R_2} \ll \beta R_E$$

which gives us the following system of equations:

$$R_1 = 5 R_2 \quad \text{and} \quad \frac{R_1 R_2}{R_1 + R_2} \ll 100\,k\Omega \quad \therefore \quad R_2 \ll \frac{6}{5} 100\,k\Omega$$

Following the rule of thumb that much larger is ten times or more, it is reasonable to choose standard value resistance $R_2 = 10\,k\Omega$, which forces $R_1 = 50\,k\Omega$. These two values give $R_{th} = 8.33\,k\Omega$, which after substituting in (6.14) results in the sensitivity of

$$S_\beta^{I_{C0}} = \frac{\dfrac{R_{Th}}{\beta} + \dfrac{R_E}{\beta}}{\dfrac{R_{Th}}{\beta} + \dfrac{(\beta+1)}{\beta} R_E} = \frac{\dfrac{8.33\,k\Omega}{100} + \dfrac{1\,k\Omega}{100}}{\dfrac{8.33\,k\Omega}{100} + \dfrac{(100+1)}{100} 1\,k\Omega} = 85.3 \times 10^{-3} \approx 9\%$$

which is to say that the variation of collector current is almost ten times lower than the β factor variation, which illustrates the validity of (6.14) and (6.15).

The introduction of R_E, depending on the emitter current, introduces a possibility that V_E voltage is raised too high and violated the condition for minimum $V_{CEmin} = 0.1$ V. We verify this possibility

$$V_{CC} = V_E + V_{CE} \Rightarrow V_{CE} \approx V_{CC} - R_E\,I_C = 10\,\text{V} - 1\,\text{k}\Omega \times 1\,\text{mA} = 9\,\text{V} > 0.1\,\text{V}$$

and confirm that the transistor still operates in the constant current mode.

5. SPICE simulations demonstrate the effect of two fundamental relations (5.8) and (5.1), with and without emitter resistor, for two values of collector current, see Fig. 6.10 (left), and for the same current but two different emitter resistors, Fig. 6.10 (right).

Adding R_E in series with the emitter resistance r_e, given biasing currents at room temperature, causes that instead of the exponential dependence, the collector current changes only moderately and almost linearly as $I_C \approx I_{C0} \pm 200\,\mu\text{A}$. In addition, using larger R_E is beneficial because the approximation $R_E + r_e \approx R_E \neq f(T)$ is better. Here, "better" means higher collector output resistance, i.e. better current source.

As a consequence, however, in order to keep "headroom" of BJT (i.e. maximal amplitude of the output signal) when larger R_E is used, the power supply voltage must also be increased. This is because, for a given V_{CC}, if V_E increases, then there is less space for V_{CE}, where the output signal is found.

This example illustrates multiple compromises and trade-offs that are possible during the design process. If even tighter voltage dependence is required by the final application, then additional more sophisticated temperature compensation techniques must be used.

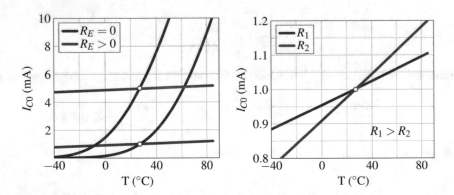

Fig. 6.10 Example 6.2-5

Again, we keep in mind that the collector current *increases* with the increase of the temperature, which is opposite to what (5.8) implies. This consequence is due to very high temperature coefficient of I_s current.

Exercise 6.3, page 147

1. By inspection of the schematic diagram, we write

$$V_{CC} = V_{R_C} + V_{CE} = I_{R_C}\,R_C + V_{CE} = I_{C0}\,R_C + V_{CE} \tag{6.17}$$

which is valid as long as $V_{CE}(min) \geq 0.1\,\text{V}$ is satisfied, in other words, $I_{C0} = \text{const}$ and $R_C < R_C(max)$. In one extreme, we set $R_C = 0$ and verify that

$$V_{CC} = I_{C0}\, \cancel{R_C}^{\,0} + V_{CE} \quad \therefore \quad V_{CE} = V_{CC} = 10\,\text{V} \geq 0.1\,\text{V}$$

Therefore, $R_C = 0$ is the minimal value of collector resistor. In the other boundary case, when $V_{CE} = V_{CE}(min)$, we write

$$V_{CC} = I_{C0}\, R_C(max) + V_{CE}(min)$$

$$\therefore$$

$$R_C(max) = \frac{V_{CC} - V_{CE}(min)}{I_{C0}} \tag{6.18}$$

Direct implementation of (6.4) gives maximum R_C as

$$R_C(max) = \frac{V_{CC} - V_{CE_{sat}}}{I_{C0}} = \frac{10\,\text{V} - 0.1\,\text{V}}{1\,\text{mA}} = 9.9\,\text{k}\Omega$$

and therefore the load resistance must be $0 \leq R_C \leq 9.9\,\text{k}\Omega$.

In conclusion, the collector voltage can fluctuate between $V_{CE}(min)$ and V_{CC}, while $I_{C0} = \text{const}$. This voltage range is where the output signal is placed. Therefore, the "optimal" resistance should set $V_{CE} \approx V_{CC}/2$ (because $V_{CE}(min) \ll V_{CC}$), that is

$$R_C = \frac{V_{CC} - V_{CC}/2}{I_{C0}} = \frac{10\,\text{V} - 5\,\text{V}}{1\,\text{mA}} = 5\,\text{k}\Omega$$

This value of R_C sets DC voltage at collector node as $V_C = V_{CC}/2$ (which is also the average value of the output signal) that allows a sinusoidal output voltage to have maximum possible amplitude to swing up and down from its average value, while the transistor is still in the constant current mode.

2. Addition of R_C does not change the biasing point (V_{BE0}, I_{C0}); however, this time, the emitter voltage is lifted from the ground, and subsequently V_B must follow. Thus, we write KVL equation in the collector–emitter branch as

$$V_{CC} = V_{R_C} + V_{CE} + V_{R_E} \approx R_C\, I_C + V_{CE} + R_E\, I_C$$

$$\therefore$$

$$R_C \approx \frac{V_{CC} - V_{CE} - R_E\, I_C}{I_C}$$

$$\therefore$$

$$R_C(max) \approx \frac{V_{CC} - V_{CE}(min) - R_E\, I_C}{I_C} = \frac{10\,\text{V} - 0.1\,\text{V} - 1\,\text{k}\Omega \times 1\,\text{mA}}{1\,\text{mA}} = 8.9\,\text{k}\Omega$$

which forces the collector voltage to

$$V_C = V_{CC} - V_{R_C} = V_{CC} - R_C\, I_C = 10\,\text{V} - 8.9\,\text{k}\Omega \times 1\,\text{mA} = 1.1\,\text{V}$$

In this case, the average collector voltage that allows for the maximum signal swing is between 10 and 1.1 V, that is when $V_{C0} = (10\,\text{V} - 1.1\,\text{V})/2 = 5.55\,\text{V}$. This "optimal" setup is achieved with collector resistance set to $R_C = (10\,\text{V} - 5.55\,\text{V})/1\,\text{mA} = 4.45\,\text{k}\Omega$.

Exercise 6.4, page 148

1. The two biasing schemes are developed as follows:

   (a) By inspection, after neglecting the base current, we write

   $$V_{EC} - V_{EB0} = V_{R_2} = I_{R_2} R_2 = \frac{V_{EC}}{R_1 + R_2} R_2 = V_{EC} \frac{1}{\dfrac{R_1}{R_2} + 1}$$

   $$\therefore$$

   $$\frac{R_1}{R_2} = \frac{V_{EC}}{V_{EC} - V_{EB0}} - 1 = \frac{\cancel{V_{EC}} - \cancel{V_{EC}} + V_{EB0}}{V_{EC} - V_{EB0}}$$

   $$\therefore$$

   $$\frac{R_2}{R_1} = \frac{V_{EC} - V_{EB0}}{V_{EB0}} = \frac{10\,\text{V} - 0.650\,\text{V}}{0.650\,\text{V}} = 14.385$$

   Without the additional constraints, we choose, for example, $R_1 = 1\,\text{k}\Omega$, which forces $R_2 = 14.385\,\text{k}\Omega$. As a result, the voltage divider generates $V_{EB0} = 650\,\text{mV}$ and $V_B = 9.35\,\text{V}$.

   (b) By inspection, we write

   $$I_{R_{BB}} = I_B = \frac{I_{C0}}{\beta} = \frac{V_B}{R_{BB}} = \frac{V_{EC} - V_{EB0}}{R_{BB}}$$

   $$\therefore$$

   $$R_{BB} = \beta \frac{V_{EC} - V_{EB0}}{I_{C0}} = 100 \frac{10\,\text{V} - 0.650\,\text{V}}{1\,\text{mA}} = 935\,\text{k}\Omega$$

   which sets $V_B = 9.35\,\text{V}$, i.e. $V_{EB0} = 650\,\text{mV}$.

Review of Basic Amplifiers

7

After a weak radio frequency (RF) signal has arrived at the antenna, it is channeled to the input terminals of the RF amplifier through a passive matching network that enables maximum power transfer of the receiving signal by equalizing the antenna impedance with the RF amplifier input impedance. Then, it is job of the RF amplifier to increase the power of the received signal and prepare it for further processing. Aside from their operating frequency, for all practical purposes, there is not much difference between the schematic diagrams of RF and IF amplifiers.

7.1 Important to Know

1. Voltage amplifier I/O resistances

$$R_i \gg R_S \quad \text{and,} \quad R_L \gg R_o \tag{7.1}$$

2. Voltage amplifier gain

$$A_v = \frac{v_{out}}{v_s} = A'_v \, A''_v = \underbrace{\frac{R_i}{R_S + R_i}}_{<1} \times \underbrace{A_v}_{\text{voltage gain}} \times \underbrace{\frac{R_L}{R_o + R_L}}_{<1} \tag{7.2}$$

3. Current amplifier I/O resistances

$$R_i \ll R_S \quad \text{and,} \quad R_L \ll R_o \tag{7.3}$$

4. Current amplifier gain

$$A_i = \frac{i_{out}}{i_s} = A'_i \, A''_i = \underbrace{\frac{R_S}{R_S + R_i}}_{<1} \times \underbrace{A_i}_{\text{current gain}} \times \underbrace{\frac{R_o}{R_o + R_L}}_{<1} \tag{7.4}$$

© Springer Nature Switzerland AG 2021
R. Sobot, *Wireless Communication Electronics by Example*,
https://doi.org/10.1007/978-3-030-59498-5_7

5. G_m amplifier I/O resistances

$$R_i \gg R_S \quad \text{and} \quad R_o \gg R_L \tag{7.5}$$

6. G_m amplifier gain

$$G_m = \frac{i_{out}}{v_s} = \underbrace{\frac{R_i}{R_i + R_S}}_{<1} \times \underbrace{G_m}_{g_m \text{ gain}} \times \underbrace{\frac{R_o}{R_o + R_L}}_{<1} \tag{7.6}$$

7. Transresistance amplifier I/O resistances

$$R_i \ll R_S \quad \text{and} \quad R_o \ll R_L \tag{7.7}$$

8. Transresistance amplifier gain

$$A_R = \frac{v_{out}}{i_s} = \underbrace{\frac{R_S}{R_i + R_S}}_{<1} \times \underbrace{A_R}_{R \text{ gain}} \times \underbrace{\frac{R_L}{R_o + R_L}}_{<1} \tag{7.8}$$

9. CE amplifier gain

$$A_V = \frac{RC}{RE} \tag{7.9}$$

("RC" is the total reassurance at collector node)

("RE" is the total reassurance at emitter node)   (7.10)

10. CE amplifier gain, without discrete "RE"

$$A_V = \frac{R}{r_e} = g_m R \tag{7.11}$$

("R" is the total reassurance at collector node)

("r_e" is the small signal emitter reassurance)   (7.12)

7.2   Exercises

7.1 *   Ideal Amplifier Models

1. Design the model of an ideal voltage amplifier whose gain is, for example, $A_V = v_{out}/v_{in}$. The ideal amplifier model is to be made as a two-stage amplifier, where the first stage must be a G_m amplifier. Propose two "back of the envelope" model solutions by assuming that both gain stages are:

1. ideal, i.e. their respective input/output resistances are assumed to be either zero or infinite as needed,
2. more realistic, i.e. their respective input/output resistances are greater than zero and less than infinity.

Comment on your two solutions and the necessary constrains resulting from the above assumptions. **Data:** $A_V = 40\,\text{dB}$

7.2 *   Basic Amplifiers: I/O Resistance

1. Given the resistance seen by looking into emitter R_out and resistance by looking into base R_in of BJT amplifier, calculate the perceived resistances R_B and R_E.
   Data: $R_\text{out} = 100\,\Omega$, $R_\text{in} = 100\,\text{k}\Omega$, $\beta = 99$.

7.3 *   CE Amplifier: Voltage Gain

1. Given a CE amplifier that includes only the NPN BJT and the external collector resistor R_C, see Fig. 7.1, calculate its voltage gain $A_V = v_{out}/v_{in}$ assuming that small AC signal source v_{in} is connected to base terminal. Without showing the biasing network, we assume that the biasing point is set to (V_{BE0}, I_{C0}), i.e. to g_m.
   Data: $\beta \gg 1$, $r_e = 25\,\Omega$, $R_C = 10\,\text{k}\Omega$, $r_o \to \infty$.

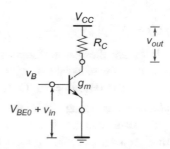

Fig. 7.1 Example 7.3-1

2. Given a simple CE amplifier as in Example 7.3-1 with non-degenerated emitter and resistive load R_C, estimate the voltage gain A_V when

   (a) $V_\text{out} = 7.5\,\text{V}$,   (b) $V_\text{out} = 5\,\text{V}$, and   (c) $V_\text{out} = 0.2\,\text{V}$.

   Comment on the calculated results. **Data:** $V_{CC} = 10\,\text{V}$, $R_C = 5.1\,\text{k}\Omega$, $V_T = 25\,\text{mV}$.

3. Given the NPN BJT and two external resistors R_C and R_E, see Fig. 7.2, calculate its voltage gain $A_V = v_{out}/v_{in}$ assuming that a small AC signal source v_{in} is connected to its base terminal. Without showing the biasing network, assume that the biasing point is set to (V_{BE0}, I_{C0}), i.e. to g_m, by means of DC voltage source V_{B0}. In addition, assume that the power supply voltage is sufficiently high to keep the transistor in the constant current mode.
   Data: $\beta \gg 1$, $r_e = 25\,\Omega$, $R_E = 1\,\text{k}\Omega$, $R_C = 10\,\text{k}\Omega$, $r_o \to \infty$.

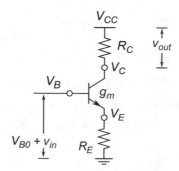

Fig. 7.2 Example 7.3-3

4. Given the NPN BJT and the external R_C and R_E resistors and a capacitor C_E, see Fig. 7.3, calculate the voltage gain A_V assuming that a small AC signal v_{in} is applied across its input terminals. Furthermore, without showing the biasing network, we assume that the transistor is already set to its biasing point (V_{BE0}, I_{C0}).
   Data: $\beta \gg 1$, $r_e = 25\,\Omega$, $R_C = 10\,\mathrm{k}\Omega$, $r_o \to \infty$, $C_E \to \infty$.

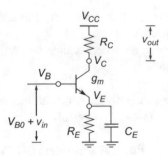

Fig. 7.3 Example 7.3-4

5. Assuming the schematic diagram of CE amplifier as in Example 7.3-4, and given that the BJT is in constant current mode of operation, estimate the voltage gain A_v and express the gain in [dB].
   Data: $R_C = 10\,\mathrm{k}\Omega$, $C_E = 0\,\mathrm{F}$ (i.e. no capacitor), $R_E = 1\,\mathrm{k}\Omega$.

6. Assuming the schematic diagram of CE amplifier as in Example 7.3-4, estimate the voltage gain A_v if:

   (a) $C_E = 1.6\,\mathrm{pF}$,   (b) $C_E = 160\,\mathrm{nF}$, and   (c) $C_E \to \infty$.

   How large are the gain differences in comparison with the gain calculated in Example 7.3-5? Comment on the calculated results.
   Data: $f = 10\,\mathrm{MHz}$, $T = 25\,°\mathrm{C}$, $I_S = 100\,\mathrm{fA}$, $V_{BE} = 650.6\,\mathrm{mV}$, $R_C = 10\,\mathrm{k}\Omega$, $R_E = 1\,\mathrm{k}\Omega$,

7. Recalculate the voltage gain A_V of CE amplifier in Example 7.3-1; however, this time the collector resistance r_o is finite.
   Data: $\beta \gg 1$, $r_e = 25\,\Omega$, $R_C = 10\,\mathrm{k}\Omega$, $r_o = 10\,\mathrm{k}\Omega$.

8. Recalculate the voltage gain A_V of CE amplifier in Example 7.3-3; however, this time the collector resistance r_o is finite.
   Data: $\beta \gg 1$, $r_e = 25\,\Omega$, $R_E = 1\,\mathrm{k}\Omega$, $R_C = 10\,\mathrm{k}\Omega$, $r_o = 10\,\mathrm{k}\Omega$.

9. Recalculate the voltage gain A_V of CE amplifier in Example 7.3-4; however, this time the collector resistance r_o is finite.
   Data: $\beta \gg 1$, $r_e = 25\,\Omega$, $R_C = 10\,\mathrm{k}\Omega$, $r_o = 10\,\mathrm{k}\Omega$, $C_E \to \infty$.

10. Given a common base amplifier, see schematic diagram in Fig. 7.4, estimate:

    (a) DC voltage at the collector,

    (b) $g_m(Q_1)$, and

    (c) AC voltage gain, $A_V = v_C/v_i$.

    Data: $\beta = \infty$, $V_T = 25\,\mathrm{mV}$, $R_C = 7.5\,\mathrm{k}\Omega$, $I = 0.5\,\mathrm{mA}$, $C = \infty$, $V_{CC} = 5\,\mathrm{V}$.

Fig. 7.4 Example 7.3-10

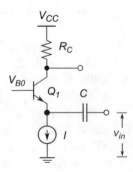

11. In the following questions, express all results both in V/V and in dB units. Note, in "midband" gain calculations, we assume all capacitors infinite. Calculate its midband AC gain A_V of CE amplifier in Fig. 7.5:

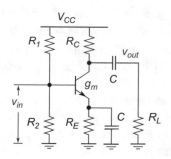

(a) if $(R_L, \beta, r_o \to \infty)$,
(b) if only $(R_L, \beta \to \infty)$,
(c) if only $(\beta \to \infty)$, and
(d) if using all given data as is.

Fig. 7.5 Example 7.3-11

Data:

$\beta = 200$, $V_{CC} = 15\,\text{V}$, $R_1 = 2.5\,\text{k}\Omega$, $R_2 = 1.0\,\text{k}\Omega$, $R_C = 1250\,\Omega$, $C = \infty$, $R_E = 500\,\Omega$, $R_L = 10\,\text{k}\Omega$, $r_o = 5\,\text{k}\Omega$, $g_m = 160\,\text{mS}$.

7.4 ** BJT Amplifier Design

1. Given the characteristics of NPN BJT, see Fig. 7.6, design a simple CE amplifier if the objective is to achieve the voltage gain $A_V = 40\,\text{dB}$ and, at the same time, to permit the amplitude of the output signal to be as large as possible.

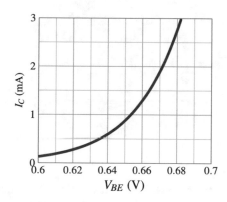 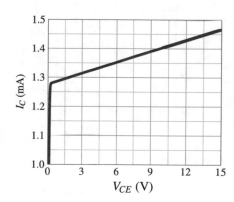

Fig. 7.6 Example 7.4-1

2. By SPICE simulations, verify the functionality of the amplifier designed in Example 7.4-1 over the temperature range specified by industrial standard, i.e. from $-40°$ to $+85°$. What the simulation results show? What do you think is the explanation for the results? Can you propose a possible technique to reduce the problem ?

3. Modify the design solution from Example 7.4-2 by adding the external emitter resistor R_E with the objective to reduce the I_C temperature dependence. Comment on your new circuit's functionality.

7.5 *   Basic Amplifiers: Biasing

1. Design a biasing circuit to support CE BJT amplifier designed in Example 7.4-1. How many possible solutions can you propose? **Data:** $\beta = 200$.

2. Design a biasing circuit to support CE BJT amplifier designed in Example 7.4-3 by providing the V_{B0} biasing voltage. How many possible solutions can you propose?

3. Given the CE amplifier schematic diagram in Fig. 7.5, assume that DC current flowing through R_1, R_2 branch is set as $I_{R_{1,2}} = 0.1\, I_E$ (for this calculation, only ignore the base current) and $V_B = 1/3\, V_{CC}$.

   At the room temperature, estimate the base voltage V_B, R_1, R_2, R_E, I_C, g_m, r_e, R_C, and the amplifier's input resistance R_{in} by looking into the input node.

   Verify that the transistor is in constant current mode, then calculate maximal R_C.

   Data: $A_V = -8$, $V_{CC} = 9\,\text{V}$, $I_E = 2\,\text{mA}$, $\beta = 100$, $V_{BE} = 0.7\,\text{V}$, $R_{\text{sig}} = 10\,\text{k}\Omega$, $T = 17\,°\text{C}$, $C = C_E = \infty$.

4. As a follow-up to Example 7.3-1, given $V_{CE}(min)$, calculate maximum resistance R_C. **Data:** $V_{CE}(min) = 200\,\text{mV}$.

7.6 **   NMOS and JFET Amplifiers

1. Given the characteristics of a general NMOS transistor, see Fig. 7.7, design an amplifier and compare it with your best solution based on the NPN BJT. What differences between the two designs you find to be important?

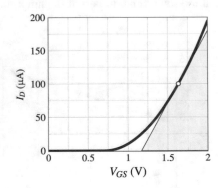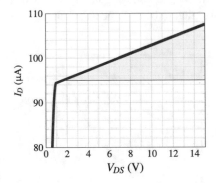

Fig. 7.7 Example 7.6-1

2. Given the voltage–current transfer characteristics in Fig. 7.8, design a cascode CS amplifiers based on JFET transistor. Later, the amplifier may be converted into an RF amplifier, that is to say, its drain resistance R_D will be equal to the dynamic resistance of a typical LC resonator. **Data:** $R_D = 1\,\text{k}\Omega$.

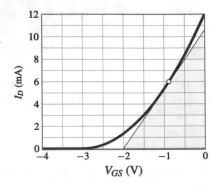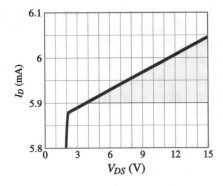

Fig. 7.8 Example 7.6-2

Solutions

Exercise 7.1, page 160

1. Given the first stage of the amplifier is g_m type (i.e. voltage-to-current), as a consequence, the input voltage v_{in} is converted into current i_C. Thus, current i_C must be converted into voltage v_{out} by means of the second-stage current-to-voltage amplifier, Fig. 7.9.

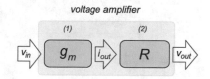

voltage amplifier

Fig. 7.9 Example 7.1-1

Given data, $A_V = 40\,\text{dB} = 100\,\text{v/v}$, we write:[1]

(a) By inspection of an ideal voltage amplifier model made of g_m and A_R amplifiers, we write

$$i_{out} = G_m v_{in} = i_{in}$$

$$\therefore$$

$$v_{out} = A_R\, i_{in} = A_R\, G_m v_{in} \qquad \therefore \qquad A_V \stackrel{\text{def}}{=} \frac{v_{out}}{v_S} = G_m\, A_R = 100$$

Therefore, in theory, any combination of G_m and A_R whose product equals A_V is correct solution. For example, if $G_m = 100\,\text{mS}$, then $A_R = 1\,\text{k}\Omega$. Simple practical implementation would be as a single transistor CE amplifier, Fig. 7.10.

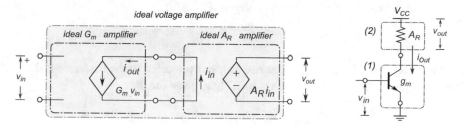

Fig. 7.10 Example 7.1-1

(b) In reality, the interface between realistic G_m and A_R amplifiers is in the form of current divider. The output resistance of a realistic current source is not infinite, instead there is some large but limited R_o internal resistance. In its literal implementation, A_R amplifier is a simple resistor R that appears in parallel with R_o, Fig. 7.11. In addition, evidently, $v_{out} = A_R\, i_{in} = R\, i_{in}$.

---

[1]Note the use of G_m and g_m; most books use capital "G_m" to indicate the gain of a circuit, while small letter version "g_m" indicates the gain of a single transistor. However, this convention is not rigorously followed.

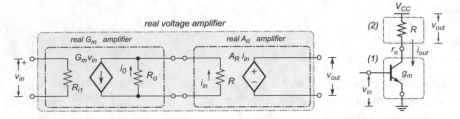

Fig. 7.11 Example 7.1-1

Thus, by inspection of this model, we write the expression for voltage across $R_o||R$ as

$$G_m \, v_{in} \times R_o||R = i_{in} \times R = \frac{v_{out}}{\cancel{R}} \cancel{R}$$

$$\therefore$$

$$A_V = \frac{v_{out}}{v_{in}} = G_m \, R_o||R = G_m \frac{R_o \, R}{R_o + R} = G_m R \frac{R_o}{R_o + R} < G_m R$$

We note that as $R_o \to \infty$ (i.e. when the current source is ideal), then the internal current $i_o \to 0$ and $i_{in} \to G_m v_{in}$, which consequently allows the voltage gain A_V to reach its theoretical maximum $A_V = G_m R$.

In practical implementation, it is the output resistance r_o of collector (or, drain if MOS is used instead of BJT) that is modelled as $r_o = R_o$. This conclusion gives us very important design guideline: in order to maximize voltage gain of realistic voltage amplifier, it is necessary to minimize the current division, which is achieved if $r_o \gg R$.

Exercise 7.2, page 161

1. Everything else being ideal, the emitter resistance is projected to the base terminal and the base resistance is projected to the emitter terminal. Thus, given the data,

$$R_{in} = (\beta + 1) R_E \quad \therefore \quad R_E = 1 \, \text{k}\Omega$$

$$R_{out} = \frac{R_B}{\beta + 1} \quad \therefore \quad R_B = 10 \, \text{k}\Omega$$

Exercise 7.3, page 161

1. When analyzing the gain of CE BJT amplifier, among the first thing to observe is the emitter's connection. In the simplest case, the emitter is connected directly to the ground, Fig. 7.12. Consequently, the biasing voltage V_{BE0} and the small signal input v_{in} are measured between the base and ground, which is exactly as same as between the base and the emitter. Therefore, the total base voltage v_B is the sum of its DC biasing V_{BE} and small signal voltage v_{in}, as

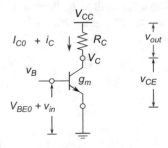

Fig. 7.12 Example 7.3-1

$$v_B = v_{BE} = V_{BE0} + v_{in}$$

In practice, we calculate DC and AC signal components separately, that is to say, DC bias V_{BE0} is assumed constant and only small plus/minus signal variations are added to it (thus, we assume g_m to be constant). Once DC point is set, we are only concerned about the signal variations alone. Small AC signal v_{in} is applied across the emitter diode whose resistance is r_e; thus, small signal emitter's current is

$$i_E = \frac{v_{in}}{r_e}$$

Due to $\beta \gg 1$, the assumption $i_E = (\beta + 1)i_C \approx i_C$ is valid, and thus

$$v_{out} = v_{R_C} = i_C R_C \approx \frac{v_{in}}{r_e} R_C$$

$$\therefore$$

$$A_V = \frac{v_{out}}{v_{in}} = \frac{R_C}{r_e} \overset{\text{def}}{=} g_m R_C = \frac{10\,\text{k}\Omega}{25\,\Omega} = 400 = 52\,\text{dB} \qquad (7.13)$$

Note that the output collector current i_C is converted into voltage by means of the loading resistance R_C that serves as a current-to-voltage amplifier.

Another way to show the same result is given by definition as

$$A_V \overset{\text{def}}{=} \frac{v_{out}}{v_{in}} = \frac{v_{out}}{v_{BE}} = \frac{i_C R_C}{i_E r_e} \approx \frac{\cancel{i_C} R_C}{\cancel{i_C} r_e} = \frac{R_C}{r_e} \overset{\text{def}}{=} g_m R_C$$

This last form or voltage gain is very useful when interpreted as follows: CE amplifier's voltage gain A_V is evaluated as a simple ratio between the *total* resistance found at the collector terminal and the *total* resistance found at the emitter node. As this statement is general, thus we use it to evaluate voltage gain "by inspection".

2. Given the data, we estimate

$$I_C = \frac{V_{CC} - V_{out}}{R_C} \quad \therefore \quad I_C = 2.5\,\text{mA}, 5\,\text{mA}, \quad \text{and} \ \ 9.8\,\text{mA}$$

$$\therefore$$

$$r_e = \frac{V_T}{I_C} \quad \therefore \quad r_e = 10\,\Omega, 5\,\Omega, \quad \text{and} \ \ 2.5\,\Omega$$

$$\therefore$$

$$|A_V| = \frac{R_C}{r_e} \quad \therefore \quad A_V = 100, 200, \quad \text{and} \ \ 400\,\text{V/v}$$

This example illustrates large g_m variations in a simple CE amplifier, and therefore its voltage gain variations. For this reason, the external $R_E \gg r_e$ is used in series so that the overall gain is function of $R_E + r_e$ instead of r_e alone.

3. Relative to Example 7.3-1, among the first thing to observe is the emitter's connection. In this case, the emitter is *not directly* connected to the ground, Fig. 7.13. Consequently, its voltage is elevated to $v_E = i_E R_E$ level. Nevertheless, the BJT biasing point must be still kept by V_{BE0} between the emitter and the base terminals, not between base and ground as in Example 7.3-1. That being the case,

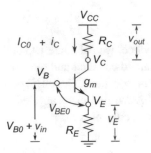

Fig. 7.13 Example 7.3-3

the base voltage must also be elevated by v_E so that V_{BE0} between the base and the emitter terminals is preserved, as

$$v_B = v_E + V_{BE0} = V_{B0} + v_{in},$$

where voltage V_{B0} is set by the external reference source. By following only the small signal, the key observations are:

(a) The total resistance at the emitter node is $R'_E = r_e + R_E$, and this is because the emitter current flows through the internal diode resistance r_e that is followed by the external R_E.

(b) Voltage variations at emitter terminal v_e are equal to and controlled by voltage variations at base terminal v_{in} (with V_{BE0} distance between the two). That is to say, small signal emitter's current is

$$i_e = \frac{v_e}{r_e + R_E} = \frac{v_{in}}{r_e + R_E} \quad \therefore \quad v_{in} = i_e (r_e + R_E)$$

To calculate the voltage gain by definition, assuming $\beta \gg 1$, i.e. $i_E \approx i_C$, we write

$$A_V \overset{\text{def}}{=} \frac{v_{out}}{v_{in}} = \frac{i_C R_C}{i_E (r_e + R_E)} \approx \frac{\not{i_C} R_C}{\not{i_C} (r_e + R_E)} = \frac{R_C}{(r_e + R_E)} \approx \frac{R_C}{R_E} \text{ if } (R_E \gg r_e)$$

In the first approximation $(r_e + R_E) \approx R_E$, CE amplifier's voltage gain A_V is evaluated "by inspection" simply as the ratio between the collector and emitter resistances. Another point to notice is that by increasing emitter resistance, the voltage gain is reduced, as

$$A_V = \frac{R_C}{(r_e + R_E)} < \frac{R_C}{R_E} \tag{7.14}$$

Finally, we calculate

$$A_V = \frac{R_C}{(r_e + R_E)} = \frac{10\,\text{k}\Omega}{(25\,\Omega + 1\,\text{k}\Omega)} = 9.76 \approx \frac{R_C}{R_E} = \frac{10\,\text{k}\Omega}{1\,\text{k}\Omega} = 20\,\text{dB},$$

which shows large gain reduction relative to Example 7.3-1, where non-degenerated CE amplifier achieved $A_V = 400 = 52\,\text{dB}$. This is one of the most important trade-offs that is used in the CE amplifier design. We recall that emitter degenerated transistor is much less sensitive to temperature variations.

4. With the experience gained in Example 7.3-1 and 3, we ask a natural question: is it possible to design a CE amplifier that is both stable (i.e. with degenerated emitter) and has high gain (as in non-degenerated emitter circuit)?

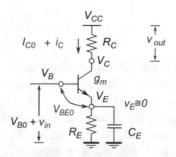

Fig. 7.14 Example 7.3-4

By comparison of (7.13) and (7.14), we conclude that R_E is needed for DC setup; however, it should be "not visible" by AC signal. With that reasoning, the addition of a "sufficiently large" C_E in parallel with R_E creates the equivalent parallel impedance Z_E, see Fig. 7.14, as

$$Z_E = R_E || Z_{C_E} \quad \therefore \quad Z_E(DC) = R_E || (\infty) = R_E \quad \text{(at DC only)}$$

$$\therefore$$

$$Z_E(\infty) = R_E || (0) = 0 \quad \text{(at AC only)}$$

That is to say, at non-DC frequencies, $Z_C \to 0$ and therefore shorts R_E to ground. For that reason, we say that the emitter node is connected to "virtual ground" for small signals and $v_E \cong 0$. Direct consequence for the gain function is

$$A_V = \frac{R_C}{(r_e + \underbrace{R_E || Z_C}_{0})} \cong \frac{R_C}{r_e} = \frac{10\,\text{k}\Omega}{25\,\Omega} = 400 = 52\,\text{dB} \tag{7.15}$$

With this technique, the maximal CE amplifier gain is restored, while its biasing current is made insensitive to temperature variations.

5. Given only the two external resistors, the best estimate is simply the ratio of collector and emitter resistors as $|A_V| \approx R_C / R_E = 10\,V/v = 20\,\text{dB}$.

6. Voltage gain estimate done by inspection is simply the ratio of *total* resistance at collector and *total* resistance at the emitter.
   In the given configuration, the base resistance is not included (i.e. assumed infinite), and therefore the total collector resistance R'_C consists of parallel resistances $r_o || R_C || R_L$, where r_o is the internal collector resistance, and R_C and R_L are the external loading resistors. Given the data, we assume that $r \to \infty$ as well as $R_L = \infty$ (i.e. not connected at all).
   Assuming ideal signal generator and/or $\beta = \infty$, the total emitter resistance R'_E consists of $r_e + R_E || Z(C_E)$, where $r_e = 1/g_m$ is the emitter resistance, R_E is the external resistor, and $Z(C_E) = 1/(2\pi f C_E)$ is impedance of the external emitter capacitor C_E at given frequency. The calculation of three $Z(C_E)$ impedances is straightforward: $Z_C = 9.95\,\text{k}\Omega, 100\,\text{m}\Omega, 0\,\Omega$.
   By definition,

$$I_C = I_S \left(e^{\frac{v_{BE}}{V_T}} - 1 \right) \approx I_S \left(e^{\frac{v_{BE}}{V_T}} \right)$$

$$\therefore$$

$$g_m = \frac{dI_C}{dV_{BE}} = \frac{I_C}{V_T} \quad \therefore \quad r_e = \frac{1}{g_m} = \frac{V_T}{I_C}$$

Given the data, $I_C = 10\,\text{mA}$ and $V_T = kT/q = 25.7\,\text{mV}$. Therefore, $r_e = 2.57\,\Omega$.
Finally, we calculate back of the envelope voltage gain as the ratio of total collector and emitter resistances. That is, $|A_V| = R_C/R'_E = 20.8\,\text{dB},\ 71.5\,\text{dB},\ 71.8\,\text{dB}$. This example illustrates how realistic C_E values are used to achieve voltage gain that is close to the theoretical maximum (i.e. when $C_E \to \infty$), while at the same time taking advantage of degenerated emitter amplifier topology with $R_E \gg r_e$ resistor.

7. From the perspective of small signal, the collector current is split between the internal collector resistance r_o and the external R_C, see Fig. 7.15. As the small signal model illustrates, the total resistance R'_C at the collector is

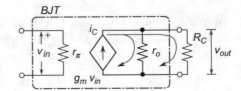

Fig. 7.15   Example 7.3-7

$$R'_C = r_o || R_C,$$

which is approximated by R_C when $r_o \to \infty$ (i.e. r_o is an open connection).
However, in the realistic case when r_o is not infinitely large, we calculate

$$\frac{1}{R'_C} = \frac{1}{r_o} + \frac{1}{R_C} \quad \therefore \quad R'_C = \frac{R_C\, r_o}{R_C + r_o} = \frac{10\,\text{k}\Omega\,10\,\text{k}\Omega}{10\,\text{k}\Omega + 10\,\text{k}\Omega} = 5\,\text{k}\Omega$$

Using the total resistance R'_C, by definition, we write

$$A_V \stackrel{\text{def}}{=} \frac{v_{out}}{v_{in}} = \frac{v_{out}}{v_{BE}} = \frac{i_C\, R'_C}{i_E\, r_e} \approx \frac{\cancel{i_C}\, R'_C}{\cancel{i_C}\, r_e} = \frac{R'_C}{r_e} \left(\stackrel{\text{def}}{=} g_m\, R'_C \right) = \frac{5\,\text{k}\Omega}{25\,\Omega} = 200 = 46\,\text{dB}$$

This results in realistic reduction in voltage gain due to finite r_o, and here the reduction is 50% or 6 dB.

8. Substitution of R'_C in (7.14) results in

$$A_V = \frac{R'_C}{(r_e + R_E)} = \frac{5\,\text{k}\Omega}{(25\,\Omega + 1\,\text{k}\Omega)} \approx 4.88 \approx 14\,\text{dB}$$

This results in realistic reduction in voltage gain due to finite r_o, and here the reduction is 50% or 6 dB.

9. Substitution of R_C' in (7.15) results in

$$A_V = \frac{R_C}{(r_e + \cancelto{0}{R_E \| Z_C})} \cong \frac{R_C}{r_e} = \frac{5\,k\Omega}{25\,\Omega} = 200 = 46\,dB$$

This results in realistic reduction in voltage gain due to finite r_o, and here the reduction is 50% or 6 dB.

10. Given the data, by inspection, we write

$$V_C = V_{CC} - R_C I_C = 1.25\,V \quad \text{and} \quad g_m = \frac{I_C}{V_T} = 20\,mS$$

and also, $v_{BE} = -v_i$, thus

$$v_C = R_C\, i_C = -R_C g_m v_{BE} = R_C g_m v_i$$

$$\therefore$$

$$A_V = \frac{v_C}{v_i} = g_m R_C = 150\,V/v = 43.5\,dB$$

11. Gradual removal of non-ideal component values results in better gain estimates, as illustrated.

(a) Assuming $R_L, \beta, r_o \to \infty$, the equivalent schematic diagram of this amplifier shows only the collector resistor and current source (Fig. 7.16).

Fig. 7.16 Example 7.3-11

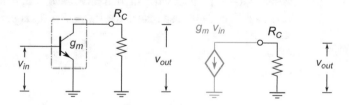

So that the voltage gain estimate is calculated simply as

$$|A_V| = g_m R_C = 160\,mS \times 1.25\,k\Omega = 200 \approx 46\,dB$$

(b) Assuming only $R_L, \beta \to \infty$, i.e. $r_o < \infty$, the equivalent schematic diagram of this amplifier shows collector's resistor R_C in parallel with internal resistance r_o, Fig. 7.17, thus

Fig. 7.17 Example 7.3-11

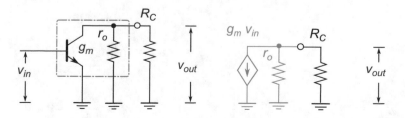

and the voltage gain estimate is calculated as

$$|A_V| = g_m(R_C||r_o) = 160\,\text{mS} \times (1.25\,\text{k}\Omega||5\,\text{k}\Omega) = 160\,\text{mS} \times 1\,\text{k}\Omega$$
$$= 160 \approx 44\,\text{dB}$$

(c) Assuming only $\beta \to \infty$, the equivalent schematic diagram of this amplifier shows collector's resistor R_C in parallel with internal resistance r_o in parallel with load resistor R_L, Fig. 7.18, thus

Fig. 7.18 Example 7.3-11

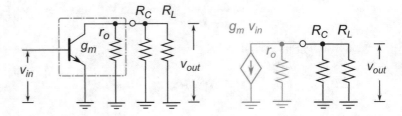

and the voltage gain estimate is calculated as

$$|A_V| = g_m(R_C||r_o||R_L) = 160\,\text{mS} \times (1.25\,\text{k}\Omega||5\,\text{k}\Omega||10\,\text{k}\Omega) = 160\,\text{mS} \times 909\,\Omega$$
$$= 145 \approx 43\,\text{dB}$$

(d) Assuming given the data, the equivalent schematic diagram of this amplifier shows collector's resistor R_C in parallel with internal resistance r_o in parallel with load resistor R_L. In addition, due to $\beta < \infty$, the base resistance $R_B = R_1||R_2$ is projected into the emitter node, Fig. 7.19, as

Fig. 7.19 Example 7.3-11

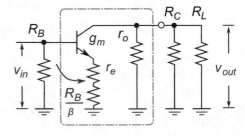

Consequently, the total emitter resistance R'_E increased as

$$r_e \overset{\text{def}}{=} \frac{1}{g_m} = 6.25\,\Omega$$

$$R_B = R_1||R_2 \approx 715\,\Omega$$

$$\therefore$$

$$R'_E = r_e + \frac{R_B}{\beta+1} \approx 9.8\,\Omega$$

and the voltage gain estimate is calculated as

$$|A_V| = \frac{R_C||r_o||R_L}{R'_E} = \frac{909\,\Omega}{9.8\,\Omega} \approx 93 \approx 39\,\text{dB}$$

This example illustrates various levels of approximations that can be used, based on the available data.

Exercise 7.4, page 163

1. On the upper side, maximal amplitude of the output signal is limited by V_{CC}, see Fig. 7.20. On the lower side, however, it is the minimal $V_{CE}(min) = 0.1\text{–}0.2\,\text{V}$ voltage that must be preserved. Due to relation $V_{CC} \gg V_{CE}(min)$, we conclude that the output signal's amplitude would have largest "space" for both positive and negative swing if its average is set close to

$$V_C \approx \frac{V_{CC}}{2}$$

Fig. 7.20 Example 7.4-1

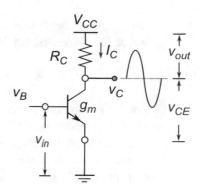

Obviously, given the collector current and the power supply voltage, the collector voltage is determined by R_C. By inspection of the schematic diagram, we calculate the required R so that

$$R = \frac{V_R}{I_C} = \frac{V_{CC} - V_C}{I_C}$$

At this point, there are two unknown variables, R and I_C. Given the transistor characteristics, we choose the collector current, for example, $I_C = 1.35\,\text{mA}$ (which is a "nice number" relatively close to $V_{CC}/2$). Consequently,

$$R = \frac{V_{CC} - V_C}{I_C} = \frac{7.5\,\text{V}}{1.35\,\text{mA}} = 5.55\,\text{k}\Omega$$

With $R = 5.55\,\text{k}\Omega$, assuming $I_C = 1.35\,\text{mA}$, we establish $V_C = V_{CC}/2$ and provide maximal space for the output signal.

On the other hand, however, the choice of $I_C = 1.35$ mA means that g_m value is also set, and we estimate g_m either by the graphical method (tangent at the biasing point) or by an analytical method (assuming the room temperature), see Fig. 7.21.

By inspection of the I_C vs. V_{BE} characteristics, the graphical method results in

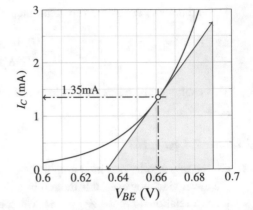

$$g_m \approx \frac{\Delta I_C}{\Delta V_{BE}} \approx \frac{2.75\,\text{mA}}{55\,\text{mV}} = 50\,\text{mS}$$

Fig. 7.21 Example 7.4-1

As a comparison, at room temperature ($T = 300$ K), the analytical calculation gives

$$V_T = \frac{kT}{q} = 25.8\,\text{mV} \qquad \therefore \qquad g_m = \frac{I_C}{V_T} = \frac{1.35\,\text{mA}}{25.8\,\text{mV}} \approx 52\,\text{mS}$$

The two g_m estimates are close, for the sake of argument, let us approximate $g_m = 50$ mS. Given g_m and R, the theoretical maximal voltage gain is

$$A_V = g_m R = 50\,\text{mS} \times 5.55\,\text{k}\Omega = 277.5 \approx 50\,\text{dB} \tag{7.16}$$

Obviously, by giving the preference to the "maximal output signal swing" specification, given transistor and power supply voltage, the achieved voltage gain $A_V \approx 50$ dB is quite a bit higher than the initial specification $A_V = 40$ dB.

Therefore, we must make some compromise and change one or more among the initial specifications: (a) power supply voltage, (b) transistor, (c) maximal output signal swing, (d) voltage gain, or even (e) amplifier topology, to name but a few. Depending on the intended application, it may not be possible to change some of the specifications; thus, it is designer's responsibility to make the final decision.

In this case, let us assume that $A_V = 40$ dB can be modified into $A_V \geq 40$ dB, in other words to say that the amplifier must have *at least* 40 dB gain. By doing so, we leave some margin to account for non-idealities and the actual loading resistance that unavoidably cause the reduction of the maximal theoretical gain.

By inspection of given I_C vs. V_{CE} characteristics and by using the graphical method, we estimate collector resistance r_o at given collector current as

$$r_o \approx \frac{\Delta V_{CE}}{\Delta I_C} \approx \frac{15\,\text{V}}{0.2\,\text{mA}} = 75\,\text{k}\Omega$$

To account for non-infinite r_o and because in the small signal model it appears in parallel with R, we must correct result (7.16) so that

$$A_V = g_m R || r_o = g_m \frac{R\,r_o}{R + r_o} = 50\,\text{mS} \times 5.17\,\text{k}\Omega = 258 \approx 48\,\text{dB} \tag{7.17}$$

where we note that due to $r_o \gg R$, the reduction in voltage gain is relatively small. In practice, this relationship between the two resistances is the preferred one, and we make our design choices accordingly.

So far, the assumption is that no external load ($R_L = \infty$) is connected to the output of this amplifier. This case is referred to as "unloaded" gain. By itself, it specifies *maximum* achievable gain of the amplifier by itself. Obviously, if no load is connected (for example, no speakers), the amplifier is not much useful. Now we can specify the *minimal* resistive load so that the specification $A_V = 40$ dB is still honoured. Thus we calculate the total resistance that corresponds to $A_V = 100$ as

$$A_V = g_m R' \quad \therefore \quad R' = \frac{A_V}{g_m} = \frac{100}{50\,\text{mS}} = 2\,\text{k}\Omega \tag{7.18}$$

Because in the small signal model, the external load R_L appears in parallel to $r_o || R = 5.17\,\text{k}\Omega$, we calculate

$$\frac{1}{R'} = \frac{1}{R} + \frac{1}{r_o} + \frac{1}{R_L} \quad \therefore \quad R_L \approx 3.41\,\text{k}\Omega,$$

which is to say that as long as $R_L \geq 3.41\,\text{k}\Omega$, then $A_V \geq 40$ dB. To compete this design, by inspection of transistor's I_C vs. V_{BE} characteristics, we estimate the biasing point as $(I_{C0}, V_{BE0}) = (1.35\,\text{mA}, 661\,\text{mV})$. In the next phase of design, we create the biasing circuit that generates $V_{BE0} = 661\,\text{mV}$ at base terminal (for example, resistive voltage divider).

2. The simulation of CE amplifier over the industrial temperature range shows very strong temperature dependence of the collector current, Fig. 7.22. We see that at low temperatures, the collector current drops almost to zero, which causes the output voltage V_C to rise to V_{CC} (because $V_R = I_C R \approx 0$). As the temperature is increased so is I_C, and consequently V_C is reduced until around 40 °C.

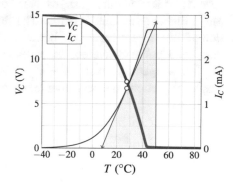

Fig. 7.22 Example 7.4-2

When $V_C \approx 200\,\text{mV}$ (that is to say, $V_{CE}(min) \approx 200\,\text{mV}$), the transistor's current is maximal at $I_C \approx 2.7\,\text{mA}$ (i.e. $V_R = 5.55\,\text{k}\Omega \times 2.7\,\text{mA} \approx 14.8$ V). We note that at room temperature, the amplifier functions as designed. Nevertheless, by the graphical method, we estimate the temperature sensitivity to be

$$TTC(I_C) \approx \frac{\Delta I_C}{\Delta T} \approx \frac{2.85\,\text{mA}}{42\,°C} \approx 68\,\mu\text{A}/°C$$

This temperature dependence is a problem that circuit designers always face. Thus, obviously, this simple circuit cannot be used not even commercially (the standard is $0°$ to $+70°$), let alone automotive applications (the standard is $-40°$ to $+125°$), or space and military applications (the standard is $-55°$ to $+125°$).

The main cause of the temperature dependence is in resistance r_e of the base–emitter pn-junction. With that in mind, adding the external temperature independent resistor $R_E \gg r_e$ is one possible technique to reduce I_C temperature dependence.

3. In reference to the CE amplifier design in Example 7.4-1, where collector current is held at $I_C = 1.35\,\text{mA}$ by $V_{BE} = 661\,\text{mV}$ and the emitter node is connected to the ground, see Fig. 7.23, the addition of, for example, $R_E = 500\,\Omega \gg r_e = 1/g_m = 20\,\Omega$ elevates the emitter voltage to

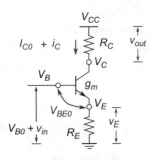

$$v_E = I_E\,R_E \approx I_C\,R_E = 1.35\,\text{mA} \times 500\,\Omega$$

$$= 675\,\text{mV}$$

Fig. 7.23 Example 7.4-3

Still, in order to generate that collector current, the base–emitter voltage must be held at $V_{BE0} = 661\,\text{mV}$, and therefore, the base voltage must be elevated to

$$v_B = v_E + V_{BE} = 675\,\text{mV} + 661\,\text{mV} = 1.336\,\text{V}$$

With the biasing base voltage set to $V_{B0} = 1.336\,\text{V}$, at room temperature, from the perspective of BJT nothing changed, thus the amplifier functions as same as before. The simulation of the temperature sweep, however, shows only mild collector current dependence over the full temperature range, see Fig. 7.24. By graph inspection, the temperature sensitivity estimate is

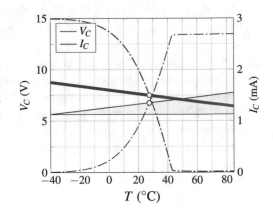

Fig. 7.24 Example 7.4-3

$$TTC(I_C) \approx \frac{\Delta I_C}{\Delta T} \approx \frac{0.425\,\text{mA}}{125\,°\text{C}} \approx 3.4\,\mu\text{A/°C}$$

Application of this technique illustrates large $TTC(I_C)$ reduction relative to the original amplifier version. What is more important, the amplifier functions across the full industrial temperature range within relatively small variation in its specifications. Using another higher value of R_E would improve this result even more. Nevertheless, there is limit in how large R_E can be: we must keep the transistor in the constant current mode, as well as the gain and the output signal swing. Again, based on the intended application, there are multiple compromises possible.

Exercise 7.5, page 163

1. The problem is to design a biasing network that, given $V_{CC} = 15\,\text{V}$, generates voltage reference so that $V_{BE0} = 661\,\text{mV}$. In principle, there are at least two possible ways to set V_{BE0} of a BJT: either by voltage or by current reference.

(a) *Voltage reference:* using voltage divider technique, Fig. 7.25, we calculate

$$V_B = I\,R_2 \quad \text{and} \quad I = \frac{V_{CC}}{R_1 + R_2}$$

$$\therefore$$

$$V_B = \frac{V_{CC}}{R_1 + R_2}\,R_2$$

$$\therefore$$

$$\frac{R_1}{R_2} = \frac{V_{CC}}{V_{BE0}} - 1 = 21.7$$

where we ignored BJT base current. The error is normally relatively small so, for example, we can round the ratio down to $R_1/R_2 = 21.5$. Without further initial constrains, we simply choose $R_1 = 21.5\,\text{k}\Omega$ and $R_2 = 1\,\text{k}\Omega$. For example, bandwidth constrains would give a second equation with R_1 and R_2.

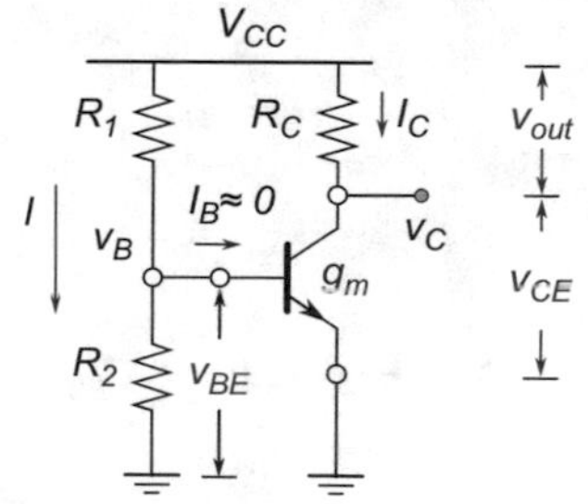

Fig. 7.25 Example 7.5-1

(b) *Current reference:* instead of using two resistors to create voltage divider, an alternative technique is to use only R_1, Fig. 7.26. In this case, however, the base voltage is created by base current and R_1. That is to say, we have to know β so that starting with $I_C = 1.35\,\text{mA}$, we calculate $I_B = I_C/\beta$ and

$$V_{BE} = V_{CC} - V_{R_1} \text{ and } V_{R_1} = I_B\,R_1$$

$$\therefore$$

$$V_{BE} = V_{CC} - \frac{I_C}{\beta}\,R_1$$

$$\therefore$$

$$R_1 = \frac{\beta(V_{CC} - V_{BE})}{I_C} \approx 2.2\,\text{M}\Omega$$

where we keep in mind that β is very dependent on temperature and, to a lesser extent, the biasing current.

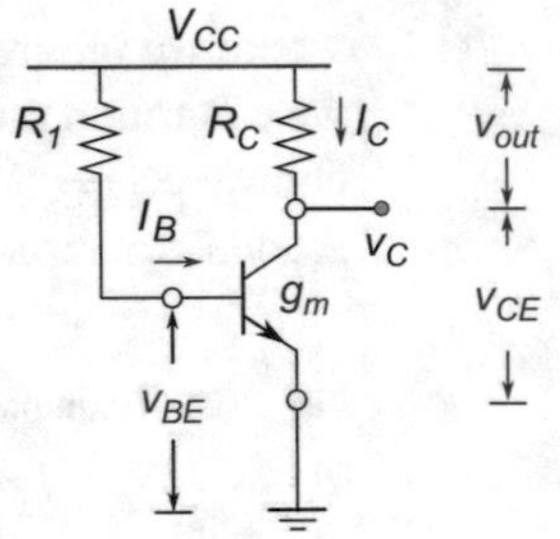

Fig. 7.26 Example 7.5-1

2. The base voltage $V_{B0} = 1.336\,\text{V}$ is set by resistive voltage divider modified as

$$V_{B0} = I\,R_2 \text{ and } I = \frac{V_{CC}}{R_1 + R_2} \quad \therefore \quad V_{B0} = \frac{V_{CC}}{R_1 + R_2}\,R_2$$

$$\therefore$$

$$\frac{R_1}{R_2} = \frac{V_{CC}}{V_{B0}} - 1 = \frac{15}{1.336} - 1 = 10.2 \tag{7.19}$$

Without any additional constrains, there is only
one equation but two unknowns R_1 and R_2.
In this case, the only option is to choose
one resistor and set the second in accordance
with (7.19). Normally, there is no reason to
allow large current through this branch, it is
the resistance ratio not the absolute values that
determine the voltage division, Fig. 7.27.

For example, we set $R_1 = 1\,\text{k}\Omega$ and calculate
$R_2 = 10.2\,\text{k}\Omega$. From the amplifier's perspective, everything is still the same: $V_{BE0} =$
$661\,\text{mV}$, $I_C = 1.35\,\text{mA}$, $g_m = 50\,\text{mS}$, and
$A_V = R_C/R_E \approx 20\,\text{dB}$.

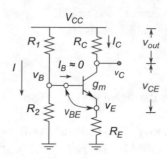

Fig. 7.27 Example 7.5-2

3. Given the data, there are multiple orders in which the calculations can be done. For example, by
   inspections and definitions, we calculate

$$V_T = \frac{kT}{q} = 25\,\text{mV} \quad \text{at} \quad 17\,°\text{C}$$

$$V_B = \frac{V_{CC}}{3} = 3\,\text{V} \quad \therefore \quad V_E = V_B - V_{BE} = 2.3\,\text{V} \quad \therefore \quad R_E = \frac{V_E}{I_E} = 1.15\,\text{k}\Omega$$

Given current through R_1, R_2 branch, we calculate

$$I_{R1,R_2} = 0.1 I_E = \frac{V_{CC}}{R_1 + R_2} \quad \therefore \quad R_1 + R_2 = 45\,\text{k}\Omega$$

Voltage divider calculation gives

$$V_B = R_2 I_{R_2} = R_2 \frac{V_{CC}}{R_1 + R_2} \quad \therefore \quad \frac{R_1}{R_2} = \frac{V_{CC}}{V_B} - 1 = \frac{3\,\cancel{V_{CC}}}{\cancel{V_{CC}}} - 1 = 2$$

Then,

$$\left.\begin{array}{r} \dfrac{R_1}{R_2} = 2 \\[2em] R_1 + R_2 = 45\,\text{k}\Omega \end{array}\right\}, \quad \text{and therefore} \quad R_1 = 30\,\text{k}\Omega, R_2 = 15\,\text{k}\Omega,$$

also

$$I_C = \frac{\beta}{\beta + 1} I_E = 1.98\,\text{mA} \quad \therefore \quad r_e = \frac{V_T}{I_C} = \frac{25\,\text{mV}}{1.98\,\text{mA}} = 12.6\,\Omega$$

$$\therefore$$

$$g_m = \frac{1}{r_e} = 79.2\,\text{mS},$$

and therefore due to β, $C_E \to \infty$ (i.e. $R_{in}/(\beta + 1) \to 0$ and R_E is shorted), we have $|A_V| = g_m R_C$, which leads into

$$R_C = |A_V| r_e = 8 \times 12.6\,\Omega = 101\,\Omega$$

From the signal source perspective, the input side resistance consists of $R_{in} = R_1 \| R_2 \| R_{sig} = 5\,k\Omega$.

We verify if the transistor is still in the constant current mode. The first condition, $V_{BE} = 0.7$ V, is assumed to give $I_E = 2$ mA, and thus it is satisfied. The second condition is to keep collector–emitter voltage above $V_{CE}(min)$. If not specified, this number is typically between 100 and 200 mV. For the sake of argument, let us assume $V_{CE}(min) = 100$ mV. Then, we calculate

$$V_C = V_{CC} - R_C I_C = 9\,V - 101\,\Omega \times 1.98\,mA = 8.77\,V$$

$$V_E = R_E I_E \quad \text{or} \quad = V_B - VBE = 1.15\,k\Omega \times 2\,mA = 2.3\,V$$

$$\therefore$$

$$V_C - V_E = 6.47\,V > V_{CE}(min)$$

Therefore, the transistor is indeed in the constant current mode.

Given constants I_C and V_{CC}, collector potential $V_C = V_{CC} - R_C I_C$ depends on R_C. When $V_C = V_E + V_{CE}(min) = 2.3\,V + 0.1\,V = 2.4\,V$, the transistor is at the edge of constant current mode. That happens when

$$V_C = V_{CC} - R_C I_C \quad \therefore \quad R_C(max) = \frac{V_{CC} - V_C(min)}{I_C} = \frac{9\,V - 2.4\,V}{1.98\,mA} = 3.33\,k\Omega$$

As a conclusion, the collector resistor can take any value $0 < R_C < 3.33\,k\Omega$ without affecting the constant current mode. Consequently, the voltage gain of CE amplifier $A_V = g_m R_C$ is also limited by $R_C(max)$. As a side note, it is a good technique to use spreadsheet and organize long calculations like this one in a logical order.

4. Given $V_{CE}(min)$, $I_{C0} = 1.35$ mA, and $V_{CC} = 15$ V, we write

$$V_{CE}(min) + R_{max} I_{C0} = V_{CC}$$

$$\therefore$$

$$R_{max} = \frac{V_{CC} - V_{CE}(min)}{I_{C0}} \approx 11\,k\Omega,$$

which forces $V_C = V_{CE}$ voltage to its minimum value, beyond that point the transistor is forced to enter the linear region (i.e. $V_{CE} < V_{CE}(min)$) and, consequently, to rapidly reduce the collector current. When that happens, the transistor does not follow exponential equation (5.2), instead the linear mode equation must be used. The simulation of amplifier with

a typical BJT shows when the collector current drops sharply, at the same time, the output voltage reduced to $V_{CE}(min)$, Fig. 7.28.

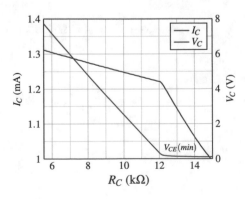

Fig. 7.28 Example 7.5-4

Exercise 7.6, page 164

1. Given the two characteristics of NMOS transistor, we can make the following initial choices:

$$V_{DD} = 15\,\text{V}$$

$$V_D \approx \frac{V_{DD}}{2} = 7.5\,\text{V}$$

$$I_{D0} = 100\,\mu\text{A} \quad \therefore \quad V_{GS0} = 1.665\,\text{V},$$

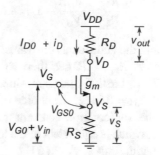

Fig. 7.29 Example 7.6-1

where V_D is chosen to permit maximum output voltage swing.

In addition, by inspection of given NMOS characteristics, we estimate

$$g_m \approx \frac{\Delta I_D}{\Delta V_{GS}} \approx \frac{180}{2 - 1.18}\,\mu\text{S} \approx 220\,\mu\text{S}$$

$$r_o \approx \frac{\Delta V_{DS}}{\Delta I_D} \approx \frac{15 - 1.5}{107 - 95}\,\text{M}\Omega \approx 1.1\,\text{M}\Omega$$

$$V_{DS}(min) \approx 0.5\,\text{V}$$

With these initial specifications and schematic diagram in Fig. 7.29, the CS amplifier design proceeds as follows.

Given the drain current and the power supply voltage, the drain voltage is determined by R_D. By inspection of the schematic diagram, we calculate the required R_D so that

$$R_D = \frac{V_{R_D}}{I_D} = \frac{V_{DD} - V_D}{I_D} = \frac{7.5\,\text{V}}{100\,\mu\text{A}} - 75\,\text{k}\Omega$$

From the signal perspective, R_D appears in parallel to drain resistance r_o, and thus

$$R_D' = R_D || r_o = 75\,\text{k}\Omega || 1.1\,\text{M}\Omega \approx 70.2\,\text{k}\Omega$$

Given g_m and R_D', the theoretical maximal voltage gain is

$$A_V = g_m\,R_D' = 220\,\mu\text{S} \times 70.2\,\text{k}\Omega \approx 15 \approx 23.5\,\text{dB}$$

In the next step, we choose R_S so that it is greater than $1/g_m$ resistance. In this example, $1/g_m \approx 4.5\,\text{k}\Omega$. Without some additional constrains, given $I_D = I_S = 100\,\mu\text{A}$, the choice of $R_S = 10\,\text{k}\Omega$ sets $V_S = R_S\,I_S = 1\,\text{V}$. As a consequence, the gate DC voltage must be set to $V_{G0} = 1\,\text{V} + 1.667\,\text{V} = 2.667\,\text{V}$.

Capacitive voltage divider is calculated as

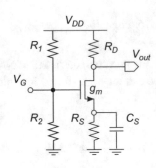

$$\frac{R_1}{R_2} = \frac{V_{DD}}{V_{G0}} - 1 = 4.62$$

Without the additional constrains, one possible choice is $R_2 = 10\,k\Omega$, then $R_1 = 46.2\,k\Omega$, Fig. 7.30. Lastly, there are three capacitors, given $R_S = 10\,k\Omega$, the choice of $C_S = 1\,\mu F$ sets the source time constant $\tau = RC = 10\,ms$, which is equivalent to the low frequency bandwidth limit at $f_\tau \approx 16\,Hz$.

Fig. 7.30 Example 7.6-1

2. Given the two characteristics of JFET transistor, we can make the following initial choices:

$$V_{DD} = 15\,V \quad \text{and} \quad I_D \approx \frac{I_{DSS}}{2} = 6\,mA$$

$$\therefore$$

$$V_{GS0} = -880\,mV$$

By inspection of the two graphs, further we estimate

$$g_m \approx \frac{\Delta I_D}{\Delta V_{GS}} \approx \frac{10.5}{2}\,mS \approx 5.25\,mS$$

$$r_o \approx \frac{\Delta V_{DS}}{\Delta I_D} \approx \frac{15 - 4}{6.045 - 5.9}\,k\Omega \approx 75\,k\Omega$$

Fig. 7.31 Example 7.6-2

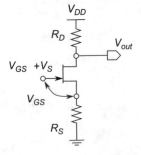

Given the drain current, the power supply voltage, and R_D, by inspection of the schematic diagram in Fig. 7.31, we calculate $V_D = V_{out}$ as

$$V_D = V_{DD} - V_{R_D} = V_{DD} - R_D I_D = 15\,V - 1\,k\Omega \times 6\,mA = 9\,V$$

From the signal perspective, R_D appears in parallel to drain resistance r_o, and thus

$$R'_D = R_D || r_o = 75\,k\Omega || 1\,k\Omega \approx 985\,\Omega$$

Given g_m and R'_D, the theoretical maximal voltage gain is

$$A_V = g_m R'_D = 5.25 \, \text{mS} \times 985 \, \Omega \approx 5.2 \approx 14.3 \, \text{dB}$$

In the next step, we choose R_S so that it is greater than $1/g_m$ resistance. In this example, $1/g_m \approx 190 \, \Omega$. Without some additional constrains, given $I_D = I_S = 6 \, \text{mA}$, the choice of $R_S = 833 \, \Omega$ sets $V_S = R_S I_S = 5 \, \text{V}$. As a consequence, the gate DC voltage must be set to $V_{G1} = 5 \, \text{V} - 0.880 \, \text{V} = 4.12 \, \text{V}$.

Resistive voltage divider needed to generate the biasing gate reference V_{G1} for J_1 is calculated as

$$\frac{R_1}{R_2} = \frac{V_{DD}}{V_{G1}} - 1 = 2.64$$

Without the additional constrains, one possible choice is $R_2 = 10 \, \text{k}\Omega$, then $R_1 = 26.4 \, \text{k}\Omega$. Again, JFET transistors require *negative* gate-source biasing voltage.

The cascode transistor: drain current versus V_{DS} characteristics shows that for this particular JFET, it is necessary to maintain $V_{DS} \geq 2 \, \text{V}$. In order to estimate biasing voltage V_{G2}, we proceed as follows. Both transistors must permit $I_D = 6 \, \text{mA}$, where J_1 is in the control (it is a CS amplifier), while J_2 serves as the "current buffer" (it is a CG amplifier), see Fig. 7.32.

Starting from the ground level, first, $V_{S1} = 5 \, \text{V}$, then there is the minimum of $V_{DS1} = 2 \, \text{V}$ required to keep J_1 in constant current mode. For the same reason, then, drain of J_2 must also be set to least $V_{DS2} = 2 \, \text{V}$ higher than its source.

Fig. 7.32 Example 7.6-2

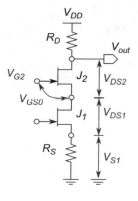

At the same time, given the drain current and R_D, the output voltage is maintained at $V_D = V_{out} = 9 \, \text{V}$. By inspection of the schematic diagram, we write the following condition:

$$V_{G2} \geq V_{S1} + V_{DS1} + V_{GS0} \geq 5 \, \text{V} + 2 \, \text{V} - 0.880 \, \text{V} \geq 6.12 \, \text{V}$$

We keep in mind that for JFET transistors, V_{GS} voltage is negative (i.e. $-V_P \leq V_{GS} = V_G - V_S \leq 0$). That is to say that at its highest level, the gate voltage should stay below the drain voltage as well. The conclusion is that J_2 gate voltage should be in the $6.12 \, \text{V} \leq V_{G2} \leq 9 \, \text{V}$ range.

Although any value of V_{G2} that is within the allowed interval is acceptable, we choose the minimum value that brings J_2 on the edge of constant current mode, in other words, $V_{G2} \geq 6.12\,\text{V}$. The reason is that higher V_{G2} results in higher $V_{DS}(\text{min})$, which subsequently reduces the overall headroom available to the output signal.

Resistive voltage divider needed to generate the biasing gate reference V_{G2} for J_2 is calculated as

$$\frac{R_3}{R_4} = \frac{V_{DD}}{V_{G2}} - 1 = 1.45$$

Without the additional constrains, one possible choice is $R_2 = 10\,\text{k}\Omega$, then $R_1 = 14.5\,\text{k}\Omega$. Again, JFET transistors require *negative* gate-source biasing voltage.

Finally, it is necessary to add source capacitor at source of J_1 to restore the maximal voltage gain, see Fig. 7.33. Without specific frequency domain constrains,[2] for example, we could set $C_S = 10\,\text{nF}$.

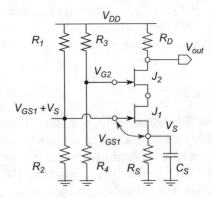

Fig. 7.33 Example 7.6-2

---

[2]See the textbook examples for more detailed discussion.

Introduction to Frequency Analysis of Amplifiers

8

Frequency independent analysis is based on a simple assumption that the circuit is capable to accept and process signals whose frequency spectrum includes all frequencies, from minus infinity to plus infinity. Nonetheless, we already know that elements capable to store energy need a finite amount of time to change their internal states. For slow changes this time delay is negligible, thus "low frequency" approximation produces acceptable results. However, as the signal frequency increases, the impedances of frequency dependent components drastically change.

8.1 Important to Know

1. Gain-bandwidth product

$$GBW \stackrel{\text{def}}{=} A_{V_{\text{mid}}} \times BW \tag{8.1}$$

2. Time constants

$$\tau = RC \ [\text{s}] \quad \text{and,} \quad \tau = \frac{L}{R} \ [\text{s}] \tag{8.2}$$

3. Dominant pole

$$f_0 \cong \frac{1}{\sqrt{\dfrac{1}{f_{p1}^2} + \dfrac{1}{f_{p2}^2} + \cdots - \dfrac{2}{f_{z1}^2} - \dfrac{2}{f_{z2}^2} - \cdots}} \tag{8.3}$$

4. Miller capacitance

$$C_M = C \ (A_v + 1) \tag{8.4}$$

© Springer Nature Switzerland AG 2021
R. Sobot, *Wireless Communication Electronics by Example*,
https://doi.org/10.1007/978-3-030-59498-5_8

8.2   Exercises

8.1 *   Transfer Functions

1. Given first order transfer function

$$H(s) = \frac{1}{1 + \frac{s}{\omega_P}}$$

where, $s = j\omega$, $\omega_P = 2\pi \times 10^6$ rad/s

(a) Sketch its Bode plots and estimate the associated bandwidth

(b) Given amplitude $V_m = 1$ V and frequency f_{in} of the input signal sinusoid, estimate the propagation time delay and amplitude of the output signal, assuming

   (1) $f_{in} = 10$ kHz,
   (2) $f_{in} = 1$ MHz,
   (3) $f_{in} = 10$ MHz.

2. · Given second order transfer function

$$H(s) = \frac{1 + \frac{s}{\omega_Z}}{\left(1 + \frac{s}{\omega_{P1}}\right)\left(1 + \frac{s}{\omega_{P2}}\right)}$$

where, $s = j\omega$ and
$\omega_Z = 6.283 \times 10^3$ rad/s,
$\omega_{P1} = 62.832 \times 10^3$ rad/s,
$\omega_{P2} = 6.283 \times 10^6$ rad/s.

(a) Sketch its Bode plots and estimate the associated bandwidth

(b) Given amplitude $V_m = 1$ V and frequency f_{in} of the input signal sinusoid, estimate the propagation time delay and amplitude of the output signal, assuming

   (1) $f_{in} = 100$ Hz,
   (2) $f_{in} = 1$ kHz,
   (3) $f_{in} = 100$ kHz,
   (4) $f_{in} = 1$ MHz,
   (5) $f_{in} = 10$ MHz.

8.2 *   RLC Networks

1. Given RC and RL networks in Fig. 8.1(a) and (b), derive and calculate module and phase of $Z_{AB}(j\omega)$ functions relative to the terminals AB.
   Data: $R = 1\,k\Omega$, $C = 1\,\mu F$, $f = 1\,kHz$, $L = 1\,\mu H$.

Fig. 8.1 Example 8.2-1

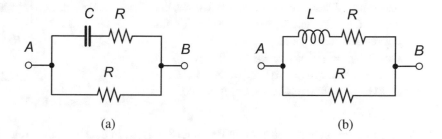

(a)   (b)

2. Given three RL networks in Fig. 8.2(a), (b) and (c), derive their respective $H(j\omega)$ functions relative to the terminals V_1, V_2 then sketch their Bode plots and estimate the bandwidths.
   Data: $R = 1\,k\Omega$, $L = 1\,\mu H$.

Fig. 8.2 Example 8.2-2

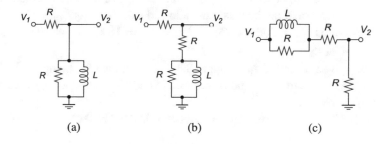

(a)   (b)   (c)

8.3 *   Miller Effect

1. Assume an ideal single stage CE amplifier whose voltage gain is A_v. In addition, there is C_{CB} capacitor connected between the transistor's collector and base, Fig. 8.3.
   Given data, estimate CE amplifier's frequency bandwidth.
   Data: $C_{CB} = 1\,pF$, $A_v = -99$, $R_S = 50\,\Omega$, $R_{in} = \infty$.

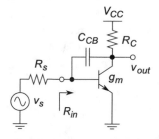

Fig. 8.3 Example 8.3-1

2. Assuming CE amplifier schematic diagram in Example 8.3-1, given data, estimate frequency of its HF pole.
   Data: $g_m = 160\,mS$, $\beta = 200$, $C_{BC} = 15\,pF$, $C_{BE} = 5\,pF$, $R_1 = 2.5\,k\Omega$, $R_2 = 1.0\,k\Omega$, $R_C = 1250\,\Omega$, $R_L = 100\,k\Omega$, $r_o = 5\,k\Omega$, $R_S = 50\,\Omega$.

3. Estimate dominant pole if transfer function of an amplifier contains two pole and one zero frequency: $f_{p1} = 10\,kHz$, $f_{p2} = 50\,kHz$, and $f_{z1} = 100\,kHz$.

4. Assuming $\beta \to \infty$ and $R_E \gg r_e$ estimate τ_1, τ_2, and τ_3 time constants of CB amplifier.

8.4 **   Amplifiers: Poles, Zeros

1. Design CE amplifier whose $-3\,\mathrm{dB}$ dominant frequency can be approximated to f_L primarily by the choice of C_E capacitor, see Fig. 8.4. Assuming the signal source resistance R_S, calculate C_E, C_1, and C_2 capacitors so that this approximation is valid.

   Data: $f_L = 100\,\mathrm{kHz}$, $R_C = R_L = 10\,\mathrm{k\Omega}$, $R_E = 1\,\mathrm{k\Omega}$, $R_B = R_1\|R_2 = 20\,\mathrm{k\Omega}$, $R_S = 10\,\mathrm{k\Omega}$, $g_m = 400\,\mathrm{mS}$, $r_o \to \infty$, $\beta = 100$.

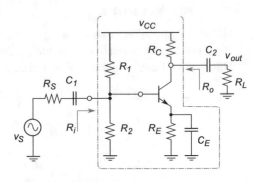

Fig. 8.4   Example 8.4-1

2. Assuming schematic diagram in Example 8.3-1, given data, calculate relevant pole/zero frequencies in a CE amplifier.

   Data: $R_B = 100\,\mathrm{k\Omega}$, $R_C = R_L = 10\,\mathrm{k\Omega}$, $r_o \to \infty$, $R_S = 1\,\mathrm{k\Omega}$, $g_m = 1/r_e = 40\,\mathrm{mS}$, $R_E = 100\,\Omega$, $\beta = 100$, $C_1 = C_2 = C_E = 100\,\mathrm{pF}$.

3. Assuming CE amplifier schematic diagram in Example 8.3-1, calculate its midband gain $A_{V_{\mathrm{mid}}}$ and its upper frequency f_H. Then, if f_H needs to be doubled and everything else being equal, what would be the maximum allowed value of C_{CB}?

   Data: $R_i = 100\,\mathrm{k\Omega}$, $R_C = R_L = R_S = 10\,\mathrm{k\Omega}$, $r_o = 20\,\mathrm{k\Omega}$, $r_e = 25\,\Omega$, $C_{BE} = 1\,\mathrm{pF}$, $C_{CB} = 0.5\,\mathrm{pF}$.

4. Given amplifier in Example 8.3-3, everything else being equal, how large is the load resistance R_L if the midband gain is reduced to half? What happened to the GBW when the gain is reduced?

8.5 **   Frequency Bandwidth, Miller Effect

1. Given data, state the necessary assumptions then estimate the Miller capacitance C_M and the frequency range where CE amplifier given in Fig. 8.5 should be used. What would be the consequence if current gain β is not infinite?

   Data: $L = 2.533\,\mathrm{\mu H}$, $R_C = 9.9\,\mathrm{k\Omega}$, $R_E = 100\,\Omega$, $\beta \to \infty$, $C_{CB} = 1\,\mathrm{pF}$.

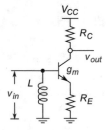

Fig. 8.5   Example 8.5-1

2. Given CE amplifier, estimate value of inductor L so that the input stage accepts a signal frequency f_0, see Fig. 8.6.

   Data: $C = 1\,\mathrm{pF}$, $f_0 = 15.915\,\mathrm{MHz}$, $A_V = -99$, $\beta = \infty$.

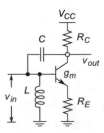

Fig. 8.6   Example 8.5-2

Solutions

Exercise 8.1, page 186

1. Bode plot is created by decomposing transfer function into its pole and zero sub-functions that simply add when in logarithmic form.

(a) By inspection of first order function $H(s)$ we conclude that pole frequency is $\omega_P = 2\pi \times 10^6 \, \text{rad/s}$, therefore $f_P = 1 \, \text{MHz}$. Being the only sub-function, this f_P frequency determines LPF function bandwidth. DC gain is calculated for $\omega = 0$ and it equals to one, i.e. $0 \, \text{dB}$. At f_P, phase of the transfer function equals $-\pi/4$ and tends to zero for $0.1 f_P$ and to $-\pi/2$ for $10 f_P$. Being at pole frequency, amplitude of the output signal is $-3 \, \text{dB} = \sqrt{2}/2$ below its highest level, therefore $V_{out} = 0.707 V_m = 0.707 \, \text{V}$.

Fig. 8.7 Example 8.1-1

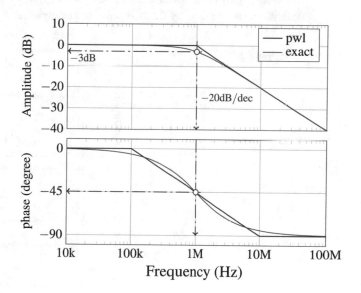

(b) By inspection of amplitude and phase Bode plots, Fig. 8.7, we calculate as follows.

(1) At $f_{in} = 10 \, \text{kHz}$ the amplitude gain graphs show $0 \, \text{dB}$, which is equivalent to gain of one, in other words $V_{out} = V_m = 1 \, \text{V}$. Similarly, the phase plot shows $0°$ at $10 \, \text{kHz}$, in other words the output signal is aligned with the input signal, the propagation delay equals zero.

(2) At $f_{in} = 1 \, \text{MHz}$ the amplitude gain graphs show $0 \, \text{dB}$ at piecewise linear approximated graph. To be more precise, since this frequency is the pole frequency, we know that the amplitude is at its $-3 \, \text{dB} = \sqrt{2}/2 \approx 0.707$ level relative to its highest level (in this case, DC level). Therefore, the output amplitude is $V_{out} = 0.707 V_m = 0.707 \, \text{V}$.

Phase plot shows $-45°$ at $1 \, \text{MHz}$, in other words the output signal is delayed by $\pi/4$ relative to the input signal. To convert angular measure into time, we use the identity for period $T \equiv 2\pi$ that results in $\pi/4 = 2\pi/8 \equiv T/8$. Given $f_{in} = 1 \, \text{MHz}$, follows that $T = 1/f = 1 \, \mu\text{s}$ and therefore $T/8 = 125 \, \text{ns}$.

(3) At $f_{in} = 10 \, \text{MHz}$ the amplitude gain graphs show $-20 \, \text{dB}$ level, which is equivalent to $1/10$ gain. Therefore, the output amplitude is $V_{out} = 0.1 V_m = 100 \, \text{mV}$.

Phase plot shows $-90°$ at $10 \, \text{MHz}$, in other words the output signal is delayed by $\pi/2$ relative to the input signal. To convert angular measure into time, we use the identity for period $T \equiv 2\pi$ that results in $\pi/2 = 2\pi/4 \equiv T/4$. Given $f_{in} = 1\,\mathrm{MHz}$, follows that $T = 1/f = 1\,\mu s$ and therefore $T/4 = 250\,\mathrm{ns}$.

In conclusion, while $10\,\mathrm{kHz}$ signal passes without delay nor attenuation, $1\,\mathrm{MHz}$ and $10\,\mathrm{MHz}$ signals are attenuated ($-3\,\mathrm{dB}$ and $-20\,\mathrm{dB}$) and delayed ($125\,\mathrm{ns}$ and $250\,\mathrm{ns}$) respectively.

2. Using piecewise linear approximations, we sketch Bode plots as follows.

(a) By inspection of second order function $H(s)$ we conclude that there is one zero at $\omega_Z = 6.283 \times 10^3\,\mathrm{rad/s}$ and two poles at $\omega_{P1} = 62.832 \times 10^3\,\mathrm{rad/s}$, and $\omega_{P2} = 6.283 \times 10^6\,\mathrm{rad/s}$. Radian frequencies are converted to Hz after division by 2π factor, i.e. $f_Z = 1\,\mathrm{kHz}$, $f_{P1} = 10\,\mathrm{kHz}$, and $f_{P2} = 1\,\mathrm{MHz}$.

DC gain equals to one (for $\omega = 0$), i.e. $0\,\mathrm{dB}$. In logarithmic scale, piecewise linear approximations of Bode plot are derived by simple addition of one zero and two pole subfunctions, both for amplitude and phase, see Fig. 8.8. Bandwidth of this BPF function is, by definition, determined by the two pole frequencies, i.e. $BW = f_{P2} - f_{P1} = 990\,\mathrm{kHz}$.

Fig. 8.8 Example 8.1-2

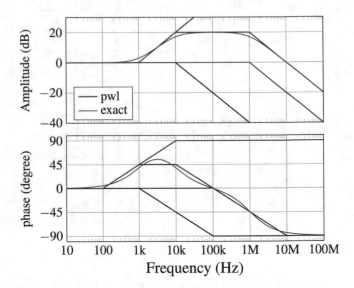

(b) By inspection of amplitude and phase Bode plots, we calculate as follows.

(1) At $f_{in} = 100\,\mathrm{Hz}$ the amplitude gain graphs show $0\,\mathrm{dB}$, which is equivalent to gain of one, in other words $V_{out} = V_m = 1\,\mathrm{V}$. Similarly, the phase plot shows $0°$ at $10\,\mathrm{kHz}$, in other words phase of the output signal is aligned with the input signal, the propagation delay equals zero.

(2) At $f_{in} = 1\,\mathrm{kHz}$ the amplitude gain graphs show $0\,\mathrm{dB}$ at piecewise linear approximated graph. To be more precise, since this frequency is the zero frequency, we know that the amplitude is at its $+3\,\mathrm{dB} = \sqrt{2} \approx 1.41$ level relative to its lowest level (in this case, DC level). Therefore, the output amplitude is $V_{out} = 1.41 V_m = 1.41\,\mathrm{V}$.

Phase plot shows $+45°$ at $1\,\mathrm{kHz}$, in other words phase of the output signal is *advanced* by $\pi/4$ relative to the input signal. To convert angular measure into time, we use the identity for period $T \equiv 2\pi$ that results in $\pi/4 = 2\pi/8 \equiv T/8$. Given $f_{in} = 1\,\mathrm{MHz}$, follows that $T = 1/f = 1\,\mu s$ and therefore $T/8 = 125\,\mathrm{ns}$.

(3) At $f_{in} = 100\,\text{kHz}$ the amplitude gain graphs show $+20\,\text{dB}$ level, which is equivalent to 10 gain, the output amplitude is $V_{out} = 10V_m = 10\,\text{V}$.

However, the phase plot shows again $0°$ at $100\,\text{kHz}$, in other words phase of the output signal is aligned with the input signal, the propagation delay equals to zero.

(4) At $f_{in} = 1\,\text{MHz}$ the amplitude gain graphs show $+20\,\text{dB}$ at piecewise linear approximated graph. To be more precise, since this frequency is the pole frequency, we know that the amplitude is at its $-3\,\text{dB} = \sqrt{2}/2 \approx 0.707$ level relative to its highest level (in this case, $20\,\text{dB}$ level). Thus, for $f_{in} = 1\,\text{MHz}$ signal gain is at $20\,\text{dB} - 3\,\text{dB} = 17\,\text{dB} \approx 7$. Consequently, amplitude of the output signal is $V_{out} = 7V_m = 7\,\text{V}$.

Phase plot shows $-45°$ at $1\,\text{MHz}$, in other words the output signal is delayed by $\pi/4$ relative to the input signal. To convert angular measure into time, we use the identity for period $T \equiv 2\pi$ that results in $\pi/4 = 2\pi/8 \equiv T/8$. Given $f_{in} = 1\,\text{MHz}$, follows that $T = 1/f = 1\,\mu\text{s}$ and therefore $T/8 = 125\,\text{ns}$.

(5) At $f_{in} = 10\,\text{MHz}$, amplitude gain graphs show $-20\,\text{dB}$ level, which is equivalent to $1/10$ gain. Consequently, amplitude of the output signal is $V_{out} = 0.1V_m = 0.1\,\text{V}$.

Phase plot shows $-90°$ at $10\,\text{MHz}$, in other words the output signal is delayed by $\pi/2$ relative to the input signal. To convert angular measure into time, we use the identity for period $T \equiv 2\pi$ that results in $\pi/2 = 2\pi/4 \equiv T/4$. Given $f_{in} = 10\,\text{MHz}$, follows that $T = 1/f = 100\,\text{ns}$ and therefore $T/4 = 25\,\text{ns}$.

Exercise 8.2, page 187

1. The equivalent impedances are found by using serial/parallel impedance transformation rules.

(a) by inspection of the given network, we write

$$Z_{AB} = R \| \left(R + \frac{1}{j\omega C} \right)$$

then, we need to find their real and imaginary parts. Complex algebra techniques give

$$\frac{1}{Z_{AB}} = \frac{1}{R} + \frac{1}{\left(R + \dfrac{1}{j\omega C} \right)} = \frac{1}{R} + \frac{j\omega C}{(1 + j\omega C R)}$$

$$= \frac{(1 + j\omega C R) + j\omega C R}{R(1 + j\omega C R)} = \frac{1 + 2j\omega C R}{R(1 + j\omega C R)}$$

therefore,

$$Z_{AB} = \frac{R(1 + j\omega C R)}{1 + 2j\omega C R} \frac{1 - 2j\omega C R}{1 - 2j\omega C R} = \frac{R[1 + 2(\omega C R)^2]}{1 + 4(\omega C R)^2} - j\frac{\omega C R^2}{1 + 4(\omega C R)^2}$$

Thus,

$$\Re(Z_{AB}) = \frac{R[1 + 2(\omega C R)^2]}{1 + 4(\omega C R)^2} \quad \text{and} \quad \Im(Z_{AB}) = -\frac{\omega C R^2}{1 + 4(\omega C R)^2}$$

Phase is calculated as

$$\angle(Z_{AB}) = \arctan\left(\frac{\Im(Z_{AB})}{\Re(Z_{AB})}\right) = \arctan\frac{-\omega C R}{1 + 2(\omega C R)^2}$$

Given data, we calculate

$$|Z_{AB}| = \sqrt{\Re(Z_{AB})^2 + \Im(Z_{AB})^2} = \sqrt{503.14^2 + (-39.5)^2}\ \Omega = 504.7\ \Omega$$

$$\angle(Z_{AB}) = -4.5°$$

(b) by inspection of the given network, we write

$$Z_{AB} = R || (R + j\omega L)$$

then, we need to find their real and imaginary parts. Complex algebra techniques give

$$\frac{1}{Z_{AB}} = \frac{1}{R} + \frac{1}{(R + j\omega L)}$$

$$= \frac{(R + j\omega L) + R}{R(R + j\omega L)} = \frac{2R + j\omega L}{R(R + j\omega L)}$$

therefore,

$$Z_{AB} = \frac{R(R + j\omega L)}{2R + j\omega L}\frac{2R - j\omega L}{2R - j\omega L} = \frac{2R^3 + R(\omega L)^2}{4R^2 + (\omega L)^2} + j\frac{\omega R^2 L}{4R^2 + (\omega L)^2}$$

Thus,

$$\Re(Z_{AB}) = \frac{2R^3 + R(\omega L)^2}{4R^2 + (\omega L)^2} \quad \text{and} \quad \Im(Z_{AB}) = \frac{\omega R^2 L}{4R^2 + (\omega L)^2}$$

Phase is calculated as

$$\angle(Z_{AB}) = \arctan\left(\frac{\Im(Z_{AB})}{\Re(Z_{AB})}\right) = \arctan\frac{\omega R^2 L}{2R^3 + R(\omega L)^2}$$

Given data, we calculate

$$|Z_{AB}| = \sqrt{\Re(Z_{AB})^2 + \Im(Z_{AB})^2} = \sqrt{500^2 + (1.57 \times 10^{-3})^2}\ \Omega \approx 500\ \Omega$$

$$\angle(Z_{AB}) \approx 0°$$

This example illustrates influence of LC components at relatively low typical frequency and large real resistance. Given different frequency and serial/parallel network, the LC influence may become dominant relative to the real resistor.

2. Transfer function is, by definition, ratio of the output and input variables. In this example, the RL networks are in fact voltage dividers. This is because the output voltage v_2 is tapped at some internal nodes.

The equivalent impedances are found by using serial/parallel impedance transformation rules.

(a) Transfer and phase functions are derived as follows:

$$H(\omega) = \frac{v_2}{v_1} = \frac{R||j\omega L}{R + R||j\omega L} = \frac{\dfrac{Rj\omega L}{R + j\omega L}}{R + \dfrac{Rj\omega L}{R + j\omega L}} = \frac{j\omega L}{R + 2j\omega L}$$

$$= \frac{1}{R} \frac{j\omega L}{1 + \dfrac{2j\omega L}{R}} = \frac{j\dfrac{\omega}{R/L}}{1 + j\dfrac{\omega}{R/(2L)}} = \frac{j\dfrac{\omega}{\omega_z}}{1 + j\dfrac{\omega}{\omega_p}}$$

where, the transfer function $H(\omega)$ has one zero at

$$\omega_z = \frac{R}{L} = \frac{1\,\mathrm{k\Omega}}{1\,\mathrm{\mu H}} = 2\pi \times 159.15\,\mathrm{MHz}$$

and one pole at

$$\omega_p = \frac{R}{2L} = \frac{1\,\mathrm{k\Omega}}{2 \times 1\,\mathrm{\mu H}} = 2\pi \times 79.58\,\mathrm{MHz}$$

Limits of gain functions can be found by inspection of the equivalent schematic diagram at DC and HF, i.e. when the inductor impedance equals zero or infinity. Obviously, at DC the inductor impedance equals zero, thus the output voltage $v_2 = 0$, therefore the gain $H(0) = v_2/v_1$ equals zero too. At the other side of the frequency spectrum, the inductor impedance is infinite, therefore the equivalent circuit contains only R, R voltage divider, thus $|H(\infty)| = 1/2 = -6\,\mathrm{dB}$.
More formal way to analyze the gain function is, for example, by using algebra techniques. First, the module $|H(\omega)|$ is derived as

$$|H(\omega)| = \frac{|j\omega L|}{|R + 2j\omega L|} = \sqrt{\frac{(\omega L)^2}{R^2 + 4(\omega L)^2}}$$

where gain function limits at DC and infinity are evaluated as

$$\lim_{\omega \to 0} |H(\omega)| = \lim_{\omega \to 0} \sqrt{\frac{(\omega L)^2}{R^2 + 4(\omega L)^2}} = 0$$

$$\lim_{\omega \to \infty} |H(\omega)| = \lim_{\omega \to \infty} \sqrt{\frac{(\omega L)^2}{R^2 + 4(\omega L)^2}} = \left(\frac{\infty}{\infty}\right)$$

$$= \lim_{\omega \to \infty} \sqrt{\frac{1}{\dfrac{R^2}{(\omega L)^2} + 4}} = \sqrt{\frac{1}{4}} = \frac{1}{2} = -6\,\mathrm{dB}$$

As usual, gain and phase function plots are derived from $H(\omega)$ in log form

$$20\log H(\omega) = 20\log \frac{j\dfrac{\omega}{\omega_z}}{1 + j\dfrac{\omega}{\omega_p}} = 20\log\left(j\frac{\omega}{\omega_z}\right) - 20\log\left(1 + j\frac{\omega}{\omega_p}\right)$$

Piecewise linear approximation and the exact numerical simulation graphs are shown in Fig. 8.9.

Fig. 8.9 Example 8.1-2

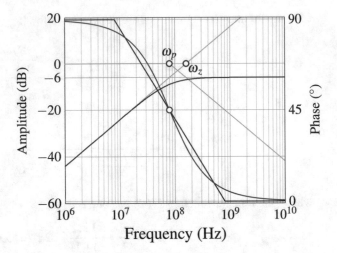

(b) By inspection of schematic diagram we write serial/parallel equivalent impedances that form voltage divider as

$$H(\omega) = \frac{v_2}{v_1} = \frac{R + R||j\omega L}{2R + R||j\omega L} = \frac{\cancel{R} + \dfrac{\cancel{R}\,j\omega L}{R + j\omega L}}{2\cancel{R} + \dfrac{\cancel{R}\,j\omega L}{R + j\omega L}} = \frac{R + 2j\omega L}{2R + 3j\omega L}$$

$$= \frac{1}{2}\frac{1 + \dfrac{\omega}{R/(2L)}}{1 + j\dfrac{\omega}{2R/(3L)}} = \frac{1}{2}\frac{1 + \dfrac{\omega}{\omega_z}}{1 + j\dfrac{\omega}{\omega_p}}$$

where, obviously $H(0) = 1/2$ (i.e. DC gain), and the transfer function $H(\omega)$ has one zero at

$$\omega_z = \frac{R}{2L} = \frac{1\,\text{k}\Omega}{2 \times 1\,\mu\text{H}} = 2\pi \times 79.58\,\text{MHz}$$

one pole at

$$\omega_p = \frac{2R}{3L} = \frac{2 \times 1\,\text{k}\Omega}{3 \times 1\,\mu\text{H}} = 2\pi \times 106.1\,\text{MHz}$$

Note that the pole and zero are less than one decade apart, which is to say that it may not be easy to do precise evaluation of the gain and phase functions within that frequency interval. We evaluate the gain function by inspection of schematic diagram.

At DC: the inductor's impedance equals zero, and the circuit transforms into a simple R, R voltage divider, thus $H(0) = R/(2R) = 1/2 = -6\,\text{dB}$. At high frequencies, the inductor's impedance tends to infinity, and the circuit transforms into a simple $R, 2R$ voltage divider, thus $H(\infty) = 2R/(3R) = 2/3 = -3.52\,\text{dB}$

Note that the difference between DC and AC gains is less than 3 dB, consequently the transfer function never reaches its $-3\,\text{dB}$ points.

Scale of the graph is too small to show details of piecewise linear approximation; however, the exact numerical simulation graphs are shown in Fig. 8.10.

Fig. 8.10 Example 8.1-2

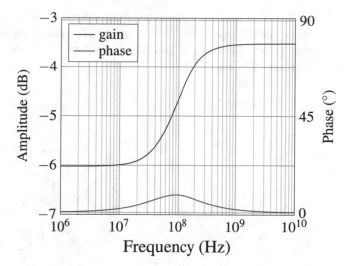

(c) By inspection of schematic diagram we write serial/parallel equivalent impedances that form voltage divider as

$$H(\omega) = \frac{v_2}{v_1} = \frac{R}{R + R + R||j\omega L} = \frac{R}{2R + \dfrac{R\,j\omega L}{R + j\omega L}} = \frac{R + j\omega L}{2R + 3j\omega L}$$

$$= \frac{1}{2}\frac{1 + \dfrac{\omega}{R/L}}{1 + j\dfrac{\omega}{2R/(3L)}} = \frac{1}{2}\frac{1 + \dfrac{\omega}{\omega_z}}{1 + j\dfrac{\omega}{\omega_p}}$$

where, obviously $H(0) = 1/2$ (i.e. DC gain), and the transfer function $H(\omega)$ has one zero at

$$\omega_z = \frac{R}{L} = \frac{1\,\text{k}\Omega}{1\,\mu\text{H}} = 2\pi \times 159.15\,\text{MHz}$$

one pole at

$$\omega_p = \frac{2R}{3L} = \frac{2 \times 1\,\text{k}\Omega}{3 \times 1\,\mu\text{H}} = 2\pi \times 106.1\,\text{MHz}$$

Note that the pole and zero are less than one decade apart, which is to say that it may not be easy to do precise evaluation of the gain and phase functions within that frequency interval. We evaluate the gain function by inspection of schematic diagram.

At DC: the inductor's impedance equals zero, and the circuit transforms into a simple $R, 2R$ voltage divider, thus $H(0) = R/(2R) = 1/2 = -6\,\text{dB}$. At high frequencies, the inductor's impedance tends to infinity, and the circuit transforms into a simple $R, 2R$ voltage divider, thus $H(\infty) = R/(3R) = 1/3 = -9.54\,\text{dB}$

Note that the difference between DC and AC gains is just a little bit more than 3 dB.

Scale of the graph is too small to show details of piecewise linear approximation; however, the exact numerical simulation graphs are shown in Fig. 8.11.

Exercise 8.3, page 187

1. CE amplifier satisfies all three conditions required for Miller effect: (a) it is an inverting amplifier; (b) its voltage gain is greater than one; and (c) it has a capacitive component that creates feedback path between the output and input terminals. Therefore, the equivalent input side schematic diagram, as in Fig. 8.12,

Fig. 8.11 Example 8.1-2

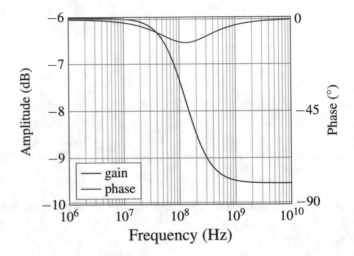

Fig. 8.12 Example 8.1-1

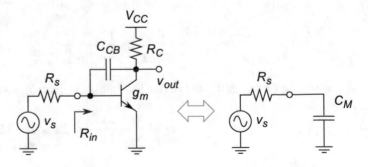

is analyzed as a voltage divider that consists of the source resistance R_S and the Miller capacitance $C_M = C_{CB}(|A_V| + 1)$. The frequency bandwidth defined by this (R_S, C_M) LP filter is in effect the "useful range" of signal frequencies, i.e.

$$f_{3\,dB} = \frac{1}{2\pi \, R_S \, C_M} = \frac{1}{2\pi \times 50\,\Omega \times 100\,pF} = 31.8\,\text{MHz}$$

Considering that a simple $(50\,\Omega, 1\,pF)$ LP filter would allow the bandwidth of $3.2\,GHz$, this example illustrates significant reduction in CE amplifier's bandwidth due to Miller effect.

2. Given data we systematically calculate

$$g_m = 160\,\text{mS} \quad \therefore \quad r_e \stackrel{\text{def}}{=} \frac{1}{g_m} = 6.25\,\Omega \quad \therefore \quad r_\pi = (\beta + 1)r_e = 1.256\,\text{k}\Omega$$

The total equivalent resistance R'_C at the collector terminal is found by network transformations, see Fig. 8.13:

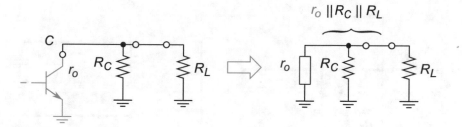

Fig. 8.13 Example 8.1-2

therefore,

$$R'_C = r_o||R_C||R_L = \cfrac{1}{\cfrac{1}{r_o} + \cfrac{1}{R_C} + \cfrac{1}{R_L}} = 990.1 \, \Omega$$

The total equivalent resistance R'_E at the emitter terminal is found by network transformations, Fig. 8.14:

Fig. 8.14 Example 8.1-2

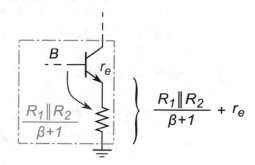

therefore,

$$R'_E = r_e + \frac{R_1||R_2}{\beta + 1} = 9.821 \, \Omega$$

Voltage gain of transistor itself and C_{CB} create Miller capacitance as

$$|A_V| = \frac{R'_C}{R'_E} = \frac{r_o||R_C||R_L}{r_e + \cfrac{R_1||R_2}{\beta}} = \frac{990.1 \, \Omega}{9.821 \, \Omega} \approx 101 \quad \therefore \quad C_M = (|A_V| + 1)C_{CB} = 1.5 \, \text{nF}$$

The total equivalent resistance at the base terminal is found by network transformations in Fig. 8.15 (decoupling capacitor C_2 is large, thus short connection):

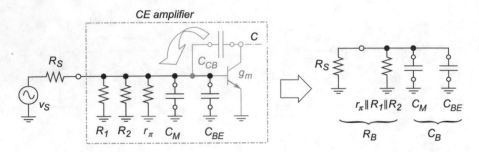

Fig. 8.15 Example 8.1-2

$$R_B = R_S || r_\pi || R_1 || R_2 = \cfrac{1}{\cfrac{1}{R_S} + \cfrac{1}{R_1} + \cfrac{1}{R_2} + \cfrac{1}{r_\pi}} = 45\,\Omega$$

By inspection of the equivalent network at base terminal we write

$$f_H = \frac{1}{2\pi\tau_H} = \frac{1}{2\pi R_B C_B} = \frac{1}{2\pi(R_S||R_1||R_2||r_\pi)(C_{BE} + C_M)}$$

$$= \frac{1}{2\pi \times 45\,\Omega \times (5\,\mathrm{pF} + 1.5\,\mathrm{nF})}$$

$$\approx 2.3\,\mathrm{MHz}$$

3. Direct application of (8.3) gives

$$f_0 = \frac{1}{\sqrt{\dfrac{1}{f_{p1}^2} + \dfrac{1}{f_{p2}^2} - \dfrac{2}{f_{z1}^2}}} = \frac{1}{\sqrt{\dfrac{1}{10\,\mathrm{kHz}^2} + \dfrac{1}{50\,\mathrm{kHz}^2} - \dfrac{2}{100\,\mathrm{kHz}^2}}} = 9.76\,\mathrm{k}\Omega$$

which is very close to $f_{p1} = 10\,\mathrm{kHz}$ because the second pole and the zero are "sufficiently" far away, thus we could make first approximation without the above calculation.

4. The CB amplifier input terminal is at emitter node, basic network transformations reveal the equivalent RC circuit, Fig. 8.16.

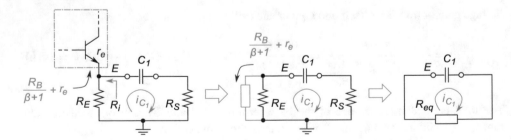

Fig. 8.16 Example 8.1-4

where, by inspection write

$$\tau_1 = C_1 \left[R_S + R_E || \left(r_e + \frac{R_B}{\beta + 1} \right) \right] \approx C_1 \left(R_S + R_E || r_e \right) \approx C_1 R_S$$

Base terminal of CB amplifier is at signal ground (virtual ground) level, basic network transformations reveal the equivalent RC circuit as in Fig. 8.17

Fig. 8.17 Example 8.1-4

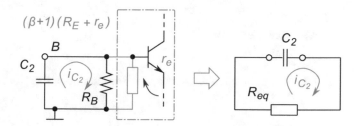

where, by inspection write

$$\tau_2 = C_2 \left[R_B || (\beta + 1)(R_E + r_e) \right] \approx C_2 R_B$$

Collector terminal of CB amplifier serves as the output node, basic network transformations reveal the equivalent RC circuit as in Fig. 8.18

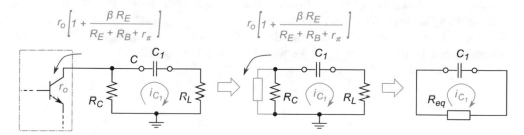

Fig. 8.18 Example 8.1-4

where, by inspection write

$$\tau_3 = C_3 \left[R_L + R_C || r_o \left(1 + \frac{\beta R_E}{R_E + R_B + r_\pi} \right) \right] \approx C_3 \left(R_L + R_C \right)$$

Exercise 8.4, page 188

1. By using basic network transformations and by inspection we derive the expression for time constant τ_E associated with C_E capacitor, see Fig. 8.19.

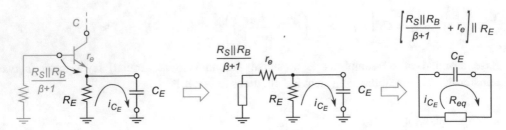

Fig. 8.19 Example 8.4-1

The total equivalent resistance at the emitter node therefore results in time constant

$$\tau_E = \left[\left(\frac{R_S||R_B}{\beta+1} + r_e\right)||R_E\right]C_E$$

Note that this resistance is very low, i.e. it is close to r_e. Given τ_E, it follows that $\omega_L = 2\pi f_L = 1/\tau_E$ and we calculate

$$C_E = \frac{1}{\left[\left(\dfrac{R_S||R_B}{\beta+1} + r_e\right)||R_E\right]2\pi f_L} = 24.8\,\text{nF}$$

To make f_L highest dominant pole, frequencies of the other two poles should be at least one decade lower, i.e. at $f \approx f_L/10 \approx 10\,\text{kHz}$, so that they start compensating for the DC zero. That is achieved by choosing C_1 and C_2 as follows.

The sequence of circuit transformations that enables us to use "by inspection" technique to determine time constant due to $R_{eq}C_1$ network is as shown in Fig. 8.20.

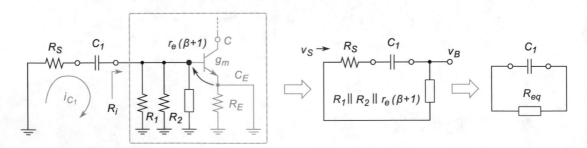

Fig. 8.20 Example 8.4-1

$$R_{eq_1} = R_S + R_1||R_2||r_e(\beta+1) \approx R_S + R_1||R_2||\beta r_e$$

$$\therefore$$

$$\tau_1 = R_{eq_1}C_1 = [R_S + R_B||(\beta+1)r_e]\,C_1, \qquad (R_B = R_1||R_2)$$

therefore,

$$\omega_{\tau_1} = \frac{1}{[R_S + R_B||(\beta + 1)r_e]\, C_1}$$

and, since $\omega_{\tau_1} = 2\pi f_{\tau_1}$ it follows that

$$C_1 = \frac{1}{[R_S + R_B||(\beta + 1)r_e]\, 2\pi f_{\tau_1}} \geq 1.55\,\mathrm{nF}$$

Similarly, by inspection and network transformations at the collector node we write, see Fig. 8.21,

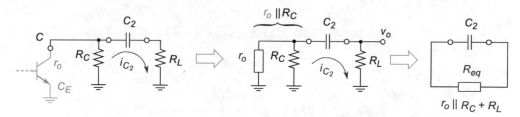

Fig. 8.21 Example 8.4-1

$$R_{eq} = r_o||R_C + R_L \approx R_C + R_L \quad \text{when} \ (r_o \gg R_C)$$

$$\therefore$$

$$\tau_2 = (r_o||R_C + R_L)\, C_2 \approx (R_C + R_L)\, C_2 \quad \text{when} \ (r_o \gg R_C)$$

which gives us pole/zero frequency at

$$\omega_{\tau_2} = \frac{1}{\tau_2} = \frac{1}{(R_C + R_L)\, C_2} \quad \text{and}$$

$$C_2 = \frac{1}{(R_C + R_L)\, 2\pi f_{\tau_2}} \geq 796\,\mathrm{pF}$$

2. We reuse derivations from Example 8.3-1 and calculate poles:

$$f_E = \frac{1}{\left[\left(\dfrac{R_S||R_B}{\beta + 1} + r_e\right)||R_E\right]\, 2\pi C_E} = 61.6\,\mathrm{MHz}$$

$$f_{\tau_1} = \frac{1}{[R_S + R_B||(\beta + 1)r_e]\, 2\pi C_1} = 460\,\mathrm{kHz}$$

$$f_{\tau_2} = \frac{1}{(R_C + R_L)\, 2\pi C_2} 79.6\,\mathrm{k\Omega}$$

and zero due to C_E is

$$f_z = \frac{1}{2\pi R_E C_E} = 15.9\,\text{MHz}$$

3. The midband gain is calculated assuming large C_1, C_2, C_E capacitances, therefore shorted. In addition, the parasitic capacitances C_{CB}, C_{BE} are assumed non-existent (i.e. an ideal BJT).

There are two "stages" of the signal amplification. First, before reaching the gate terminal, the signal amplitude is reduced by gain A'_V of resistive voltage divider formed by the source resistance R_S and the input resistance R_i. Second, assuming large C_E, BJT amplification A''_V itself is calculated as $A''_V = g_m R'_C$, where R'_C is the total collector resistance.

First, we calculate

$$R_i = R_1 || R_2 = 100\,\text{k}\Omega \quad \therefore \quad A'_V = \frac{R_i}{R_i + R_S} = 0.909$$

Second, we calculate

$$R'_C = r_o || R_C || R_L = \frac{1}{\frac{1}{20\,\text{k}\Omega} + \frac{1}{10\,\text{k}\Omega} + \frac{1}{10\,\text{k}\Omega}} = 4\,\text{k}\Omega$$

$$\therefore$$

$$A''_V = -g_m R'_C = -\frac{R'_C}{r_e} = -\frac{4\,\text{k}\Omega}{25\,\Omega} = -160$$

The total voltage gain A_V is therefore

$$A_V = A'_V \times A''_V = 0.909 \times (-160) = -145.44 \approx 43\,\text{dB}$$

where, CE is an inverting amplifier (noted by "$-$" sign for the gain) while gain given in decibel assumes the phase inversion. Reminder: gain of this *inverting* amplifier is still 43 dB where the negative phase is assumed. Writing -43 dB would mean the gain of $(1/160)$, which would be *attenuation* of the signal's amplitude.

Taking into account the two parasitic capacitances, C_{CB}, C_{BE} results in limited frequency bandwidth, where the upper frequency pole frequency ω_H is calculated as follows.

CE amplifier is inverting, thus it suffers from Miller effect, as

$$C_M = (A''_V + 1) C_{CB} = 161 \times 0.5\,\text{pF} = 80.5\,\text{pF}$$

This Miller capacitance C_M appears between the base and ground nodes. At the same time there is C_{BE} connected from the base to the virtual ground at the emitter node. Reminder: C_E is assumed large capacitance (i.e. zero AC resistance), thus from the signal's perspective the emitter node is at the ground level ("virtual ground"). Consequently, the time constant at base terminal is created with $R_S || R_i$ and $(C_M + C_{BE})$, as the simple network transformations show, see Fig. 8.22.

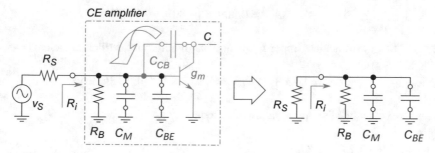

Fig. 8.22 Example 8.4-3

We calculate the upper bandwidth frequency as

$$\tau_H = (R_S || R_i)(C_M + C_{BE})$$

$$\therefore$$

$$f_H = \frac{1}{2\pi \tau_H} = \frac{1}{2\pi (R_S || R_i)(C_M + C_{BE})} = \frac{1}{2\pi \times 9.1\,\text{k}\Omega \times 81.5\,\text{pF}} = 214.8\,\text{kHz}$$

Doubling f_H can be achieved as

$$2 \times f_H = \frac{1}{2\pi (R_S || R_i)(C_M' + C_{BE})}$$

$$\therefore$$

$$C_M' = \frac{1}{2\pi \times (R_S || R_i) \times 2 \times f_H} - C_{BE} = 40.75\,\text{pF} - 1\,\text{pF} = 39.5\,\text{pF}$$

$$\therefore$$

$$(A_V + 1)C_{CB}' = 39.5\,\text{pF} \quad \therefore \quad C_{CB}' = \frac{39.5\,\text{pF}}{101} \approx 0.39\,\text{pF} = 390\,\text{fF}$$

In conclusion, a HF IC manufacturing technology should be used to design BJT whose parasitic C_{CB} reduced by two orders of magnitude.

4. Reduction of gain A_V by half is due to R_L being part of R_C' (the input side voltage divider stays the same) thus by reusing results already found in Example 8.3-3, we calculate

$$0.5 \times |A_V''|r_e = r_o || R_C || R_L' = \frac{1}{\dfrac{1}{r_o} + \dfrac{1}{R_C} + \dfrac{1}{R_L'}}$$

$$\therefore$$

$$R_L' = \frac{1}{\dfrac{2}{|A_V''|r_e} - \dfrac{1}{r_o} - \dfrac{1}{R_C}} \approx 2.9\,\text{k}\Omega$$

Therefore, the total voltage gain is also reduced by 6 dB (i.e. by factor of two), as

$$A_V = A'_V \times A''_V = 0.909 \times (-80) = -72.72 \approx 37\,\text{dB}$$

At the same time we calculate upper frequency bandwidth, first, Miller capacitance as

$$C_M = (A_V + 1)\,C_{CB} = 73.72 \times 0.5\,\text{pF} = 36.36\,\text{pF}$$

Therefore, we calculate the upper bandwidth frequency as

$$\tau_H = (R_S\|R_i)(C_M + C_{BE})$$

$$\therefore$$

$$f_H = \frac{1}{2\pi\,\tau_H} = \frac{1}{2\pi\,(R_S\|R_i)(C_M + C_{BE})} = \frac{1}{2\pi \times 9.1\,\text{k}\Omega \times 37.36\,\text{pF}} = 468.13\,\text{kHz}$$

We compare the two gain-bandwidth (GBW) products as

$$GBW_1 = 145.44 \times 214.8\,\text{kHz} \approx 31\,\text{MHz}$$

$$GBW_2 = 72.72 \times 468.1\,\text{kHz} \approx 34\,\text{MHz}$$

Gain-bandwidth product (ideally) stays the same, here the two results are slightly different due to approximations used along the way. That is to say, depending on the application, we trade gain for bandwidth and vice versa.

Exercise 8.5, page 188

1. The existence of C_{CB} and voltage gain A_V in the case of CE amplifier enables "projection" of real C_{CB} into perceived Miller capacitance $C_M = (A_V + 1)C_{CB}$ at the base terminal, Fig. 8.23.

   Independently from Miller effect, due to β factor the real emitter resistance R_E is also "projected" to the base terminal as $R'_B = (\beta + 1)R_E$.

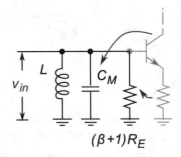

Fig. 8.23 Example 8.5-1

Therefore, the base terminal perceives three components connected in parallel: L, C_M, and R'_B. The three components, RLC, by definition create BPF[1] that shapes frequency profile and selects frequency range of the input signal to be amplified. Voltage gain is estimated as

$$|A_V| \approx R_C/R_E = 99 \quad \therefore \quad C_M = (A_V + 1)\,C_{CB} = 100\,\text{pF}$$

In the first approximation, given data and $\beta \to \infty$, therefore $R'_B \to \infty$. That leaves inductor L to form an ideal LC band-pass filter with C_M. Being ideal BPF, there is only one frequency that passes through and is subsequently amplified,

---

[1] See chapter on RLC resonance.

$$f_0 = \frac{1}{2\pi\sqrt{LC_M}} = \frac{1}{2\pi\sqrt{2.533\,\mu\text{H} \times 100\,\text{pF}}} = 10\,\text{MHz}$$

If $\beta \ll \infty$, the consequence would be that a finite $R_B' = (\beta + 1)R_E$ resistance is projected in parallel with the LC resonator, thus causing the associated BPF bandwidth to be greater than zero, i.e. there would be a range of frequencies centred around f_0 that would be amplified to a certain degree, not only the central $f_0 = 10\,\text{MHz}$ frequency.

2. Miller capacitance is $C_M = (A_V + 1)C = 100\,\text{pF}$ and it forms an ideal LC band-pass filter with C_M. Therefore

$$f_0 = \frac{1}{2\pi\sqrt{LC_M}} \quad \therefore \quad L = \frac{1}{(2\pi f_0)^2 C_M} = 2.533\,\mu\text{H}$$

Electrical Noise

9

Any electrical signal that makes recovery of the information signal more difficult is considered noise. For example, "white snow" on a TV picture and "hum" in an audio signal are typical electrical noise manifestations. Noise mainly affects receiving systems, where it sets the minimum signal level that it is possible to recover before it becomes swamped by the noise. We note that amplifying a signal already mixed with noise does not help the signal recovery process at all. Once it enters the amplifier, noise is also amplified, which is to say that the ratio of signal to noise (S/N) power does not improve and that is what matters. When the power of the noise signal becomes too large relative to the power of the information signal, information content may be irreversibly lost.

9.1 Important to Know

1. Noise spectrum density and power

$$S_n(f) = k\,T \;\; [\text{W}/\text{Hz}] \tag{9.1}$$

$$P_n = \int_{f_1}^{f_1+\Delta f} S_n(f)\,df = S_n(f) \int_{f_1}^{f_1+\Delta f} df = k\,T\,\Delta f \;\; [\text{W}] \tag{9.2}$$

2. Resistor's thermal noise

$$P_{Lmax} = \frac{(V_n/2)^2}{R_S} \quad \therefore \quad k\,T\,\Delta f = \frac{V_n^2}{4\,R} \quad \therefore \quad V_n = \sqrt{4\,R\,k\,T\,\Delta f} \tag{9.3}$$

3. The equivalent noise in RC network

$$V_n^2 = \frac{k\,T}{C} \tag{9.4}$$

4. The equivalent noise in RLC network

$$V_n^2 = 4\,r\,k\,T\,Q^2\,\Delta f = 4\,R_D\,k\,T\,\Delta f \tag{9.5}$$

© Springer Nature Switzerland AG 2021
R. Sobot, *Wireless Communication Electronics by Example*,
https://doi.org/10.1007/978-3-030-59498-5_9

5. SNR definition

$$\text{SNR} = \frac{P_s}{P_n} \tag{9.6}$$

$$\text{SNR} = 10 \log \frac{P_s}{P_n} \quad [\text{dB}] \tag{9.7}$$

$$= 10 \log \frac{V_s^2/\cancel{R}}{V_n^2/\cancel{R}} = 10 \log \left(\frac{V_s}{V_n}\right)^2 = 20 \log \frac{V_s}{V_n} \quad [\text{dB}] \tag{9.8}$$

6. The "dBm" definition

$$P_{\text{dBm}} = 10 \log \frac{P_1}{1\,\text{mW}} \quad [\text{dBm}] \tag{9.9}$$

7. Noise factor and figure

$$\text{SNR}_{in} = \frac{P_{si}}{P_{ni}} \quad \text{and} \quad \text{SNR}_{out} = \frac{P_{so}}{P_{no}} \tag{9.10}$$

$$\therefore$$

$$F = \frac{\text{SNR}_{in}}{\text{SNR}_{out}} = \frac{P_{si}}{P_{ni}} \frac{P_{no}}{P_{so}} = \frac{P_{no}}{A_P\, P_{ni}} \tag{9.11}$$

$$NF = 10 \log F \;[\text{dB}] \tag{9.12}$$

8. Noise temperature

$$T_n = (F-1)\,T \quad \text{or} \quad F = 1 + \frac{T_n}{T} \tag{9.13}$$

9. Friis's formula

$$F_{(tot)} = F_1 + \frac{F_2-1}{A_{P1}} + \frac{F_3-1}{A_{P1}A_{P2}} + \cdots + \frac{F_n-1}{A_{P1}A_{P2}\cdots A_{P(n-1)}} \tag{9.14}$$

9.2   Exercises

9.1 Definitions

1. Briefly explain the reason to use BP filter in front of amplifiers, then

(a) Sketch frequency-domain graphs that show: (a) white noise power spectral density spectrum; (b) ideal "brick–wall" bandpass (BP) filter transfer profile; and (c) the white noise spectrum after being filtered by the BP filter.

(b) Show how the filtered white noise power may be estimated graphically. Assume that the white noise $PSD = 1\,\mu\text{W/Hz}$, and the audio signal power $P_S = 100\,\text{mW}$. Calculate SNR for both cases, i.e. without and with BP filter applied.

2. Power spectrum graph of a signal and the noise floor power levels at the input side of an amplifier, Fig. 9.1.

Data: $G = 10\,\text{dB}$, NF $= 3\,\text{dB}$.

(a) Find the input side SNR_I?

(b) Assuming this signal is amplified by an amplifier with gain G and NF, redraw the power spectrum graph and find the output side SNR_o?

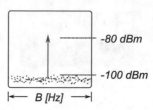

Fig. 9.1 Example 9.1-2

3. Calculate:

(a) spectrum density for thermal noise at room temperature ($T = 300\,\text{K}$);

(b) available noise power within a bandwidth of $BW = 1\,\text{MHz}$;

(c) available signal power if ($v_S = 1\,\mu\text{V}$, $R_S = 50\,\Omega$) source is connected to the matched load;

(d) SNR for noise in (b) and signal in (c).

4. Given $BW = 20\,\text{kHz}$, calculate the noise voltage v_n generated by a resistor at room temperature 290 K if

(a) $R = 50\,\Omega$, (b) $R = 1\,\text{k}\Omega$, (c) $R = 1\,\text{M}\Omega$.

5. Given two resistors at room temperature, within frequency bandwidth, calculate the thermal noise voltage for:

(a) each resistor separately; (b) their serial combination;

(c) their parallel combination.

Data: $R_1 = 20\,\text{k}\Omega$, $R_2 = 50\,\text{k}\Omega$, $T = 290\,\text{K}$, $BW = 100\,\text{kHz}$.

6. First, convert signal power levels into dBm units, then calculate SNR of the same signals if the noise power is $P_n = 1\,\text{mW}$ and:

(a) $P_S = 1\,\text{mW}$, (b) $P_S = 100\,\text{mW}$, (c) $P_S = 1\,\text{W}$, (d) $P_S = 10\,\text{W}$.

9.2 * Definitions: Noise Figure

1. Calculate F and NF for an amplifier whose output signal to noise ratio is $\text{SNR}(out) = 5$ and its input signal to noise ratio is $\text{SNR}(in) = 10$.

2. Given power spectrum measurement results at the input and output ports of an amplifier, see Fig. 9.2, estimate its

(a) power gain, (b) SNR_{in}, SNR_{out}, (c) N and NF.

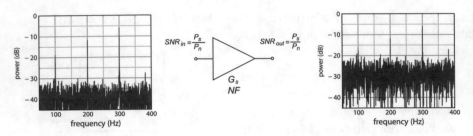

Fig. 9.2 Example 9.2-2

3. Calculate the amplifier's noise factor F at room temperature $T = 300\,\text{K}$, given that the equivalent noise temperature of an amplifier is:

(a) $T_n = 300\,\text{K}$,   (b) $T_n = 50\,\text{K}$,   (c) $T_n = 0\,\text{K}$.

4. An amplifier with the input signal power of P_{sig}, noise input power P_{ni}, has output signal power of P_{out} and the output noise power P_{no}. Calculate the noise factor F and the nose figure NF of this amplifier.
   Data: $P_{sig} = 5\,\mu\text{W}$, $P_{ni} = 1\,\mu\text{W}$, $P_{out} = 50\,\text{mW}$, $P_{no} = 40\,\text{mW}$.

9.3 * RLC Network: Bandwidth Noise

1. Calculate the equivalent noise voltage V_n generated by a resistor R in series with a capacitor C at room temperature. **Data:** $C = 100\,\text{pF}$, $T = 300\,\text{K}$.

2. Knowing that one of the forms for dynamic resistance of a tuned parallel LC tank is $R_D = Q/\omega_0 C$, calculate the effective noise voltage of the LC tank at room temperature within the given bandwidth.
   Data: $f_0 = 120\,\text{MHz}$, $C = 25\,\text{pF}$, $Q = 30$, $T = 300\,\text{K}$, $BW = 10\,\text{kHz}$.

9.4 * Friis Formula

1. Calculate the overall noise figure NF of a three stage amplifier that consists of:

   - 1st stage: $A_{P1} = 14\,\text{dB}$, $NF_1 = 3\,\text{dB}$,
   - 2nd stage: $A_{P2} = 20\,\text{dB}$, $NF_2 = 8\,\text{dB}$,
   - 3rd stage is identical to the second stage.

2. An amplifier model shows separately its equivalent internal noise resistance R_n and its input resistance R_i, as well as the source voltage V_s and its resistance R_s.
   Calculate the signal voltage V_{in} and the equivalent noise voltage V_n that appear at the amplifier's input terminals, Fig. 9.3.

   Data: $R_s = 50\,\Omega$, $V_s = 1\,\mu\text{V}$. $BW = 10\,\text{kHz}$, $T = 290\,\text{K}$, $R_n = 400\,\Omega$, $R_i = 600\,\Omega$.

   Fig. 9.3 Example 9.4-2

   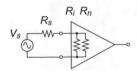

3. Front-end RF amplifier whose gain is A_{RF} and noise temperature NT provide signal to receiver whose noise figure is NF. Calculate the noise temperature of the receiver by itself, and the overall noise temperature of the amplifier plus the receiver system at the room temperature.
Data: $A_{RF} = 50\,\text{dB}$, $NT = 90\,\text{K}$, NF $= 12\,\text{dB}$, $T = 300\,\text{K}$.

4. A television set consists of the following chain of sub-blocks: two RF amplifiers with 20 dB gain and 3 dB noise figure each, a mixer with a gain of $-6\,\text{dB}$ and a noise figure of 8 dB, two additional amplifiers with 20 dB gain and a noise figure of 10 dB each. Assuming room temperature, calculate the system's:

(a) noise figure;   (b) noise temperature.

9.5 * Device Noise

1. Calculate noise current i_n and equivalent noise voltage v_n for a diode biased with I_{DC} within the bandwidth of BW. **Data:** $I_{DC} = 1\,\text{mA}$, $BW = 1\,\text{MHz}$, $T = 300\,\text{K}$.

2. A voltage source v_S with the internal resistance of R_S provides a signal to amplifier whose input resistance is R_{in}. At room temperature, the equivalent DC shot noise current is I_{DCn}. Given data, calculate the input SNR(in).
Data: $v_S = 100\mu\,V_{rms}$, $R_S = 50\,\Omega$, $R_{in} = 1\,\text{k}\Omega$, $I_{DCn} = 1\,\mu\text{A}$, $BW = 10\,\text{MHz}$, $T = 300\,\text{K}$.

9.6 *** Op-Amplifier Noise

1. Given inverting amplifier model and assuming that the only noise source in the system is due to the total thermal noise voltage e_n at its input terminals, see Fig. 9.4. The two resistors R_1, R_2 are assumed ideal, i.e. noiseless, and amplifier bandwidth is BW.

Data: $R_1 = 1\,\text{k}\Omega$, $R_2 = 10\,\text{k}\Omega$, $e_n = 5\,\text{nV}/\sqrt{\text{Hz}}$, $BW = 400\,\text{Hz}$ to $19\,\text{kHz}$.

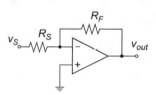

(a) Derive expression for rms value of the output noise voltage e_{out}.

(b) Calculate rms value of the amplifier's noise voltage e_{out}.

Fig. 9.4 Example 9.6-1

2. Given the same amplifier model as in Example 9.6-1 where the two resistors R_1, R_2 are assumed ideal, i.e. noiseless. The noisy operational amplifier is specified in terms of: (a) its total input noise voltage e_n and (b) its total input noise currents i_{n+} and i_{n-}, at the positive and negative input nodes of the operational amplifier. Noise currents i_{n+} and i_{n-} are equal, and within the total frequency bandwidth of the system, they are assumed to be non-correlated.

(a) Derive expression for rms value of the total output noise voltage e_{out}.

(b) Calculate e_{out}.

Data: $e_n = 5.082\,\text{nV}/\sqrt{\text{Hz}}$, $i_n = 5\text{pA}/\sqrt{\text{Hz}}$, $R_1 = 1\,\text{k}\Omega$, $R_2 = 10\,\text{k}\Omega$.

3. Given the same amplifier model as in Example 9.6-1. However, this time assume that the two resistors R_1 and R_2 are not ideal, i.e. they also contribute to the total noise power in the system.

(a) Derive expression for rms value of the total output noise voltage e_{out}.

(b) Calculate numerical value of e_{out}.

Data: $e_n = 5\,\text{nV}/\sqrt{\text{Hz}}$, $i_n = 5\,\text{pA}/\sqrt{\text{Hz}}$, $R_1 = 1\,\text{k}\Omega$, $R_2 = 10\,\text{k}\Omega$, $T = 3.75\,°\text{C}$, $BW = 20\,\text{Hz}$ to $20\,\text{kHz}$.

4. Given the amplifier model that is connected to the input signal source v_S through node node①so that the source resistance R_S is matched to the amplifier's input resistance R_3, see Fig. 9.5.

(a) Derive expression for the noise figure NF of this amplifier.

(b) Calculate NF.

Data: $R_1 = 1\,\text{k}\Omega$, $R_2 = 10\,\text{k}\Omega$, $e_n = 5\,\text{nV}/\sqrt{\text{Hz}}$, $i_n = 5\,\text{pA}/\sqrt{\text{Hz}}$, $R_3 = 1\,\text{k}\Omega$, $T = 95\,°\text{C}$.

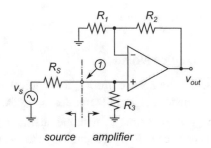

Fig. 9.5 Example 9.6-4

5. Given the amplifier model that is connected to the input signal source v_S through node node①so that the source resistance R_S is matched to the amplifier's input resistance R_4, see Fig. 9.6.

(a) Derive expression for the noise figure NF of this amplifier.

(b) Calculate NF.

Data: $e_n = 5\,\text{nV}/\sqrt{\text{Hz}}$, $i_n = 5\text{pA}/\sqrt{\text{Hz}}$, $R_S = 800\,\Omega$, $R_1 = 1\,\text{k}\Omega$, $R_2 = 10\,\text{k}\Omega$, $R_3 = 1\,\text{k}\Omega$, $T = 80.8\,°\text{C}$.

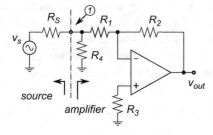

Fig. 9.6 Example 9.6-5

Solutions

Exercise 9.1, page 208

1. The total amount of energy contained within white noise is, by definition, *infinite*. That is because there are infinitely non-zero tones between zero and infinite frequency; therefore, each tone has some non-zero energy, and thus, the sum of all energies in the spectrum must be infinite. Practical consequence is that if the complete white noise is allowed to enter an amplifier, then that amplifier would need to draw infinite amount of energy from the energy source (e.g. the battery) in order to amplify the noise signal.

   Naturally, this is not realistic engineering solution. Therefore, we must apply BP filtering operation before signal amplification, where BW of the BP filter is set wide enough to let accept the signal itself. Unavoidably, white noise also enters, nevertheless limited to the BW range of frequencies. By using the BP filter, we now calculate SNR by comparing complete signal energy with the amount of noise energy that was allowed to enter (which is now finite).

   (a) To the first approximation, power spectral density (PSD) function of a white noise is assumed flat across all frequencies in the (DC to $f \to \infty$) range, that is, the PSD function is constant P. In other words, when comparing any two 1 Hz wide frequency windows, as illustrated by red boxes in Fig. 9.7a, the measured power levels are always the same. For example, the same power is contained in (10–20 Hz) as is in (1000–1010 Hz) frequency window.

   The ideal brick–wall type bandpass (BP) filter transfer function is defined as being centred around frequency ω_0, to have gain $A = 1$ within the frequency range BW, and gain $A = 0$ for any other frequency, Fig. 9.7b. Therefore, the consequence of applying BP filter to the white noise (i.e. applying the multiplication operation in frequency domain) is that all frequencies outside of the BW frequency range are completely suppressed (because amplitudes of those tones are multiplied by zero gain), while the power level inside the BW frequency range stays unchanged (because amplitudes of those tones are multiplied by the gain of one), Fig. 9.7c.

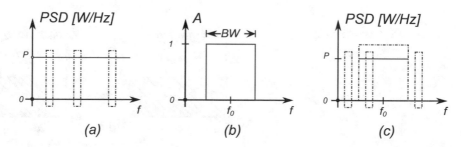

(a)   (b)   (c)

Fig. 9.7 Example 9.1-1

(b) By definition, for the known PSD, the total power P contained within a given frequency
     (f_1, f_2) window is calculated by solving integral

$$P = \int_{f_1}^{f_2} PSD \, df \qquad (9.15)$$

which, for the trivial case of $PSD = \text{const}$ and $(f_1, f_2) = (0, \infty)$, reduces to

$$P = PSD \int_{0}^{\infty} df = PSD \times \infty = \infty \; [\text{W}] \qquad (9.16)$$

for any positive value of $(PSD > 0)$. Geometrical interpretation of limited integral (9.15)
is that it represents rectangular area under PSD function, i.e. greyed areas in Fig. 9.8.
Obviously, if at least one side of a rectangle is not bound its area must be infinite. We
calculate noise power within audio bandwidth (i.e. $BW = 20\,\text{Hz}, 20\,\text{kHz}$) while holding
PSD constant as

$$P_n = PSD \int_{20}^{20 \times 10^3} df = 1\,\mu\text{W/Hz} \times 19\,980\,\text{Hz} = 19\,980\,\mu\text{W} \qquad (9.17)$$

Thus, the total noise power is equivalent to the area under PSD function in Fig. 9.8 (right).
For instance, given $P_S = 100\,\text{mW}$ voice signal, we calculate two SNR levels as

$$\text{SNR} = \frac{P_S}{P_n} = \frac{100\,\text{mW}}{\infty} = -\infty \; \text{dB}$$

and

$$\text{SNR} = \frac{P_S}{P_n} = \frac{100\,\text{mW}}{19,980\,\mu\text{W}} \approx 7 \; \text{dB}$$

Fig. 9.8 Example 9.1-1

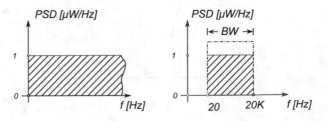

In other words, without the application of a BPF to help us limit the power of surrounding
noise, wanted signal is completely swamped by the noise and it would be impossible to
recover it again.

2. Frequency spectrum at the input and output nodes is redrawn in Fig. 9.9.

(a) Given the bandwidth B and the signal power before amplification (narrow arrow) in decibel scale, we find SNR_I as the difference between the signal $P_{\text{sig}}(in)$ and noise $P_{\text{n}}(in)$ powers, i.e.

$$\text{SNR}_I = P_{\text{sig}}(in) - P_{\text{n}}(in) = -80\,\text{dBm} - (-100\,\text{dBm}) = 20\,\text{dB}$$

Fig. 9.9 Example 9.1-2

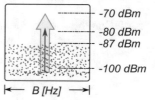

(b) Amplifier increases power of both noise and signal by $G = 10\,\text{dB}$. In addition, it also generates the internal noise whose power equals to $\text{NF} = 3\,\text{dB}$. As a consequence, the output noise power is increased. That is to say, the signal $P_{\text{sig}}(out)$ and noise $P_{\text{n}}(out)$ powers at the output node are

$$P_{\text{sig}}(out) = P_{\text{sig}}(in) + G = -80\,\text{dBm} + 10\,\text{dB} = -70\,\text{dBm}$$

and,

$$P_{\text{sig}}(n) = P_{\text{n}}(in) + G + \text{NF} = -100\,\text{dBm} + 10\,\text{dB} + 3\,\text{dB} = -87\,\text{dBm}$$

Therefore, at the output node, we find

$$\text{SNR}_o = P_{\text{sig}}(out) - P_{\text{n}}(out) = -70\,\text{dBm} - (-87\,\text{dBm}) = 17\,\text{dB} \qquad (9.18)$$

which reflects the amplifier's noise figure $\text{NF} = 3\,\text{dB}$ as expected.

3. By definitions, we write

(a) $S_n = kT = 1.38 \times 10^{-23} \times 300 = 4.14 \times 10^{-21}$ W/Hz

(b) $P_n = kT\,\Delta f = 4.14 \times 10^{-21} \times 10^6 = 4.14 \times 10^{-15}$ W

(c) $P_s = \dfrac{(v_s/2)^2}{R_s} = \dfrac{\left(1 \times 10^{-6}/2\right)^2}{50} = 5 \times 10^{-15}$ W

(d) $\text{SNR} = \dfrac{P_s}{P_n} = \dfrac{5 \times 10^{-15}}{4.14 \times 10^{-15}} = 0.82\,\text{dB}$

4. Given data, by definition

$$V_n = \sqrt{4kTR\Delta f}$$

we calculate

(a) $R = 50\,\Omega$ $\therefore$ $V_n^2 = 1.601 \times 10^{-14}\,\text{V}^2$ $\therefore$ $V_n = 126\,\text{nV}$

(b) $R = 1\,\text{k}\Omega$ $\therefore$ $V_n^2 = 3.202 \times 10^{-13}\,\text{V}^2$ $\therefore$ $V_n = 565\,\text{nV}$

(c) $R = 1\,\text{M}\Omega$ $\therefore$ $V_n^2 = 3.202 \times 10^{-10}\,\text{V}^2$ $\therefore$ $V_n = 17.9\,\mu\text{V}$

5. Given the data, by definition

$$V_n = \sqrt{4kTR\Delta f}$$

(a)   For the two resistors separately
$$V_n^2(R_1) = 4 \times 20\,\text{k}\Omega \times 1.38 \times 10^{-23} \times 290\,\text{K} \times 100\,\text{kHz} = 32 \times 10^{-12}\,\text{V}^2$$
$$V_n^2(R_2) = 4 \times 50\,\text{k}\Omega \times 1.38 \times 10^{-23} \times 290\,\text{K} \times 100\,\text{kHz} = 80 \times 10^{-12}\,\text{V}^2$$
$$\therefore$$
$$V_n(R_1) = 5.658\,\mu\text{V} \quad \text{and} \quad V_n(R_2) = 8.946\,\mu\text{V}$$

(b)   Serial resistance is $R_s = R_1 + R_2 = 70\,\text{k}\Omega$ $\therefore$ $V_n(R_s) \approx 10\,\mu\text{V}$.

(c)   Parallel resistance is $R_p = R_1 || R_2 = 14.286\,\text{k}\Omega$ $\therefore$ $V_n(R_p) = 4.78\,\mu\text{V}$.

6. By definition, power in units of W is converted into units of dBm as

$$P_{\text{dBm}} = 10\log\frac{P_1}{1\,\text{mW}} \quad [\text{dBm}] \tag{9.19}$$

and signal to noise ratio for power and voltage signals is defined as

$$\text{SNR} = 10\log\frac{P_s}{P_n} \quad [\text{dB}] \tag{9.20}$$

$$= 10\log\frac{V_s^2/R}{V_n^2/R} = 10\log\left(\frac{V_s}{V_n}\right)^2 = 20\log\frac{V_s}{V_n} \quad [\text{dB}] \tag{9.21}$$

Therefore, in accordance with (9.9), we calculate:

(a) $P_S = 1\,\text{mW}$ $\therefore$ $P_S = 10\log\dfrac{1\,\text{mW}}{1\,\text{mW}} = 10\log(1) = 0\,\text{dB}$

(b) $P_S = 100\,\text{mW}$ $\therefore$ $P_S = 10\log\dfrac{100\,\text{mW}}{1\,\text{mW}} = 10\log(100) = 20\,\text{dB}$

(c) $P_S = 1\,\text{W}$ $\therefore$ $P_S = 10\log\dfrac{1000\,\text{mW}}{1\,\text{mW}} = 10\log(1000) = 30\,\text{dB}$

(d) $P_S = 10\,\text{W}$ $\therefore$ $P_S = 10\log\dfrac{10,000\,\text{mW}}{1\,\text{mW}} = 10\log(10,000) = 40\,\text{dB}$

Similarly, in accordance with (9.20), we calculate:

(a) $P_S = 1\,\text{mW}$ $\therefore$ $\text{SNR} = 10\log \dfrac{1\,\text{mW}}{1\,\text{mW}} = 10\log(1) = 0\,\text{dBm}$

(b) $P_S = 100\,\text{mW}$ $\therefore$ $\text{SNR} = 10\log \dfrac{100\,\text{mW}}{1\,\text{mW}} = 10\log(100) = 20\,\text{dBm}$

(c) $P_S = 1\,\text{W}$ $\therefore$ $\text{SNR} = 10\log \dfrac{1000\,\text{mW}}{1\,\text{mW}} = 10\log(1000) = 30\,\text{dBm}$

(d) $P_S = 10\,\text{W}$ $\therefore$ $\text{SNR} = 10\log \dfrac{10{,}000\,\text{mW}}{1\,\text{mW}} = 10\log(10{,}000) = 40\,\text{dBm}$

Note that in the first calculations the signal power levels are the *absolute* values in dBm, while SNR calculations are in *relative* units of dB.

Exercise 9.2, page 209

1. By definition, SNR is the ratio of signal P_s and noise P_n powers calculated separately at input and output terminals

$$\text{SNR}_{in} = \frac{P_{si}}{P_{ni}} \quad \text{and} \quad \text{SNR}_{out} = \frac{P_{so}}{P_{no}} \tag{9.22}$$

which is used to define noise factor F of a given amplifier as

$$F = \frac{\text{SNR}_{in}}{\text{SNR}_{out}} = \frac{P_{si}\,P_{no}}{P_{ni}\,P_{so}} = \frac{P_{no}}{A_P\,P_{ni}} \tag{9.23}$$

where $A_P = P_{so}/P_{si}$ is the signal power gain. Given the data, we calculate

$$F = \frac{\text{SNR}(in)}{\text{SNR}(out)} = \frac{10}{5} = 2 \quad \therefore \quad \text{NF} \overset{\text{def}}{=} 10\log F = 10\log 2 = 3\,\text{dB}$$

that is, due to its internal heating and operations, this amplifier itself contributes 3 dB to the total noise budget of the overall system.

2. By inspection of power spectrum plots, Fig. 9.10, we conclude that the input signal consists of three single tones: 100, 200, and 300 Hz.

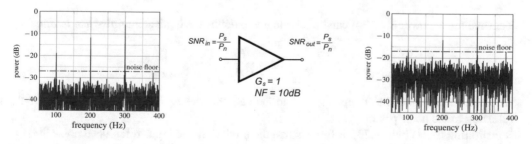

Fig. 9.10 Example 9.2-2

(a) At the input side, the power levels of these three tones are approximately $-18\,\mathrm{dB}$, $-12\,\mathrm{dB}$, and $-6\,\mathrm{dB}$, respectively. By inspection of the output side spectrum, we find that the power levels stayed approximately the same, i.e. power level difference between output and input signals is $0\,\mathrm{dB}$ or $G = 1$.

(b) At the input side, we estimate that the noise level is approximately $-27\,\mathrm{dB}$. The weakest of the three tones is at $-18\,\mathrm{dB}$; therefore, SNR is a simple difference of powers measured in dB, i.e. $\mathrm{SNR}_{in} \approx -18\,\mathrm{dB} - (-27\,\mathrm{dB}) = +9\,\mathrm{dB}$. By the same criteria, although the signal powers did not change, the output side noise power increased to $-18\,\mathrm{dB}$ level. As a consequence, the weakest tone in signal spectrum is indistinguishable from the noise, i.e. power level difference between the noise and the signal's component equals $\mathrm{SNR}_{out} \approx -18\,\mathrm{dB} - (-18\,\mathrm{dB}) = 0\,\mathrm{dB}$. The original signal is permanently distorted due to loss of one of the three components.

(c) We recall that notion of "noise" and "signal" is based strictly on our desire to suppress and ignore the "noise" while at the same time to amplify and receive the "signal". Nevertheless, our "noise" may be desired signal for someone else, while our "signal" may be noise for some other receiver. That being said, an amplifier does not distinguish between "noise" and "signal", and thus, both are valid and amplified equally. We deduced that in this example $G = 1$; however, the noise level is increased by $10\,\mathrm{dB}$. The question is where this difference is coming from?

Unless the operational temperature is $0\,\mathrm{K}$, every electronic component and circuit generates the internal thermal noise. Unavoidably, this noise is added to the signal being processed. Qualitative measure of this internally generated noise (thus, the component's "quality") is *noise figure* NF.

In this example, $G = 1$, therefore, NF is measured as the output/input noise level difference $\mathrm{NF} \approx -17\,\mathrm{dB} - (-27\,\mathrm{dB}) = 10\,\mathrm{dB}$. By definition,

$$\mathrm{NF} \stackrel{\mathrm{def}}{=} 10 \log F \quad \therefore \quad F = 10^{\frac{\mathrm{NF}}{10}} = 10^{\frac{10}{10}} = 10$$

3. The noise power generated within frequency bandwidth Δf is, by definition,

$$P_n = \int_{f_1}^{f_1 + \Delta f} S_n(f)\, df = S_n(f) \int_{f_1}^{f_1 + \Delta f} df = k\, T\, \Delta f \quad [\mathrm{W}] \qquad (9.24)$$

Thermal noise power in (9.24) can be rearranged to define the equivalent *noise temperature* T_n as

$$T_n = \frac{P_n}{k\, \Delta f} \qquad (9.25)$$

where index n is added to temperature T to indicate that the equivalent noise temperature T_n is referring to the noise power P_n.

Substituting (9.24) into (9.23), it follows that the total available input-referenced noise is

$$F = \frac{P_{no}}{A_P\, k\, T\, \Delta f} = \frac{P_{ni}}{k\, T\, \Delta f} \quad \therefore \quad P_{ni} = F\, k\, T\, \Delta f \qquad (9.26)$$

where amplifier cannot distinguish between "signal" and "noise" amplification, that is to say, $P_{no} = A_P P_{ni}$ ∴ $P_{ni} = P_{no}/A_P$. Therefore, the amplifier's noise contribution P_{na} is simply the difference between the output and input noise powers

$$P_{na} = F k T \Delta f - k T \Delta f = (F - 1) k T \Delta f \tag{9.27}$$

Substituting (9.27) into (9.25) (in the case of an amplifier for which $P_n = P_{na}$), we write

$$T_n = \frac{P_n}{k \Delta f} = \frac{(F - 1) T k \Delta f}{k \Delta f} = (F - 1) T \quad \text{or} \quad F = 1 + \frac{T_n}{T} \tag{9.28}$$

where T_n is the noise temperature and T is the ambient temperature. With this formula, we express noise in units of temperature.

(a) Given $T_n = 300\,\text{K}$:

$$300\,\text{K} = (F - 1) \times 300\,\text{K} \quad \therefore \quad F = \frac{300\,\text{K}}{300\,\text{K}} + 1 = 2$$

NF $= 10 \log 2 = 3\,\text{dB}$

(b) Given $T_n = 50\,\text{K}$:

$$50\,\text{K} = (F - 1) \times 300\,\text{K} \quad \therefore \quad F = \frac{50\,\text{K}}{300\,\text{K}} + 1 = 1.167$$

NF $= 10 \log 1.167 = 0.669\,\text{dB}$

(c) Given $T_n = 0\,\text{K}$:

$$0\,\text{K} = (F - 1) \times 300\,\text{K} \quad \therefore \quad F = \frac{0\,\text{K}}{300\,\text{K}} + 1 = 1$$

NF $= 10 \log 1 = 0\,\text{dB}$

which is to say that noiseless amplifier is equivalent to $T_n = 0\,\text{K}$ (it does not add noise to the input signal), while when T_n is equal to the environment temperature the noise of this amplifier equals to the noise of the input signal.

4. By definitions, we write

$$F = \frac{\text{SNR}_{in}}{\text{SNR}_{out}} = \frac{P_{si} P_{no}}{P_{ni} P_{so}} = \frac{5\,\mu\text{W} \times 40\,\text{mW}}{1\,\mu\text{W} \times 50\,\text{mW}} = 4 \quad \therefore \quad \text{NF} = 10 \log 4 = 6\,\text{dB}$$

Exercise 9.3, page 210

1. Noise voltage within RC bandwidth is calculated as

$$V_n^2 = \frac{k T}{C} = \frac{1.38 \times 10^{-23} \times 300\,\text{K}}{100\,\text{pF}} = 4.14 \times 10^{-11}\,V^2 \quad \therefore \quad V_n = 6.434\,\mu\text{V}$$

2. Given an expression for dynamic resistance of LC resonator, we write

$$R_D = \frac{Q}{\omega_0 C} = \frac{30}{2\pi \times 120\,\text{MHz} \times 25\,\text{pF}} = 1.59\,\text{k}\Omega$$

If the noise calculation is limited to narrow bandwidth $\Delta f \ll f_0$ around the resonant frequency f_0, then the RLC network's transfer function $H(\omega_0)$ is approximated as $H(\omega_0) \approx Q$, where one of Q definitions is in the form of

$$Q = \frac{\omega_0}{BW}, \quad (\omega_0 = 2\pi f_0) \tag{9.29}$$

where BW is defined strictly by the -3 dB points. Solving the spectrum density integral in the case of $H(\omega_0) \approx Q$, thus $S_{no} = Q^2 kT$, the noise power generated within frequency bandwidth $BW = \Delta f$ gives

$$P_{no} = \int_0^\infty S_{no}\, df = \int_{f_0 - \Delta f/2}^{f_0 + \Delta f/2} Q^2 kT\, df = Q^2 kT\, \Delta f \stackrel{\text{def}}{=} kT\, \Delta f_{\text{eff}}$$

$$\therefore$$

$$\Delta f_{\text{eff}} = Q^2\, \Delta f \tag{9.30}$$

which gives the equivalent noise voltage V_n as

$$V_n^2 = 4\,r\,k\,T\,Q^2\,\Delta f = 4\,R_D\,k\,T\,\Delta f \tag{9.31}$$

where $R_D = Q^2\, r$ is the "dynamic resistance" of the RLC circuit at resonance. This result is very important for practical calculations because the noise bandwidth in RLC tuned networks is indeed limited to a narrow bandwidth around the resonant frequency. Therefore, the thermal noise within bandwidth is

$$V_n^2 = 4R_D\,k\,T\,\Delta f = 0.254 \times 10^{-12}\, \text{V}^2 \quad \therefore \quad V_n = 0.50\,\mu\text{V}$$

Exercise 9.4, page 210

1. Using Friis's formula, we write

$$A_{P1} = 14\,\text{dB} = 25.1, \quad A_{P2} = A_{P3} = 20\,\text{dB} = 100$$

$$\text{NF}_1 = 3\,\text{dB} \quad \therefore \quad F_1 = 2, \quad \text{NF}_2 = NF_3 = 8\,\text{dB} \quad \therefore \quad F_2 = F_3 = 6.31$$

therefore,

$$F_{(tot)} = 2 + \frac{6.31 - 1}{25.1} + \frac{6.31 - 1}{25.1 \times 100} = 2.212 \quad \therefore \quad \text{NF} = 10\log 2.212 = 3.448\,\text{dB}$$

This example illustrates that NF of the first stage is the one that is most important, even if the subsequent stages have terrible NF, the overall NF of the system is close to the NF of the first stage. For that reason, most of the effort in design of RF front-end circuits is given to design the first stage amplifier, also known as "low-noise-amplifier" (LNA).

2. Application of Thévenin theorem on the input side network V_s, R_s, and R_i results in the equivalent Thévenin generator

$$R_t = \frac{R_s\, R_i}{R_s + R_i} = 46.15\,\Omega \quad \text{and} \quad V_t = V_s\, \frac{R_i}{R_s + R_i} = 0.923\,\mu V$$

The equivalent noise voltage at the amplifier input is then calculated for the case of serial resistors as

$$R = R_t + R_n = 400\,\Omega + 46.15\,\Omega = 446.15\,\Omega$$

which generates thermal voltage of

$$V_n = \sqrt{4\,R\,k\,T\,\Delta f} = \sqrt{4 \times 446.15\,\Omega \times 1.38 \times 10^{-23} \times 290 \times 10\,\text{kHz}}$$
$$= 0.267\,\mu V$$

3. First, we convert $12\,\text{dB} = 15.85$ and $50\,\text{dB} = 1 \times 10^5$.
   By definition of noise temperature, we write

$$T\text{rec} = (F - 1)T = (15.85 - 1) \times 300\,\text{K} = 4455\,\text{K}$$

and the total noise temperature is $T\text{sys} = 90\,\text{K} + 4455\,\text{K}/1 \times 10^5 \approx 90\,\text{K}$.

4. There are five stages in total, and we summarize gains and noise figures as

	1	2	3	4	5
	amp1	amp2	mixer	amp3	amp4
A [dB]	20	20	−6	20	20
NF [dB]	3	3	8	10	10
A_P	100	100	0.25	100	100
F	2	2	6.31	10	10

Using Friis's formula in the case a five stage system, we calculate

$$F_{(tot)} = F_1 + \frac{F_2 - 1}{A_{P1}} + \frac{F_3 - 1}{A_{P1}A_{P2}} + \frac{F_4 - 1}{A_{P1}A_{P2}A_{P3}} + \frac{F_5 - 1}{A_{P1}A_{P2}A_{P3}A_{P4}}$$
$$= 2 + \frac{2 - 1}{100} + \frac{6.31 - 1}{100 \times 100} + \frac{10 - 1}{100 \times 100 \times 0.25} + \frac{10 - 1}{100 \times 100 \times 0.25 \times 100}$$
$$= 2.014$$

Therefore, by definition: (a) $NF = 10\log(2.014) = 3.04\,\text{dB}$ and (b) $T_n = (F - 1)\,T = 1.014 \times 300\,\text{K} = 304\,\text{K}$.

Exercise 9.5, page 211

1. Shot noise current i_{sn} generated in a diode is

$$i_{sn}^2 = 2q\, I_D\, B \quad \therefore \quad i_{sn} = 18\text{nA}$$

and

$$V_T \approx 26\,\text{mV} \quad \therefore \quad r_D \approx 26\,\Omega \quad \therefore \quad V_n = i_{sn}\,r_D = 463\,\text{nV}$$

2. Shot noise current is found as

$$I_{ns} = \sqrt{2q\,I_{DCn}\,BW} = 1.79\,\text{nA}$$

which generates noise voltage in the source resistance as

$$V_{ns}(R_S) = I_{ns}\,R_S = 85.9\,\text{nV}$$

Noise voltages generated by two resistors are $V_n(R_S) = 2.88\,\mu\text{V}$ and $V_n(R_{in}) = 12.87\,\mu\text{V}$. Therefore, the total noise voltage is found as

$$V_n = \sqrt{V_{ns}(R_S)^2 + V_n(R_S)^2 + V_n(R_{in})^2} = 13.19\,\mu\text{V}$$

Relative to v_S, the total noise voltage causes SNR $= 17.6\,\text{dB}$.

Exercise 9.6, page 211

1. Noise analysis of amplifiers follows already known principles:

1. each noisy component is replaced with its equivalent noiseless component model plus the ideal voltage/current source to model the associated noise voltage/current itself; and
2. networks containing multiple sources are solved by applying the superposition principle.

Thus, this noisy operational amplifier is replaced with its equivalent noiseless version plus the thermal input-referred voltage noise source e_n at the positive input node, see Fig. 9.11. Dotted line represents boundaries of the original noisy operational amplifier. We note that polarity of ideal thermal noise sources is meaningless because the noise signal is random and we use the squares of noise voltages.

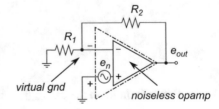

Fig. 9.11 Example 9.6-1

(a) *Contribution of voltage noise source e_n* is found by shorting the input signal source v_S, and after recalling that both terminals of an opamp are at the same potential, thus relative to the non-inverting input node, we write

$$e_{out}(e_n) = i_{R1}\,(R_1 + R_2) = \frac{e_n}{R_1}\,(R_1 + R_2) \tag{9.32}$$

(b) Given the numerical data, we write

$$e_{out} = \left(1 + \frac{10\,\text{k}\Omega}{1\,\text{k}\Omega}\right) 5\,\text{nV}/\sqrt{\text{Hz}} = 55\,\text{nV}/\sqrt{\text{Hz}}$$

To calculate the total noise voltage within a certain BW, we simply multiply the last result by $\sqrt{BW}$. The amplifier bandwidth is specified as $BW = 400\,\text{Hz}$ to $19\,\text{kHz}$, and the total measured noise voltage at the output terminal is

$$e_{out}(\texttt{total}) = 55\,\text{nV}/\sqrt{\text{Hz}} \times \sqrt{(19k - 400)\text{Hz}} = 7.5\,\mu\text{V} \qquad (9.33)$$

2. The noisy operational amplifier is replaced with its equivalent noiseless version plus the thermal voltage noise source e_n at the positive input node, and the thermal current noise sources i_{n-} and i_{n+} at its input terminals. Dotted line represents the original noisy operational amplifier, Fig. 9.12.

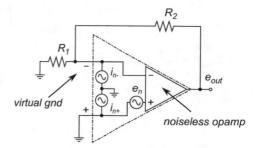

Fig. 9.12 Example 9.6-2

In this case, there is more than one noise source, and we have to apply the superposition principle. Each of the noise sources is considered on its own as follows.

(a) *Contribution of voltage noise source e_n:* we reused result in (9.32) to calculate

$$e_{out}(e_n) = \left(1 + \frac{10k}{1k}\right) 5.082\,\text{nV}/\sqrt{\text{Hz}} = 55.902\,\text{nV}/\sqrt{\text{Hz}}$$

$$\therefore$$

$$[e_{out}(e_n)]^2 = 3.125 \times 10^{-15}\ \text{V}^2/\text{Hz} \qquad (9.34)$$

(b) *Contribution of current noise source i_{n-}* is found after shorting the noise voltage source and after the current noise source i_{n+} is left open, as illustrated in Figs. 9.13 and 9.14. Under these circumstances, we can reason that there is no noise current flowing through R_1 because both of its nodes are at the same potential (i.e. gnd). Therefore, the complete noise current i_{n-} is forced through R_2, which generates noise voltage at the output node (one node of R_2 is also at the virtual gnd potential), and we write

$$e_{in-} = i_{n-}\,R_2 \quad \therefore \quad e_{in-}^2 = (i_{n-}\,R_2)^2 \qquad (9.35)$$

Given the numerical data, we calculate

$$e_{in-} = i_{n-}\,R_2 = 5\text{pA}/\sqrt{\text{Hz}} \times 10\,\text{k}\Omega = 50\,\text{nV}/\sqrt{\text{Hz}}$$

$$\therefore$$

$$e_{in-}^2 = 2.5 \times 10^{-15}\ \text{V}^2/\text{Hz} \qquad (9.36)$$

Fig. 9.13 Example 9.6-2

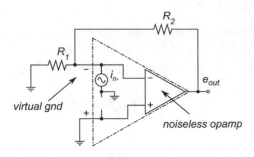

Fig. 9.14 Example 9.6-2

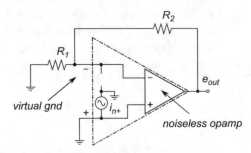

(c) *Contribution of current noise source* i_{n+} is found after shorting the noise voltage source, and after the current noise source i_{n-} is left open. We reason that there is no resistor for i_{n+} to develop voltage, i.e. $(i_{n+} \times 0\,\Omega) = 0$, because the current source is shorted by ideal wire where the whole dissipated energy is contained.

Total noise contribution of two combined noise sources is found by statistical addition (i.e. by adding the noise powers) of (9.34) and (9.36) as

$$e_{out}^2 = e_{en}^2 + e_{in-}^2 = \left[\left(1 + \frac{R_2}{R_1} \right) e_n \right]^2 + (i_{n-} R_2)^2$$

$$= (3.125 + 2.5) \times 10^{-15}\ \mathrm{V^2/Hz} = 5.625 \times 10^{-15}\ \mathrm{V^2/Hz}$$

$$\therefore$$

$$e_{out} = 75\,\mathrm{nV}/\sqrt{\mathrm{Hz}} \tag{9.37}$$

3. The noisy operational amplifier and noisy resistors are replaced with their equivalent noiseless versions plus the respective thermal voltage noise sources, Fig. 9.15. Important point to keep in mind is that the thermal noise voltage appears across the noisy resistor terminal. That is, in our model, it includes both the noiseless resistor and the associated thermal voltage generator, and the thermal noise appears between the same circuit nodes where the noisy resistor was connected.

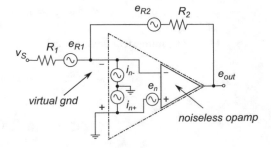

Fig. 9.15 Example 9.6-3

Thus, in addition to the already found solutions Example 9.6-2, we need to add contribution of the two noisy resistors.

(a) *Contribution due to noise source* e_{R1} is found from the equivalent circuit network as follows. The analysis can be done in a number of ways, and here we use reasoning to say that voltage e_{R1} appears at the negative input of the operational amplifier, which is at the virtual ground potential, Fig. 9.16. Therefore, the noise current through R_1 must be $i_{R1} = e_{R1}/R_1$.

Fig. 9.16 Example 9.6-3

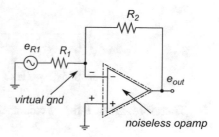

This current has nowhere else to go but through R_2, and thus, we write

$$e_{out}(R_1) = i_{R1}\, R_2 = \frac{e_{R1}}{R_1}\, R_2$$

$$\therefore$$

$$e_{our}^2(R_1) = e_{R1}^2 \left(\frac{R_2}{R_1}\right)^2 = 4kTR_1\Delta f \left(\frac{R_2}{R_1}\right)^2 \; \mathrm{V}^2 = 4kTR_1 \left(\frac{R_2}{R_1}\right)^2 \; \mathrm{V}^2/\mathrm{Hz}$$

$$= \left[4 \times 1.3806488 \times 10^{-23} \times 276.9 \times 1\mathrm{k} \times \left(\frac{10\mathrm{k}}{1\mathrm{k}}\right)^2 \right] \; \mathrm{V}^2/\mathrm{Hz}$$

$$= 1.529 \times 10^{-15} \; \mathrm{V}^2/\mathrm{Hz} \tag{9.38}$$

(b) *Contribution due to noise source* e_{R2} is found from the equivalent circuit network as follows. The noise voltage is due to the *noisy* R_2 resistor that is connected between the amplifier output node and the virtual ground, see Fig. 9.17.

Fig. 9.17 Example 9.6-3

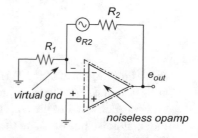

Therefore, by inspection, we simply write

$$e_{out}(R_2) = e_{R2} \quad \therefore \quad e_{our}^2(R_2) = 4kTR_2\Delta f \; \mathrm{V}^2 = 4kTR_2 \; \mathrm{V}^2/\mathrm{Hz}$$

$$e_{our}^2(R_2) = \left(4 \times 1.3806488 \times 10^{-23} \times 276.9 \times 10k\right) \ \text{V}^2/\text{Hz}$$

$$= 0.153 \times 10^{-15} \ \text{V}^2/\text{Hz} \tag{9.39}$$

Therefore, after combining (9.37) to (9.39), the complete solution is

$$e_{out}^2 = \left\{ \left[\left(1 + \frac{R_2}{R_1}\right) e_n \right]^2 + (i_{n-} R_2)^2 + 4kTR_1 \left(\frac{R_2}{R_1}\right)^2 + 4kTR_2 \right\} \ \text{V}^2/\text{Hz}$$

$$\therefore$$

$$e_{out} = \sqrt{\left((3.025 + 2.5 + 1.529 + 0.153) \times 10^{-15}\right)} \ \text{V}/\sqrt{\text{Hz}}$$

$$= 84.895 \ \text{nV}/\sqrt{\text{Hz}} \tag{9.40}$$

Therefore, given BW, we calculate the total noise rms voltage as

$$e_{out}(total) = e_{out} \times \sqrt{BW} = 84.895 \ \text{nV}/\sqrt{\text{Hz}} \times \sqrt{(20\,\text{kHz} - 20\,\text{Hz})}$$

$$= 12\,\mu\text{V} \tag{9.41}$$

The above analysis is valid under assumption that the noise spectrum density function is constant within the given bandwidth.

4. Noise figure NF of a device quantifies how much noise power is actually created internally within the device itself. Complete schematic diagram includes all thermal noise sources along with their respective noiseless components, Fig. 9.18.

Fig. 9.18 Example 9.6-4

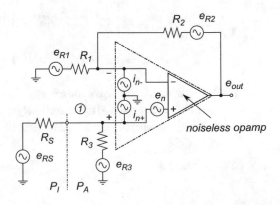

By definition, noise factor F is a ratio of SNR_i at the input of a device, and signal to noise ratio at its output SNR_o. Thus, in order to calculate F, we have to find the ratio of the noise power P_I that was delivered to the amplifier by the source (i.e. presented at the input node node①), and the total noise power e_{out} at the output of the amplifier.

The fact that e_{out} is measured at the amplifier's output node complicates the analysis because the source noise power is measured at the input node. Additionally, the output noise power e_{out} is the sum of the amplified input noise power plus the internally generated noise power. It is important to realize, therefore, that in order to compare the input noise power P_I, as delivered by the source to the input node, with the internally generated noise power, the amplifier noise contribution needs to be referenced relative to the same node. Thus, we introduce the term *input-referred* voltage/current noise power P_A. In accordance with this definition, P_A is equivalent to the complete internally generated noise power that is integrated at the input node, i.e. before the amplification.

Naturally, if an amplifier is ideal (i.e. noiseless), the input-referred noise power $P_A = 0$, which is equivalent to $F = 1$. In that case, the output noise power is due to only the source noise power amplified by the amplifier.

Assuming that the amplifier gain is A, the total output noise power is then $A(P_I + P_A)$, and the input signal power is P_S, and accordingly, we rewrite noise figure F as

$$F = \frac{SNR_i}{SNR_o} = \frac{\dfrac{P_S}{P_I}}{\dfrac{A\,P_S}{A\,(P_I + P_A)}} = 1 + \frac{P_A}{P_I} \tag{9.42}$$

which illustrates the point. We now derive input-referred noise and the input noise powers as follows.

(a) *Contribution due to noise source e_{RS}*: Thermal noise voltage generated by the source resistance R_S is delivered to the input node through the resistive voltage divider (R_S, R_3), and thus, at node node①, we write

$$e_I = \frac{e_{RS}}{R_S + R_3}\,R_3 = \sqrt{4kTR_S}\,\frac{R_3}{R_S + R_3} \quad \therefore \quad e_I^2 = 4kTR_S\left(\frac{R_3}{R_S + R_3}\right)^2$$

When the source resistance is matched to the amplifier input impedance, i.e. $R_S = R_3$, we have

$$P_I = e_I^2 = kTR_3 \ \ \text{V}^2/\text{Hz} = 5.083 \times 10^{-18} \ \ \text{V}^2/\text{Hz} \tag{9.43}$$

(b) *Contribution due to noise source e_{R3}*: Thermal noise voltage generated by the source resistance R_3 is delivered to the input node through the resistive voltage divider (R_3, R_S), and thus, at node node①, we write

$$e_3 = \frac{e_{R3}}{R_S + R_3}\,R_S = \sqrt{4kTR_3}\,\frac{R_S}{R_S + R_3} \quad \therefore \quad e_3^2 = 4kTR_3\left(\frac{R_S}{R_S + R_3}\right)^2$$

When the source resistance is matched to the amplifier input impedance, i.e. $R_S = R_3$, we write

$$e_3^2 = kTR_3 \ \ \text{V}^2/\text{Hz} \tag{9.44}$$

(c) *Contribution due to noise source e_{R1}:* Thermal noise voltage generated by the source resistance R_1 is delivered to the input node through the resistive voltage divider (R_1, R_2), and thus, at node node①, we write

$$e_1 = \frac{e_{R1}}{R_1 + R_2} \, R_2 = \sqrt{4kTR_1} \, \frac{R_2}{R_1 + R_2} \quad \therefore \quad e_1^2 = 4kTR_1 \left(\frac{R_2}{R_1 + R_2}\right)^2$$

Noise voltage source e_{R2} is shorted, and we remember that an operational amplifier is also a voltage source, and thus its output impedance equals to zero. That is, for purposes of this analysis, R_2 is effectively connected to the gnd node through the operational amplifier's output.

(d) *Contribution due to noise source e_{R2}:* Thermal noise voltage generated by the source resistance R_2 is delivered to the input node through the resistive voltage divider (R_2, R_1), and thus at node node①, we write

$$e_1 = \frac{e_{R2}}{R_1 + R_2} \, R_1 = \sqrt{4kTR_2} \, \frac{R_1}{R_1 + R_2} \quad \therefore \quad e_1^2 = 4kTR_2 \left(\frac{R_1}{R_1 + R_2}\right)^2$$

(e) *Contribution due to noise voltage source e_n:* The internal thermal noise voltage generated by operational amplifier is directly connected to the input node, and thus, at node node①, we write

$$e_{n1} = e_n \quad \therefore \quad e_{n1}^2 = e_n^2$$

(f) *Contribution due to noise current source i_{n+}:* This current source forces noise current through parallel connection $R_3 \| R_S$, which generates voltage directly at the input node,

$$e_{in+} = i_{n+} \, R_3 \| R_S \quad \therefore \quad e_{in+}^2 = \left[i_{n+} \left(\frac{R_3 R_S}{R_3 + R_S}\right)\right]^2 = \left[i_{n+} \, \frac{R_3}{2}\right]^2$$

(g) *Contribution due to noise current source i_{n-}:* This current source forces noise current through parallel connection $R_1 \| R_2$, which generates voltage directly at the negative input node of operational amplifier that appears also at the positive input node, and thus,

$$e_{in-} = i_{n-} \, R_1 \| R_2 \quad \therefore \quad e_{in-}^2 = \left[i_{n-} \left(\frac{R_1 R_2}{R_1 + R_2}\right)\right]^2$$

Therefore, after combining all six amplifier thermal noise power terms, the complete solution for the internal noise power spectrum density is

$$P_A = e_{out}^2 = \left[kTR_3 + 4kTR_1 \left(\frac{R_2}{R_1 + R_2}\right)^2 + 4kTR_2 \left(\frac{R_1}{R_1 + R_2}\right)^2 + \right.$$

$$e_n^2 + i_{n+}^2 \left(\frac{R_3 R_S}{R_3 + R_S}\right)^2 + i_{n-}^2 \left(\frac{R_1 R_2}{R_1 + R_2}\right)^2 \Bigg] \quad \text{V}^2/\text{Hz}$$

$$\therefore$$

$$P_A = (5.083 + 16.803 + 1.680 + 25 + 20.661 + 6.25) \times 10^{-18} \quad \text{V}^2/\text{Hz}$$

$$= 75.447 \times 10^{-18} \quad \text{V}^2/\text{Hz} \tag{9.45}$$

Substituting (9.45) and (9.43) into (9.42), we calculate

$$F = 1 + \frac{75.477 \times 10^{-18} \; \text{V}^2/\text{Hz}}{5.083 \times 10^{-18} \; \text{V}^2/\text{Hz}} = 15.849$$

$$\therefore$$

$$NF = 10 \, \log(15.849) = 12 \, \text{dB} \tag{9.46}$$

5. Complete schematic diagram includes all thermal noise sources along with their respective noiseless components, see Fig. 9.19. By the superposition principle, we write.

Fig. 9.19 Example 9.6-5

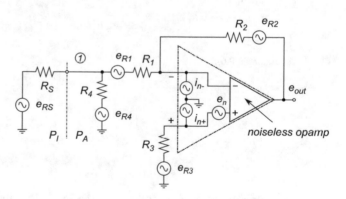

(a) *Contribution due to noise source e_{RS}*: Thermal noise voltage generated by resistance R_S is delivered to the input node through the resistive voltage divider $(R_S, R_1 || R_4)$, and thus, at node node①, we write

$$e'_{RS} = \frac{e_{RS}}{R_S + R_1 || R_4} \; (R_1 || R_4) = \sqrt{4kTR_S} \; \frac{R_1 || R_4}{R_S + R_1 || R_4}$$

$$\therefore$$

$$(e'_{RS})^2 = 4kTR_S \left(\frac{R_1 || R_4}{R_S + R_1 || R_4}\right)^2 \quad \text{V}^2/\text{Hz} \tag{9.47}$$

In order to match the source resistance to the amplifier input impedance, first we calculate R_4 so that

$$R_S = R_1 || R_4 = \frac{R_1 R_4}{R_1 + R_4} = \frac{R_1}{\dfrac{R_1}{R_4} + 1} \quad \therefore \quad R_4 = \frac{R_1 R_S}{R_1 - R_S} = 4\,\text{k}\Omega$$

then from (9.47), we write

$$P_I = (e'_{RS})^2 = kT R_S \quad \text{V}^2/\text{Hz} = 3.909 \times 10^{-18} \quad \text{V}^2/\text{Hz} \tag{9.48}$$

(b) *Contribution due to noise source e_{R3}*: Thermal noise voltage due to resistance R_3 is directly connected to the positive input node of the operational amplifier, Figs. 9.20 and 9.21. Thus, $e_{R3} = \sqrt{4kT R_3}$ V/$\sqrt{\text{Hz}}$ is also measured at the negative input node of the operational amplifier.

Fig. 9.20 Example 9.6-5

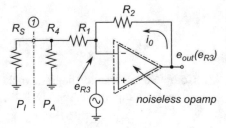

Fig. 9.21 Example 9.6-5

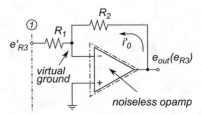

Our reasoning goes as following. We know that thermal voltage e_{R3} is amplified and is therefore measured at the output node as $e_{out}(e_{R3})$. This output voltage due to e_{R3} is then found by the following current i_0 through the $(R_2 + R_1 + R_S || R_4)$ resistive path. The feedback current i_0 is set by e_{R3} potential at the negative input terminal of the operational amplifier that forces i_0 down through $(R_1 + R_S || R_4)$ resistor path, i.e. the output voltage is

$$e_{out}(e_{R3}) = i_0 \, (R_2 + R_1 + R_S || R_4)$$

$$= \frac{e_{R3}}{R_1 + R_S || R_4} \, (R_2 + R_1 + R_S || R_4) \tag{9.49}$$

Nevertheless, our goal is to find the equivalent input voltage source e'_{R3} that, if connected to node①, would generate the output $e_{out}(e_{R3})$. We note that under conditions of this particular setup where e'_{R3} is the only signal source there is no current flow through R_3. This is because R_3 is connected between the ground and the opamp's input terminal, and thus, no current can flow into the opamp's input node. Thereafter, we write

$$e_{out}(e_{R3}) = -i_0' R_2 = -\frac{e_{R3}'}{R_1} R_2 \quad \therefore \quad (e_{R3}')^2 = \left[\frac{R_1}{R_2} e_{out}(e_{R3})\right]^2 \quad (9.50)$$

From (9.49) and (9.50), it follows that

$$(e_{R3}')^2 = \left[\frac{R_1}{R_2} \frac{R_2 + R_1 + R_S||R_4}{R_1 + R_S||R_4} e_{R3}\right]^2$$

$$= \left[\frac{R_1}{R_2} \frac{R_2 + R_1 + R_S||R_4}{R_1 + R_S||R_4}\right]^2 4kT R_3 \ \text{V}^2/\text{Hz} \quad (9.51)$$

which, given the data, results in

$$(e_{R3}')^2 =$$

$$\left[\left(\frac{1k}{10k} \frac{10k + 1k + 666.667}{1k + 666.667}\right)^2 \times 4 \times 1.38 \times 10^{-23} \times 353.95 \ \text{K} \times 1k\right] \ \text{V}^2/\text{Hz}$$

$$= 9.578 \times 10^{-18} \ \text{V}^2/\text{Hz} \quad (9.52)$$

(c) *Contribution due to noise source e_{R1}:* In order to derive thermal noise voltage generated by resistance R_1, as seen from the input node node①　through the resistive voltage divider $(R_1, R_4||R_S)$, we use similar reasoning as for e_{R3}' above, and thus, we write

$$e_{R1}' = i_{R1} R_1 = \frac{e_{R1}}{R_1 + R_4||R_S} R_1 = \sqrt{4kT R_1} \frac{R_1}{R_1 + R_4||R_S}$$

$$\therefore$$

$$(e_{R1}')^2 = 4kT R_1 \left(\frac{R_1}{R_1 + R_4||R_S}\right)^2 \quad (9.53)$$

which, given the data, results in

$$(e_{R1}')^2 = 4 \times 1.38 \times 10^{-23} \times 353.95 \ \text{K} \times 1k \left(\frac{1k}{1k + 666.667}\right)^2 \ \text{V}^2/\text{Hz}$$

$$= 7.037 \times 10^{-18} \ \text{V}^2/\text{Hz} \quad (9.54)$$

(d) *Contribution due to noise source e_{R2}:* Thermal noise voltage generated by resistance R_2 is delivered to the input node through the resistive network (R_2, R_1). Current through R_1 and R_2 is set by e_{R2}, i.e. $i_{R1} = i_{R2}$, and thus, at node node①, we write

$$e_{R2}' = i_{R1} R_1 = \frac{e_{R2}}{R_2} R_1 = \sqrt{4kT R_2} \frac{R_1}{R_2}$$

$$\therefore$$

$$(e_{R2}')^2 = 4kT R_2 \left(\frac{R_1}{R_2}\right)^2 \ \text{V}^2/\text{Hz} \quad (9.55)$$

which, given the data, results in

$$(e'_{R2})^2 = 4 \times 1.38 \times 10^{-23} \times 353.95 \, \text{K} \times 10\text{k} \left(\frac{1\text{k}}{10\text{k}}\right)^2 \, \text{V}^2/\text{Hz}$$

$$s = 1.955 \times 10^{-18} \, \text{V}^2/\text{Hz} \tag{9.56}$$

(e) *Contribution due to noise source e_{R4}:* Thermal noise voltage generated by resistance R_4 is delivered to the input node through the resistive divider $(R_4 + R_1 || R_S)$, Fig. 9.22. Thus, we write

$$e'_{R4} = i_{R4} \, R_S||R_1 = \frac{e_{R4}}{R_4 + R_S||R_1} \, R_S||R_1$$

$$\therefore$$

$$(e'_{R4})^2 = 4kT R_4 \left(\frac{R_S||R_1}{R_4 + R_S||R_1}\right)^2 \, \text{V}^2/\text{Hz} \tag{9.57}$$

which, given the data, results in

$$(e'_{R4})^2 = 4 \times 1.38 \times 10^{-23} \times 353.95 \, \text{K} \times 4\text{k} \left(\frac{444.444}{4\text{k} + 444.444}\right)^2 \, \text{V}^2/\text{Hz}$$

$$= 0.782 \times 10^{-18} \, \text{V}^2/\text{Hz} \tag{9.58}$$

Fig. 9.22 Example 9.6-5

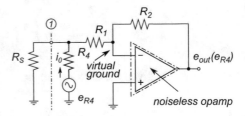

(f) *Contribution due to noise voltage source e_n:* The internal thermal noise voltage generated by operational amplifier is directly connected to the positive input node of operational amplifier, and thus, it is at the same position as e_{R3} in the network. Thus, we reuse (9.51) and write

$$(e'_n)^2 = \left[\frac{R_1}{R_2} \frac{R_2 + R_1 + R_S||R_4}{R_1 + R_S||R_4}\right] e_n^2 \, \text{V}^2/\text{Hz} \tag{9.59}$$

which, given the data, results in

$$(e'_n)^2 = 17.5 \times 10^{-18} \, \text{V}^2/\text{Hz} \tag{9.60}$$

(g) *Contribution due to noise current source i_{n+}:* This current source forces noise current through resistance R_3, which generates voltage directly at the positive input node of the operational amplifier $e_{in+} = i_{n+} \, R_3$. Again, we reuse (9.51) and write

$$(e'_{in+})^2 = \left[\frac{R_1}{R_2} \frac{R_2 + R_1 + R_S||R_4}{R_1 + R_S||R_4}\right]^2 (i_{n+} R_3)^2 \; \text{v}^2/\text{Hz} \tag{9.61}$$

which, given the data, results in

$$(e'_{in+})^2 = 12.25 \; \text{v}^2/\text{Hz} \tag{9.62}$$

(h) *Contribution due to noise current source* i_{n-}: This noise source forces a current through R_2, because due to the virtual ground both terminals of $(R_1 + R_S||R_4)$ resistive network is at ground potential, which forces that i_{n-} flows through R_2, Fig. 9.23. That being said, $e_{out}(i_{n-}) = i_{n-} R_2$ is seen at the output node. Thus, referencing the output voltage back to node node①through voltage divider R_1, R_2, we write

$$e'_{in-} = \frac{e_{out}(i_{n-})}{R_2} R_1 == \frac{i_{n-} R_2}{R_2} R_1 \quad \therefore \quad (e'_{in-})^2 = (i_{n-} R_1)^2 \tag{9.63}$$

which, given the data, results in

$$(e'_{in-})^2 = (5\text{p} \times 1\text{k})^2 \; \text{v}^2/\text{Hz} = 25 \times 10^{-18} \; \text{v}^2/\text{Hz} \tag{9.64}$$

Fig. 9.23 Example 9.6-5

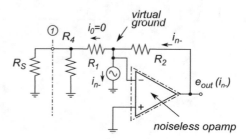

Therefore, after combining the seven amplifier thermal noise power terms (9.51) to (9.64), the complete solution for the internal noise power spectrum density is

$$P_A = e^2_{out} = (e'_{R1})^2 + (e'_{R2})^2 + (e'_{R3})^2 + (e'_{R4})^2 + (e'_n)^2 + (e'_{in+})^2 + (e'_{in-})^2 \; \text{v}^2/\text{Hz}$$

$$\therefore$$

$$P_A = (7.037 + 1.955 + 9.578 + 0.782 + 17.5 + 12.25 + 25) \times 10^{-18} \, \text{V}^2/\text{Hz}$$

$$= 74.102 \times 10^{-18} \, \text{V}^2/\text{Hz} \tag{9.65}$$

After substituting (9.65) and (9.48) into (9.42), as shown in (9.42), we calculate

$$F = 1 + \frac{P_A}{P_I} = 1 + \frac{74.102}{3.909} = 19.955 \quad \therefore \quad \text{NF} = 10 \log(19.955) = 13 \, \text{dB}$$

Part II

Radio Receiver Circuit

Radio Receiver Architecture

10

Wireless communication systems are result of multidisciplinary research that exploits various mathematical, scientific, and engineering principles in a very creative manner. The inner structure of signal waveforms (such as a voice, for example) is revealed by Fourier transformations, which enables us to design appropriate filters and amplifiers. By using Fourier theory we are able to both synthesize and decompose waveform that are continuous (i.e. analog) or sampled (i.e. digital). By using Maxwell's theory we explain the creation and propagation of EM waves. By using mathematical theorems, such as basic trigonometry identities, for example, we are able to manipulate signals at the system level. By using circuit theory and techniques, we are able to practically implement the underlying theoretical equations and therefore create "wireless communication system".

10.1   Important to Know

1. "Heterodyne" architecture: one mixer, i.e. frequency shifting step.
2. "Super heterodyne" architecture: two or more mixers, i.e. multiple frequency shifting steps.
3. A simple "half-wavelength" dipole antenna.

$$L = \frac{\lambda}{2} = \frac{c\,T}{2} = \frac{c}{2\,f} \approx \frac{300 \times 10^6 \mathrm{m/s}}{2}\,\frac{1}{f\,[^1\!/\mathrm{s}]} \approx \frac{143}{f\,[\mathrm{MHz}]}\,[\mathrm{m}] \tag{10.1}$$

© Springer Nature Switzerland AG 2021
R. Sobot, *Wireless Communication Electronics by Example*,
https://doi.org/10.1007/978-3-030-59498-5_10

10.2   Exercises

10.1   RF Antenna

1. Sketch a rough drawing of EM wave as being generated by a dipole antenna.
2. Using information in textbook or some other source that lists the allocation of frequency bands, estimate length of either real or superficial dipole antenna that would have to be used by:

    (a)   wireless transmission of audio frequencies;

    (b)   amateur radios;

    (c)   GSM-850 cell phones (carrier frequency $f_c = 850\,\text{MHz}$), UMTS-FDD cell phones (carrier frequency $f_c = 2.1\,\text{GHz}$);

    (d)   radio astronomy systems.

10.2 *   RF Signal Operations

1. Given two sine waveforms whose frequencies are relatively close to each other, $s_1(t) = \sin(2\pi \times 2\,\text{kHz} \times t)$ and $s_2(t) = \sin(2\pi \times 3\,\text{kHz} \times t)$ using a simulation software show frequency spectrum of:

    (a) $s_1(t)$ and $s_2(t)$ as standalone waveforms;

    (b) $s(t) = s_1(t) + s_2(t)$;

    (c) $p(t) = s_1(t) \times s_2(t)$. In addition, for this product, derive the analytical form of $p(t)$ waveform.

    Compare these plots and comment on the results.

2. Given $\omega = 2\pi \times 10^3\,\text{Hz}$ and a sine waveform

$$p(t) = \frac{4}{\pi} \left[\sin(\omega t) + \frac{1}{3} \sin(3\omega t) + \frac{1}{5} \sin(5\omega t) + \cdots + \frac{1}{33} \sin(33\omega t) \right] \qquad (10.2)$$

    using a simulation software show their time and frequency domain plots. Then,

    (a)   randomly choose and completely remove one or more of the harmonics in (10.2), or change amplitude and/or phase of one or more of the harmonics. Compare both time and frequency domain plots of the perturbed waveforms relative to the original form in (10.2);

    (b)   plot $f(t) = (1 + p(t))/2$; $f(t) = 2 + p(t)$; $f(t) = -3 + p(t)$;

    (d)   add a simple RC filter and observe the $p(t)$ waveforms at its output if:

(a) $\tau = 1.59\,\mu\text{s}$,	(b) $\tau = 15.9\,\mu\text{s}$,
(c) $\tau = 159\,\mu\text{s}$,	(d) $\tau = 1.59\,\text{ms}$.

    Comment on the observed waveforms.

3. Given $\omega = 2\pi \times 10^3$ Hz and a sine waveform

$$w(t) = \frac{2}{\pi}\left[\sin(\omega t) - \frac{1}{2}\sin(2\omega t) + \frac{1}{3}\sin(3\omega t) - \cdots + \frac{1}{17}\sin(17\omega t)\right] \qquad (10.3)$$

using a simulation software show their time and frequency domain plots. Then, repeat same exercises as in Example 10.1–2.

10.3 ***   RF Transmission Principle

1. The objective is to create model of an artificial transmission system that is based only on frequency multiplying operation. The signal to be transmitted consists only of one 1 kHz sinusoid. In the transmitter model there are two local oscillators: one at 455 kHz and one at 10 MHz. Show the analytical equations as the signal progresses through the transmitter, transmitting media, and receiver until it is recovered. By using the arbitrary voltage controlled source from SPICE library (B element) create and simulate the analytical equations, then show the frequency spectrum at each stage of the transmission system.

Solutions

Exercise 10.1, page 238

1. A dipole antenna is simplest geometrical form of antenna that is derived from a capacitive structure. Therefore, it consists of two symmetrical conductive elements, for example, two simple metallic rods. Once the antenna is stimulated with an AC signal generator source, it creates EM filed that keeps propagating through the space at the speed of light even when the source is turned off.

A simple capacitor stimulated by AC signal generator creates changing electric field in between the two capacitive electrodes, Fig. 10.1. However, even if the two electrodes are physically separated at the end that is further away from the source, the electric field does not break. Instead, once the two electrodes are vertically aligned with each other, the field morphs into radial shape. In other words, we can imagine that a dipole antenna is only a fully "open" capacitor.

As the AM current $i(t)$ is continuously changing its direction along the antenna electrodes, the electric and magnetic fields also keep changing their respective directions, Fig. 10.1(right). In this graphical quasi 3D representation of EM field moving radially away from the dipole antenna, we should actually imagine a torus shaped electrical field lines orthogonally wrapped around circular magnetic field lines. In reality, the whole space is filled with the alternating fields, not only the indicated vector lines.

Fig. 10.1 Example 10.1-1

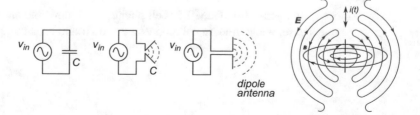

2. A half-wave dipole antenna consists of two identical wire sections whose total length is approximately equal to half-wavelength of the signal that is to be transmitted, see Fig. 10.2. Practical formula is

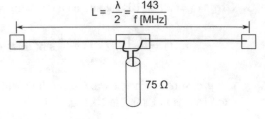

$$L = \frac{\lambda}{2} = \frac{143}{f \,[\text{MHz}]}$$

Fig. 10.2 Example 10.1-2

Given data, we calculate:

(a) Audio frequencies occupy bandwidth from $20\,\text{Hz} = 20 \times 10^{-6}\,\text{MHz}$ to $20\,\text{kHz} = 20 \times 10^{-3}\,\text{MHz}$. Straightforward calculations yield

$$L(min) = \frac{143}{20 \times 10^{-3}\,\text{MHz}} = 7.15\,\text{km}$$

$$L(max) = \frac{143}{20 \times 10^{-6}\,\text{MHz}} = 7\,150\,\text{km}$$

This example illustrates why, even after ignoring all power level and other practical issues, it is not practical to wirelessly transmit audio frequencies. The required antenna is much longer than a typical telephone cable section (which is about 1.6 km long). Compare and visualize these antenna sizes relative to Earth.

(b) Amateur radios are licensed to use 3 MHz–30 MHz HF bandwidth, therefore practical antenna lengths are

$$L(min) = \frac{143}{30\,\text{MHz}} = 4.7\,\text{m}$$

$$L(max) = \frac{143}{3\,\text{MHz}} = 47\,\text{m}$$

These lengths are acceptable to radio amateurs, for example, the popular 10 MHz radio transceiver requires around 14 m long wire antenna, which is easily stretched around houses.

(c) Similarly, cell phone systems require antennas

$$L(\text{GSM-850}) = \frac{143}{850\,\text{MHz}} = 16.8\,\text{cm}$$

$$L(\text{UMTS-FDD}) = \frac{143}{2\,100\,\text{MHz}} = 68\,\text{mm}$$

Indeed, first generation GSM-850 cell phones used wire antenna that was pulled out of the phone's case, while modern cell phones have their antennas completely embedded into the phone case.

(d) Frequencies used in radio astronomy, theoretically, would need dipole antennas as

$$L(min) = \frac{143}{30 \times 10^3 \text{ MHz}} = 4.7 \text{ mm}$$

$$L(max) = \frac{143}{300 \times 10^3 \text{ MHz}} = 470 \, \mu\text{m}$$

Of course, requirements for radio astronomy are constrained by the available signal power, thus large dish antennas are used.

Exercise 10.2, page 238

1. Transient simulations followed by FFT transformation are used to generate the following plots.

(a) Fourier transformation of each sinusoid in time domain results in its corresponding Dirac pulse in the frequency domain. Each pulse is located at the frequency of its corresponding sinusoid, at 2 kHz and 3 kHz, Fig. 10.3.

(b) Simple addition of multiple sinusoids is a linear operation, thus the Fourier transformation of their sum does not change total frequency spectrum, i.e. Dirac pulses whose amplitudes are unchanged are found at the same locations, at 2 kHz and 3 kHz, Fig. 10.4.

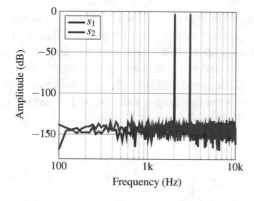

Fig. 10.3 Example 10.2-1

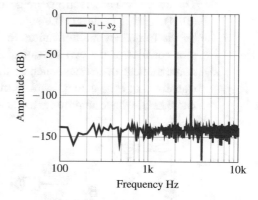

Fig. 10.4 Example 10.2-1

(c) Multiplication of two sinusoid functions obeys the trigonometric identity

$$\sin\alpha \sin\beta = \frac{1}{2} \left[\cos(\alpha - \beta) - \cos(\alpha + \beta) \right] \tag{10.4}$$

where the resulting function consists of two sinusoid whose arguments are *difference* and *sum* of the two original arguments. In addition, the output amplitude is reduced by $1/2$ factor. After substituting two abstract mathematical arguments of angle α and β with two physical arguments of frequency ω_1 and ω_2, we write

$$p(t) = s_1(t) \times s_2(t) = \sin(2\,\text{kHz}) \times \sin(3\,\text{kHz})$$

$$= \frac{1}{2}\left(\cos(2\,\text{kHz} - 3\,\text{kHz}) - \cos(2\,\text{kHz} + 3\,\text{kHz})\right)$$

$$= \frac{1}{2}\left(\cos(1\,\text{kHz}) - \cos(5\,\text{kHz})\right)$$

Note that using Hz instead of radian units does not change the multiplication operation. In addition, we use other trigonometric identities, i.e. $\cos(-x) = \cos(x)$. The negative sign in front of a sinusoid indicates its phase inversion, which does not show in the amplitude graphs. This frequency multiplication operation is fundamental to RF transmission. We note the following:

(a)   The input signal consists of *two separate* sinusoid functions, while the output signal consists of *one composite* waveform,

(b)   Frequency spectrum of the output waveform also consists of two sinusoids, however, two input side frequencies are "shifted" to another two output side frequencies: one that corresponds to the *sum* and the other that corresponds to the *difference* of the input frequencies. The two input frequencies are *not found* in the spectrum of the resulting equation,

(c)   There is $1/2$ factor in the output of (10.4), which is to say that amplitudes of both sinusoid components in the output waveform are reduced by half, i.e. $-6\,\text{dB}$.

The above two observations are general and direct consequence of the mathematical identity (10.4), thus it is always the sum and difference of the input frequencies that are produced by the frequency multiplication. Simulation setup used to generate the frequency spectrum plot consists of the two sinusoidal sources and one arbitrary voltage source that performs mathematical operations, such as the addition and multiplication of this example. Fourier transform of time domain waveform shows the two output tones in $p(t)$ waveform at $1\,\text{kHz}$ and $5\,\text{kHz}$, whose amplitudes are at $-6\,\text{dB}$, Fig. 10.5.

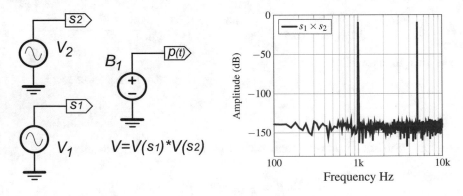

Fig. 10.5 Example 10.2-1

2.   Equation (10.2) is result of Fourier series of the square signal whose base frequency is $1\,\text{kHz}$. By using one sinusoid signal generator for each tone, and one arbitrary voltage source to do the addition and multiplication with the $4/\pi$ factor, we generate time and frequency domain plots,

Fig. 10.6. The Gibbs phenomenon is visible in the time domain while the frequency domain plot shows all odd frequency harmonics that are used to synthesize the composite signal $p(t)$, i.e. 1 kHz, 3 kHz, 5 kHz, ..., as well as their relative amplitudes.

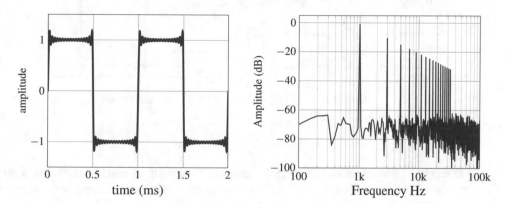

Fig. 10.6 Example 10.2-2

(a) Fourier series is a very delicate sum of infinitely many harmonics. Any perturbation of either amplitude, phase or frequency of any of the harmonics results in waveform distortion that is visible in the time domain. Frequency domain plot shows what exactly is perturbed.

For example, shape of the square waveform s_2 is distorted relative to ideal shape of s_1 waveform, Fig. 10.7. By looking at the frequency domain plot we find that 3 kHz, 13 kHz and 29 kHz harmonics are for some reason missing from the ideal square waveform spectrum, therefore causing the signal distortion. Similar distortions would be if result of partial amplitude or phase perturbations. Recall that phase perturbation simply means that particular harmonic is being delayed during the transmission relative to the rest of the harmonics in the packet.

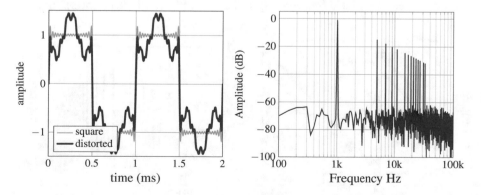

Fig. 10.7 Example 10.2-2

(b) Addition of a constant to a waveform is equivalent to addition of DC voltage to a signal, thus moves it average value. Multiplying by a constant factor (which is equivalent to the amplifier gain) obviously changes the signal's amplitude. Simulated waveforms illustrate these types of signal processing, Fig. 10.8.

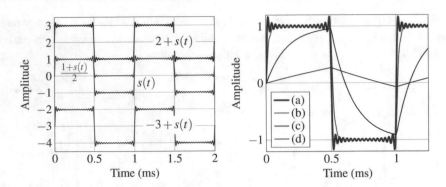

Fig. 10.8 Example 10.2-2

(c) LPF suppresses amplitudes of harmonics that are outside of its bandwidth. What is more, LPF non-uniformly reduces amplitudes of HF harmonics—higher frequency, larger amplitude reduction (recall, -20 dB/decade). Final effect is that the signal's delicate Fourier sum is perturbed therefore, depending how many of the harmonics are affected, the signal is distorted to various extends. This LPF effect sets fundamental upper limit on $p(t)$ signal's frequency content, because the distortion may be almost unnoticeable (case (a)), may be very mild (case (b)), very strong (case (c)), or so big that the square signal is converted into a triangular waveform (case (d)).

3. Equation (10.3) is result of Fourier series of the sawtooth signal whose base frequency is 1 kHz. By using one sinusoid signal generator for each tone, and one arbitrary voltage source to do the addition and multiplication with the $2/\pi$ factor, we generate time and frequency domain plots. The Gibbs phenomenon is visible in the time domain while the frequency domain plot shows all frequency harmonics that are used to synthesize the composite signal $p(t)$, i.e. 1 kHz, 2 kHz, 3 kHz, ..., as well as their relative amplitudes. Same as already shown for the square waveform in Example 10.2-2, perturbations of Fourier sum are the cause of distortions, and LPF attenuation is one practical example of this effect. Simulated plots illustrate distorted waveform as well as the ideal frequency spectrum (only the first 16 harmonics), Fig. 10.9.

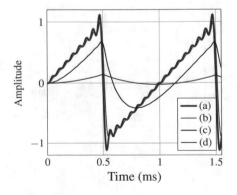
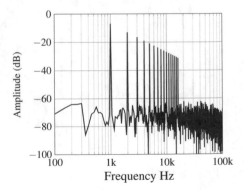

Fig. 10.9 Example 10.2-3

Exercise 10.3, page 239

1. Given a simple signal $s(t) = \sin(\omega_s t)$, where $\omega_s = 2\pi \times 1\,\text{kHz}$, the wireless signal transmission steps are as follows:

Transmitter:

(a) Multiplications of signal $s(t)$ and $s_{IF}(t) = \sin(\omega_{IF}(t))$ generated by first Tx oscillator, where $\omega_{IF} = 2\pi \times 455\,\text{kHz}$ results in

$$s_1 = s(t) \times s_{IF}(t) = \sin(\omega_s t) \times \sin(\omega_{IF} t)$$

$$= \frac{1}{2}\left[\cos(\omega_s - \omega_{IF})t - \cos(\omega_s + \omega_{IF})t\right]$$

$$= \frac{1}{2}\left[\cos(\omega_{DIF})t - \cos(\omega_{SUM})t\right]$$

$$= \frac{1}{2}\left[\cos(2\pi \times 1\,\text{kHz} - 2\pi \times 455\,\text{kHz})t\right.$$

$$\left. - \cos(2\pi \times 1\,\text{kHz} + 2\pi \times 455\,\text{kHz})t\right]$$

$$= \frac{1}{2}\left[\cos(2\pi \times 454\,\text{kHz}\,t) - \cos(2\pi \times 456\,\text{kHz}\,t)\right]$$

The composite output signal s_1 contains of two tones whose frequencies are sum and difference of the two input signal frequencies ω_s and ω_{IF}. Signal and IF frequencies are widely apart, for that reason the full frequency spectrum graph appears to show only one signal around 455 kHz. Zoom around that region in the graph shows that there are indeed two tones in the spectrum, one at $f_{DIF} = 454\,\text{kHz}$ and one at $f_{SUM} = 456\,\text{kHz}$, see Fig. 10.10, as predicted analytically.

Two points to note: first, there is no LF component $s(t)$ present in s_1 signal spectrum, both the sum and the difference tones are close to $s_{IF}(t)$, which is in HF range; second, the two output tones are separated by $2 \times f_s$, which is important specification of circuit BW that is required by this operation.

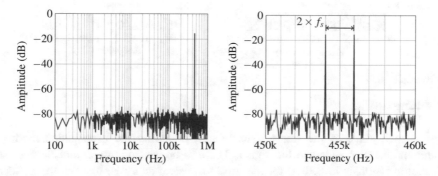

Fig. 10.10 Example 10.3-1

(b) Multiplications of signal $s_1(t)$ and $s_{RF} = \sin(\omega_{RF}(t))$ generated by the second Tx oscillator, where $\omega_{RF} = 2\pi \times 10\,\text{MHz}$ produces

$$
\begin{aligned}
s_2(t) =& s_1(t) \times s_{RF}(t) \\
=& [\cos(\omega_{DIF}\,t) - \cos(\omega_{SUM}\,t)] \times \sin(\omega_{RF}t) \\
=& \cos(\omega_{DIF}\,t) \times \sin(\omega_{RF}t) - \cos(\omega_{SUM}\,t) \times \sin(\omega_{RF}t) \\
=& \frac{1}{2}\,[\sin(\omega_{DIF} + \omega_{RF})\,t - \sin(\omega_{DIF} - \omega_{RF})\,t] \\
& - \frac{1}{2}\,[\sin(\omega_{SUM} + \omega_{RF})\,t - \sin(\omega_{SUM} - \omega_{RF})\,t] \\
=& \frac{1}{2}\,[\sin(2\pi(454\,\text{kHz} + 10\,\text{MHz}))\,t - \sin(2\pi(454\,\text{kHz} - 10\,\text{MHz}))\,t] \\
& - \frac{1}{2}\,[\sin(2\pi(456\,\text{kHz} + 10\,\text{MHz}))\,t - \sin(2\pi(456\,\text{kHz} - 10\,\text{MHz}))\,t] \\
=& \frac{1}{2}\,[\sin(2\pi \times 10.454\,\text{MHz} \times t) + \sin(2\pi \times 9.546\,\text{MHz} \times t) \\
& + \sin(2\pi \times 10.456\,\text{MHz} \times t) + \sin(2\pi \times 9.544\,\text{MHz} \times t)]
\end{aligned}
$$

In other words, after the second multiplication, in total there are two pairs, i.e. four tones, in close proximity of ω_{RF}, one pair below 10 MHz and pair above, Fig. 10.11. Separation between the two pairs is $2 \times f_{IF} = 910\,\text{kHz}$, while two paired frequencies are separated by $2 \times f_s = 2\,\text{kHz}$. It is to be noted that all four tones are very close to each other and they are all HF signals, thus suitable for RF transmission with one antenna sized for 10 MHz carrier frequency.

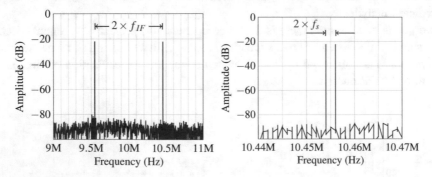

Fig. 10.11 Example 10.3-1

Transmitting media:

(d) Along each stage of the transmission chain, including the propagation media, there is incremental noise added to the signal. There is no real non-deterministic Gaussian noise generator available in SPICE for the use in transient analysis. Instead, the arbitrary signal generator B element has option "WHITE" that generates a quasi-random signal that can be used as the noise model in SPICE simulation. Syntax of the B element is, for example, `V=a*WHITE(b*time)`, where "a" is a number to control amplitude, and "b" is a number that effectively increases the noise BW. It is necessary to experiment a little bit for each specific simulation to generate desired power and spectrum of this noise.

Receiver:

After the receiver's antenna, in order to recover the original 1 kHz signal, the frequency multiplications must be done in the opposite order relative to the ones in the transmitter.

(c) Multiplication of RF signal $s_2(t)$ with the receiver's local sinusoid reference $f_{\text{LO}} = 10\,\text{MHz}$, gives

$$
\begin{aligned}
s_3(t) =\ & s_2(t) \times s_{\text{LO}}(t) \\
=\ & \frac{1}{2} \Bigg[\sin(2\pi \times 10.454\,\text{MHz} \times t) + \sin(2\pi \times 9.546\,\text{MHz} \times t) \\
& + \sin(2\pi \times 10.456\,\text{MHz} \times t) + \sin(2\pi \times 9.544\,\text{MHz} \times t) \Bigg] \\
& \times \sin(\omega_{\text{LO}}\, t)
\end{aligned}
$$

Obviously, there are in total eight tones generated by this multiplication. Four sums are all located in the proximity of 20 MHz and may be removed immediately by filtering. For example, $10\,\text{MHz} + 9.544\,\text{MHz} - 19.544\,\text{MHz}$. Four differences, however, produce the following frequencies:

$$
d_1 = 10\,\text{MHz} - 9.544\,\text{MHz} = 456\,\text{kHz}
$$

$$
d_2 = 10\,\text{MHz} - 9.546\,\text{MHz} = 454\,\text{kHz}
$$

$$
d_3 = 10.456\,\text{MHz} - 10\,\text{MHz} = 456\,\text{kHz}
$$

$$
d_4 = 10.454\,\text{MHz} - 10\,\text{MHz} = 454\,\text{kHz}
$$

which shifts the received RF signal to the same state as after multiplication with s_{IF}, albeit with duplications.

(d) Multiplication of $d(t) = \sin(2\pi \times 456\,\text{kHz} \times t) + \sin(2\pi \times 454\,\text{kHz} \times t)$ with $f_{\text{LO2}} = 455\,\text{kHz}$ generated by receiver's second oscillator results in

$$
\begin{aligned}
s_4(t) =\ & d_1(t) \times s_{\text{LO2}} \\
=\ & \frac{1}{2} \Bigg[\sin(2\pi \times 456\,\text{kHz}\, t) + \sin(2\pi \times 454\,\text{kHz}\, t) \Bigg] \times \sin(\omega_{\text{LO}} t)
\end{aligned}
$$

Obviously, there are in total four tones generated by this multiplication at the following frequencies:

$$
e_1 = 456\,\text{kHz} - 455\,\text{kHz} = 1\,\text{kHz}
$$

$$
e_2 = 455\,\text{kHz} - 454\,\text{kHz} = 1\,\text{kHz}
$$

$$
e_3 = 456\,\text{kHz} + 455\,\text{kHz} = 911\,\text{kHz}
$$

$$
e_4 = 455\,\text{kHz} + 454\,\text{kHz} = 909\,\text{kHz}
$$

therefore, the received signal is

$$e(t) = \sin(2\pi \times 1\,\text{kHz} \times t)$$
$$+ \sin(2\pi \times 909\,\text{kHz} \times t)$$
$$+ \sin(2\pi \times 911\,\text{kHz} \times t)$$

where the two summing frequencies are in HF range relative to the 1 kHz recovered signal (which is duplicated), therefore they can be removed by filtering. Comparative time domain plot shows that it is the *envelope* of $e(t)$ that corresponds to the original $s(t)$ signal, Fig. 10.12. In consequence, some filtering (i.e. peak detecting) technique must be applied to $e(t)$ signal in order to remove its HF components and faithfully recover only $s(t)$ signal.

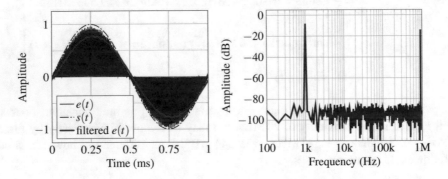

Fig. 10.12 Example 10.3-1

Electrical Resonance

11

In the most familiar form of mechanical oscillations, the pendulum, the total system energy constantly bounces back and forth between the kinetic and potential forms. In the absence of friction (i.e. energy dissipation), a pendulum would oscillate forever. Similarly, after two ideal electrical elements capable of storing energy (a capacitor and an inductor) are connected in parallel then the total initial energy of the system bounces back and forth between the electric and magnetic energy forms. This process is observed as electrical oscillations and the parallel LC circuit is said to be in resonance. The phenomenon of electrical resonance is essential to wireless radio communications technology because without it, simply put, there would be no modern communications.

11.1 Important to Know

1. LC resonant frequency (ideal)

$$\omega_0 = \frac{1}{\sqrt{LC}} \tag{11.1}$$

2. LC resonant frequency (non-ideal)

$$\omega_{p0} = \sqrt{\frac{1}{LC} - \frac{r^2}{L^2}} \quad (r \text{ is wire resistance}) \tag{11.2}$$

3. Q factor

$$Q = 2\pi \times \frac{\text{Energy Stored}}{\text{Energy dissipated per cycle}} = \omega_0 \times \frac{\text{Energy Stored}}{\text{Power Loss}} \tag{11.3}$$

4. Q factor

$$Q = \frac{\omega_0}{\Delta\omega} = \frac{\omega_0}{BW} \tag{11.4}$$

© Springer Nature Switzerland AG 2021
R. Sobot, *Wireless Communication Electronics by Example*,
https://doi.org/10.1007/978-3-030-59498-5_11

5. Q factor, series RLC network

$$Q_S = \frac{\omega_0 L}{R} = \frac{1}{\omega_0 RC} = \frac{1}{R}\sqrt{\frac{L}{C}} \tag{11.5}$$

6. Q factor, parallel RLC network

$$Q_P = \frac{R}{\omega_0 L} = \omega_0 RC = R\sqrt{\frac{C}{L}} \tag{11.6}$$

7. Dynamic resistance

$$R_D = \frac{L}{RC} = \omega_0 LQ = \frac{Q}{\omega_0 C} = Q^2 R \tag{11.7}$$

11.2   Exercises

11.1 **   Definitions

1. Amplitude of a decaying cosine function at approximately $t = 6.5$ periods from $t = 0$ is e times smaller than the initial amplitude value A_0. Calculate the Q factor of this resonator.

2. The amplitude of a decaying oscillation is $A(t) = A_0 \exp(-t/\tau)$, where A_0 is the initial amplitude and τ is the decay time. For a guitar string that produces a tone at $f_0 = 334\,\mathrm{Hz}$, the sound decays by factor 2 after 4 s. Estimate the decaying time τ and the Q factor.

3. An AC voltage source V is connected across a serial LC connection. Find the capacitive X_C and inductive X_L reactances as well as voltages V_C and V_L across their respective terminals.
   Data : $V = 5\,\mathrm{V}$, $f = 10\,\mathrm{MHz}$, $C = 1\,\mathrm{nF}$, and $L = 1\,\mu\mathrm{H}$.

4. RLC circuit is connected to a realistic voltage source V_{in} whose internal resistance is R_S, Fig. 11.1. Find:

   (a) output voltage $V_{out} = V_R$ as measured across the resistor R, at the resonant frequency $f = f_0$;

   (b) output voltage $V_{out} = V_C$ as measured across the capacitor C at $f = 1\,\mathrm{kHz}$.

   Data: $C = 1\,\mathrm{nF}$, $L = 1\,\mu\mathrm{H}$, $R = 1\,\mathrm{m}\Omega$, $R_S = 50\,\Omega$, $V_{in} = 1\,\mathrm{mV}$.

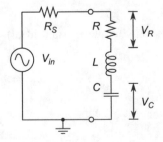

Fig. 11.1   Example 11.1-4

11.2 **   RLC Resonator

1. For RLC circuit in Example 11.1-4, find the resonant frequency f_0 and calculate the total impedance Z_{tot} at

   (a) 1 kHz,   (b) 7.335 MHz,   (c) 1 GHz.

   Data: $R_S = 0\,\Omega$, $R = 1\,\mathrm{m}\Omega$, $L = 4.708\,\mathrm{nH}$, and $C = 100\,\mathrm{nF}$.

2. Given parallel RLC resonator, calculate the total current I_{tot} that is supplied by the voltage source and the circuit's impedance Z_{tot}, Fig. 11.2.

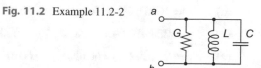

Fig. 11.2 Example 11.2-2

Data: $R_S = 0\,\Omega$, $V_{in} = 12\,V$, $G = 400\,\Omega$, $X_L = 500\,\Omega$, and $X_C = 200\,\Omega$.

11.3 *** Q Factor

1. Given data and the total resistance at the resonant frequency R, estimate the resonant frequency and the Q factor for a typical serial RLC network.
   Data: $L = 1\,mH$, $C = 25.33\,pF$, $R = 15\,\Omega$.

2. Assuming same component values, compare the resonant frequencies of an ideal RLC resonator relative to realistic RLC resonator.
   Data: $L = 1\,mH$, $C = 25.33\,pF$, $R = 15\,\Omega$.

3. Given a parallel LC tank, calculate its:

   (a) resonant frequency f_0,   (b) Q factor at resonance,

   (c) resistance at resonance R_D,   (d) bandwidth BW.

   Data: $L = 2.533\,nH$, $R_L = 1\,m\Omega$, $C = 100\,nF$.

4. For a given coil, calculate:

   (a) its equivalent series resistance,   (b) its parallel resistance

   (c) value of the resonating capacitor,   (d) parallel resistance which, when added, increases bandwidth to BW'.

   Data: $L = 2\,\mu H$, $Q = 200$, $f_0 = 10\,MHz$, $BW' = 200\,kHz$.

5. Calculate the Q-factor of a serial RLC network if inductor $L = 2.5\,nH$ and the lumped wire resistance $r = (\pi)m\Omega$, at: (a) $f_1 = 10\,MHz$; and (b) $f_2 = 100\,MHz$.

11.4 ** BP Filter

1. The objective is to design a BPF whose bandwidth is $BW = 10\,kHz$, which is equivalent to $Q = 100$. The filter should allow a $f = 1\,MHz$ signal to pass.

   (a) Use one low-pass and one high-pass RC filter in series so that the two individual transfer functions added together create BPF function and calculate their R and C values. What are practical problems with this approach?

   (b) Design RLC BPF with $Q = 100$ and compare with RC version in Example 11.4-1, (a). What are differences between these two design solutions?

2. Given data, calculate the tuning range ($\Delta f = f_{max} - f_{min}$) of tuneable LC resonator. **Data:** $L = 2.533\,nH$, $C = 80\,nF$ to $120\,nF$.

3. Design LC resonator whose resonant frequency is $f_0 = 10\,MHz$, given inductor $L = 2.533\,nH$ and only the following capacitors:

(a) $C_1 = 10\,\text{nF}, C_2 = 40\,\text{nF}, C_3 = 50\,\text{nF}$

(b) $C_1 = 200\,\text{nF}, C_2 = 300\,\text{nF}, C_3 = 600\,\text{nF}$

(c) $C_1 = 70\,\text{nF}, C_2 = 60\,\text{nF}, C_3 = 60\,\text{nF}$

(d) $C_1 = 200\,\text{nF}, C_2 = 200\,\text{nF}, C_3 = 50\,\text{nF}$

4. For an RLC resonator, given its resonant frequency ω_0 and Q factor, derive expression for its bandwidth BW. What is the conclusion?

5. Realistic LC resonator model includes both the inductive coil's wire resistance r and its parasitic capacitance C_L. The LC resonator is used at frequency f_0. Calculate LC resonator's effective inductance L_{eff} and effective Q_{eff} factor.
   Data: $f_0 = 25\,\text{MHz}, L = 1\,\mu\text{H}, r = 5\,\Omega, C_L = 5\,\text{pF}$.

11.5 ***   Frequency Characteristics

1. Calculate resonant frequency f_0 of a serial RLC network. Then, calculate its impedance at given frequency f that is close to f_0.
   Data: $R = 30\,\Omega, L = 3\,\text{mH}, C = 100\,\text{nF}. f = 10\,\text{kHz}$.

2. Given a frequency response curve of an LC resonator, Fig. 11.3, determine the resonator bandwidth, Q factor, inductance L, and the total internal circuit resistance R.
   Data: $f_1 = 450\,\text{kHz}, f_2 = 460\,\text{kHz}, f_0 = 455\,\text{kHz}, C = 1\,\text{nF}$.

Fig. 11.3 Example 11.5-2

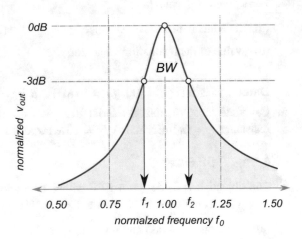

3. A parallel LC tank consists of inductor L whose wire resistance is r, and capacitor C. Determine, resonant frequency f_0, the Q factor, dynamic resistance R_D, and bandwidth BW of this resonator.
   Data: $L = 1\,\text{mH}, r = 1\,\Omega, C = 100\,\text{nF}$.

Solutions

Exercise 11.1, page 250

1. Non-ideal oscillator is modelled by second order differential equation

$$\frac{d^2x}{dt^2} + \gamma \frac{dx}{dt} + \omega_0^2 x = 0 \qquad (11.8)$$

where γ is a constant and ω_0 is the natural frequency of damped harmonic oscillator. Solution to (11.8) is assumed to be in the form of

$$x = \exp\left(-\frac{t}{\tau}\right) A_0 \cos \omega t \qquad (11.9)$$

where the exponential term is used to model the energy loss, and therefore the gradual reduction of sinusoidal amplitude. Coefficient τ controls the *rate of amplitude decay*. For example, if $\tau = \infty$ then there is no reduction in the amplitude A_0 because the exponential term becomes equal to one at all times. At the other extreme, if $\tau = 0$ then the exponential term becomes zero, that is, the cosine function is completely suppressed. For any other value of τ, there will be natural decay in the initial amplitude A_0.

Assuming solution of (11.8) in the form (11.9), first and second derivative of (11.9) are,

$$\frac{dx}{dt} = -A_0 \exp\left(-\frac{t}{\tau}\right) \left(\omega \sin \omega t + \frac{1}{\tau} \cos \omega t\right) \qquad (11.10)$$

$$\frac{d^2x}{dt^2} = A_0 \exp\left(-\frac{t}{\tau}\right) \left[\frac{2\omega}{\tau} \sin \omega t + \left(\frac{1}{\tau^2} - \omega^2\right) \cos \omega t\right] \qquad (11.11)$$

After substituting (11.9)–(11.11) into (11.8), we have

$$A_0 \exp\left(-\frac{t}{\tau}\right) \left[\left(\frac{2\omega}{\tau} - \gamma\omega\right) \sin \omega t + \left(\frac{1}{\tau^2} - \omega^2 - \frac{\gamma}{\tau} + \omega_0^2\right) \cos \omega t\right] = 0 \qquad (11.12)$$

Equation (11.12) is possible at all times only if either of the two main product terms is equal to zero. The exponential term equals zero if $(t \to \infty)$, therefore, it is the second term that must be equal to zero at all times, i.e.

$$\left[\left(\frac{2\omega}{\tau} - \gamma\omega\right) \sin \omega t + \left(\frac{1}{\tau^2} - \omega^2 - \frac{\gamma}{\tau} + \omega_0^2\right) \cos \omega t\right] = 0 \qquad (11.13)$$

We conclude that both summing terms in (11.13) are equal to zero if the multiplying constants of both sine and cosine terms equal zero, i.e.

$$\frac{2\omega}{\tau} - \gamma\omega = 0 \qquad \therefore \qquad \tau = \frac{2}{\gamma} \qquad (11.14)$$

$$\therefore$$

$$\frac{1}{\tau^2} - \omega^2 - \frac{\gamma}{\tau} + \omega_0^2 = 0 \qquad \Rightarrow \qquad \omega = \sqrt{\omega_0^2 - \left(\frac{\gamma}{2}\right)^2} \qquad (11.15)$$

after substituting τ that is calculated in (11.14). Now, we can rewrite solution (11.9) for the case a lightly dumped oscillator as

$$x = A_0 \exp\left(-\frac{\gamma}{2} t\right) \cos \omega t \tag{11.16}$$

The solution (11.16) is valid for ω as found in (11.15) and represents oscillatory motion if ω is not complex, i.e. if

$$\omega_0^2 > \frac{\gamma^2}{4} \tag{11.17}$$

which is the condition for lightly damped harmonic oscillations. In addition, the frequency of a lightly damped oscillator is close to its natural resonant frequency if

$$\omega_0^2 \gg \frac{\gamma^2}{4} \quad \therefore \quad \omega = \sqrt{\omega_0^2 - \left(\frac{\gamma}{2}\right)^2} \approx \omega_0 \tag{11.18}$$

In order to answer the question of this problem, we write (11.16) at time $t = t_0$ and at $t = t_0 + nT$, where T is the cosine period

$$\omega \stackrel{\text{def}}{=} 2\pi f \stackrel{\text{def}}{=} \frac{2\pi}{T} \quad \therefore \quad T = \frac{2\pi}{\omega} \tag{11.19}$$

and n represents the index of the n–th maxima away from the one at t_0. Hence, this system of two equations results in

$$A_k = x(t_0) = A_0 \exp\left(-\frac{\gamma}{2} t_0\right) \cos \omega t_0 \tag{11.20}$$

$$A_{k+n} = x(t_0 + nT) = A_0 \exp\left[-\frac{\gamma}{2} (t_0 + nT)\right] \cos \omega(t_0 + nT)$$

$$= A_0 \exp\left(-\frac{\gamma}{2} t_0\right) \exp\left(-\frac{\gamma}{2} nT\right) \cos \omega t_0 \tag{11.21}$$

Calculation of the ratio (11.20) and (11.21), after substituting (11.19), yields

$$\frac{A_k}{A_{k+n}} = \frac{1}{\exp\left(-\frac{\gamma}{2} nT\right)} = \exp\left(\frac{\gamma}{2} nT\right)$$

$$\therefore$$

$$\ln\left(\frac{A_k}{A_{k+n}}\right) = \frac{\gamma \, nT}{2} = \frac{\gamma \, n \, 2\pi}{2\omega} = \frac{\gamma \, n \, \pi}{\omega} \approx \frac{\gamma \, n \, \pi}{\omega_0} = \frac{n \, \pi}{Q} \tag{11.22}$$

because $\cos \omega(t_0 + nT) = \cos \omega t_0$. In (11.22), we define the ratio of the natural resonant frequency and γ as the figure of merit for energy loss during the oscillations, a.k.a. the Q factor, see Fig. 11.4,

$$Q = \frac{\omega_0}{\gamma} \tag{11.23}$$

Given data, from (11.22) we write

$$\ln(e) = \frac{6.5\,\pi}{Q} \qquad \therefore \qquad Q = 6.5\,\pi \approx 20.4$$

Fig. 11.4 Example 11.1-1

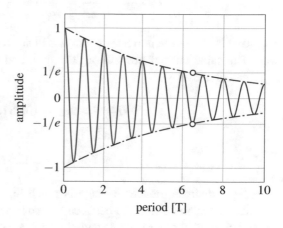

2. Given conditions, we write,

$$A(t) = A_0 \exp\left(-\frac{t}{\tau}\right) \qquad \therefore \qquad \tau = \frac{t}{\ln\left(\dfrac{A_0}{A(t)}\right)}$$

$$\tau = \frac{4s}{\ln(2)} = 5.77s$$

and, from (11.14) and (11.23) we write

$$Q = \frac{\omega_0}{\gamma} = \frac{\omega_0\,\tau}{2} = \frac{2\pi \times 334\text{Hz} \times 5.77\text{s}}{2} \approx 6 \times 10^3$$

This relatively high Q indicates that the guitar string produces very narrowband[1] tone, i.e. very clean sustainable sinusoid, which is what is expected from a musical instrument.

3. Given data, the two reactances are calculated as

$$X_L \stackrel{\text{def}}{=} j\omega L = j\,2\pi f L = +j\,62.832\,\Omega$$

$$X_C \stackrel{\text{def}}{=} \frac{1}{j\omega C}\,\frac{j}{j} = -j\frac{1}{2\pi f C} = -j\,15.915\,\Omega$$

$$\therefore$$

$$X_{LC} = X_L + X_C = +j\,62.832\,\Omega - j\,15.915\,\Omega = +j\,46.917\,\Omega$$

Phase of the total equivalent reactance X_{LC} equals to $(+\pi/2)$ phase (recall the phases of j and $-j$), which is the defining characteristics of an ideal inductor. Therefore, from the perspective of the AC voltage source, there would be no difference if this LC network is replaced with

---

[1] Recall that also $Q = f_0/\Delta f$.

$$X_{eq} = X_{LC} = j\omega L_{eq} \quad \therefore \quad L_{eq} = \frac{|X_{LC}|}{2\pi f} = 746.7\,\text{nH}$$

The total branch current is, therefore,

$$I = \frac{V}{Z} = \frac{V}{X_{LC}} = \frac{V}{j\omega L_{eq}} \frac{j}{j} = -j\frac{5\,\text{V}}{46.917\,\Omega} = -j106.571\,\text{mA}$$

that is to say, the current waveform is delayed by $-90°$ relative to the voltage.

The same current flows through both L and C, therefore voltages across each component are calculated as

$$V_L = I \times X_L = 106.573\,\text{mA} \times 62.832\,\Omega = 6.696\,\text{V}$$

and,

$$V_C = I \times X_C = 106.573\,\text{mA} \times 15.915\,\Omega = 1.696\,\text{V}$$

Note that the inductor voltage is much higher than the one provided by the voltage source. However, the difference between the voltages is $V_L - V_C = 6.696\,\text{V} - 1.696\,\text{V} = 5\,\text{V}$, as it should be in order to agree with the applied voltage. The two components are complex, therefore the resulting voltages and currents are delayed (phase different) relative to each other. Thus, we must apply the vector addition rules not the simple arithmetic addition.

As a consequence, we must be very careful about the operational range of components used to build high-Q RLC resonators—the internal voltages perceived by LC components may be much higher than the power supply voltage, which could damage the LC circuit.

4. (a) At the resonant frequency f_0, the ideal capacitive and inductive reactances are equal $Z_L = Z_C$ and having the opposite signs their sum equals zero, which is to say that there is only real resistance left $Z_{RLC}(\omega_0) = R$. Therefore, after applying the voltage-divider rule, it follows:

$$\frac{V_{out}}{V_{in}} = \frac{V_R}{V_{in}} = \frac{R}{R + R_S} = \frac{1\,\text{m}\Omega}{1\,\text{m}\Omega + 50\,\Omega} \approx 20 \times 10^{-6}\,\text{V/v}$$

$$\therefore$$

$$V_{out} = 20 \times 10^{-6}\,\text{V/v} \times 1\,\text{mV} = 20\text{n}\,\text{V}$$

(b) At the frequency $f = 1\,\text{kHz}$, inductor and capacitor reactances are

$$X_L = j\omega L = j2\pi \times 1\,\text{kHz} \times 1\,\mu\text{H} = j6.283\,\text{m}\Omega$$

$$X_C = \frac{1}{j\omega C} = -j\frac{1}{2\pi \times 1\,\text{kHz} \times 1\,\text{nF}} = -j159.2\,\text{k}\Omega$$

The total impedance of the RLC branch is, including the source resistance

$$Z_{tot} = R_S + R + Z_C + Z_L = R_S + R + j\omega L + \frac{1}{j\omega C} = (R_S + R) + j(X_L - X_C)$$

The total impedance seen by the ideal voltage source is

$$|Z_{tot}| = \sqrt{(R_S + R)^2 + (X_L - X_C)^2}$$
$$= \sqrt{(50\,\Omega + 1\,m\Omega)^2 + (6.283\,m\Omega - 159.2\,k\Omega)^2}$$
$$\approx 159.2\,k\Omega$$

Voltage across the capacitor V_C is found by voltage division rule as

$$\frac{V_{out}}{V_{in}} = \frac{V_C}{V_{in}} = \frac{X_C}{Z_{tot}} \approx \frac{159.2\,k\Omega}{159.2\,k\Omega} \approx 1\,{}^V\!/\!v \quad \therefore \quad V_{out} = 1\,{}^V\!/\!v \times 1\,mV = 1\,mV$$

which illustrates how RLC impedances, thus voltages behave widely different at resonance and at non-resonant frequencies.

Exercise 11.2, page 250

1. The resonant frequency is

$$f_0 = \frac{1}{2\pi\sqrt{LC}} = \frac{1}{2\pi\sqrt{4.708\,nH\,100\,nF}} \approx 7.335\,MHz$$

(a) at $f = 1\,kHz$:

$$|X_L| = 2\pi f L = 2\pi \times 1\,kHz \times 4.708\,nH = 29.581\mu\Omega$$

$$|X_C| = \frac{1}{2\pi f C} = \frac{1}{2\pi \times 1\,kHz \times 100\,nF} = 1.592\,k\Omega$$

therefore,

$$Z_{tot} = \sqrt{R^2 + (X_L - X_C)^2} = \sqrt{1\,m\Omega^2 + (29.581\mu\Omega - 1.592\,k\Omega)^2}$$
$$\approx 1.592\,k\Omega$$

that is to say, at $f = 1\,kHz$ this RLC network is dominated by the capacitor's reactance.

(b) at $f = 7.335\,MHz$: this is the resonant frequency, hence $Z_{tot} = R = 1\,m\Omega$, i.e. only the real resistance R is "visible" to the source.

(c) at $f = 1\,GHz$: at this frequency capacitor reactance is $X_C \approx 1.6\,m\Omega$, while $X_L = 29.581\,\Omega$, thus $Z_{tot} \approx X_L = 29.581\,\Omega$, i.e. dominated by the inductor's reactance.

2. There are three branches, each current is

$$I_R = \frac{V_{in}}{R} = \frac{12\,\text{V}}{400\,\Omega} = 30\,\text{mA}$$

$$I_L = \frac{V_{in}}{X_L} = \frac{12\,\text{V}}{500\,\Omega} = 24\,\text{mA}$$

$$I_C = \frac{V_{in}}{X_C} = \frac{12\,\text{V}}{200\,\Omega} = 60\,\text{mA}$$

Then, the total current is the sum of three currents, real and complex as

$$I_{tot} = I_R + jI_L - jI_C = I_R + j(I_L - I_C)$$

therefore, module of the total current is

$$I_{tot} = \sqrt{I_R^2 + (I_L - I_C)^2} = \sqrt{(30\,\text{mA})^2 + (24\,\text{mA} - 60\,\text{mA})^2} = 46.862\,\text{mA}$$

$$\therefore$$

$$Z_{tot} = \frac{V_{in}}{I_{tot}} = \frac{12\,\text{V}}{46.862\,\text{mA}} = 256.074\,\Omega$$

Note that the current through the capacitive branch is greater than the total current I_{tot} provided by the signal generator. Recall that we must use vector algebra for complex components, which means that module must be greater than either real or complex vector projection alone.

Exercise 11.3, page 251

1. The resonant frequency is

$$f_0 = \frac{1}{2\pi\sqrt{(LC)}} \approx 1\,\text{MHz}$$

and the Q factor of serial RLC network (see textbook for derivations) is

$$Q = \frac{1}{R}\sqrt{\frac{L}{C}} = \frac{1}{15\,\Omega}\sqrt{\frac{1\,\text{mH}}{25.33\,\text{pF}}} \approx 420$$

which are typical numbers in the current state of the art.

2. The resonant frequency of an ideal resonator (i.e. $R = 0$) is simply

$$\omega_0 = \frac{1}{\sqrt{LC}} = \frac{1}{\sqrt{1\,\text{mH} \times 25.33\,\text{pF}}} = 1.00000584\,\text{MHz}$$

while the resonant frequency of a realistic resonator (i.e. $R \neq 0$, see textbook for derivation) is calculated as

$$\omega_0 = \sqrt{\frac{1}{LC} - \frac{R^2}{L^2}} = \sqrt{\frac{1}{1\,\text{mH} \times 25\,\text{pF}} - \frac{(15\,\Omega)^2}{(1\,\text{mH})^2}} = 1.00000299\,\text{MHz}$$

therefore, except for very high precision circuits, the difference of 2.85 Hz relative to 1 MHz is negligible for most practical purposes.

3. By definitions for LC network we calculate

(a) resonant frequency   (b) Q factor

$$f_0 = \frac{1}{2\pi\sqrt{LC}} = 10\,\text{MHz}$$
$$Q = \frac{X_L}{R_L} = \frac{2\pi f_0 L}{R} = 159.153$$

(c) dynamic resistance   (d) bandwidth

$$R_D = Q^2 R_L = 29.330\,\Omega$$
$$BW = \frac{f_0}{Q} = \frac{R_L}{2\pi L} = 62.833\,\text{kHz}$$

4. Given data, by definitions of $Q = Q_S \approx Q_P$ (assuming $Q \gg 1$),

$$\omega_0 = \frac{1}{\sqrt{LC}}$$

$$Q_S = \frac{\omega_0 L}{R} = \frac{1}{\omega_0 RC} = \frac{1}{R}\sqrt{\frac{L}{C}}$$

$$Q_P = \frac{R}{\omega_0 L} = \omega_0 RC = R\sqrt{\frac{C}{L}}$$

Given f_0 and Q, by definition we can immediately calculate $Q = f_0/BW$, therefore $BW = f_0/Q = 50\,\text{kHz}$. Furthermore, by definitions we find

(a) $Q_S = \dfrac{\omega_0 L}{R_S} \quad \therefore \quad R_S = \dfrac{\omega_0 L}{Q} = \dfrac{2\pi f_0 L}{Q} = 628\,\text{m}\Omega$

(b) $Q_P = \dfrac{R_P}{\omega_0 L} \quad \therefore \quad R_P = \omega_0 Q L = 2\pi f_0 Q L = 25\,\text{k}\Omega$

(c) $\omega_0 = \dfrac{1}{\sqrt{LC}} \quad \therefore \quad C = \dfrac{1}{\omega_0^2 L} = 126\,\text{pF}$

(d) Given resonant frequency f_0 and new bandwidth BW', first we determine that the new Q' factor must be

$$Q' = \frac{f_0}{BW'} = \frac{10\,\text{MHz}}{200\,\text{kHz}} = 50 \quad \therefore \quad R'_P = \omega_0 Q' L = 2\pi f_0 Q L = 6.28\,\text{k}\Omega$$

where, R'_P is the total parallel resistance that sets $Q = 50$. Resistance R_x should be added in parallel to the internal $R_P = 25\,\text{k}\Omega$ so that

$$\frac{1}{R'_P} = \frac{1}{R_P} + \frac{1}{R_x} \quad \therefore \quad \frac{1}{R_x} = \frac{1}{R'_P} - \frac{1}{R_P} \quad \therefore \quad R_x = 8.39\,\text{k}\Omega$$

5. By definition,

(a) $Q_S = \dfrac{\omega_0 L}{R} = \dfrac{2\pi \times 10\,\text{MHz} \times 2.5\,\text{nH}}{\pi \times 10^{-3}\,\Omega} = 50$

(b) $Q_S = \dfrac{\omega_0 L}{R} = \dfrac{2\pi \times 100\,\text{MHz} \times 2.5\,\text{nH}}{\pi \times 10^{-3}\,\Omega} = 500$

Exercise 11.4, page 251

1. Indeed, BPF frequency profile is created by HP and LP filters. That is because the two filters together create the second order function: one pole and one zero, where pole frequency is superior to the zero frequency, $\omega_P > \omega_Z$. Similarly, RLC transfer function is also second order function, thus BPF transfer function.

(a) Starting with the $BW = 10\,\text{kHz}$ and $f_0 = 10\,\text{MHz}$ requirements and knowing that BW of BPF is determined by two $-3\,\text{dB}$ frequencies that are themselves determined by time constants $\tau = RC = 1/\omega$, the calculations of the lower (HP) and upper (LP) frequencies are as follows:

$$f_L = \frac{\omega_L}{2\pi} = f_0 - \frac{BW}{2} = 0.995\,\text{MHz}$$

$$f_U = \frac{\omega_U}{2\pi} = f_0 + \frac{BW}{2} = 10.005\,\text{MHz}$$

After choosing, for example, $R = 1\,\text{k}\Omega$ we calculate two capacitances as

$$\omega_L = \frac{1}{\tau_L} = \frac{1}{RC_L} \quad \therefore \quad C_L \approx 159.955\,\text{pF}$$

$$\omega_U = \frac{1}{\tau_U} = \frac{1}{RC_L} \quad \therefore \quad C_U \approx 158.363\,\text{pF}$$

First practical problem is that by simply connecting HP to LP passive filters the overall transfer function is difficult to derive. The reason is that in order to calculate the time constant there are two Rs and two Cs directly connected, which changes the overall calculation. Consequently, it is necessary to "isolate" HP function from interfering with the LP function. In other words, we use a voltage buffer amplifier between the two filters, Fig. 11.5. In addition, passive networks are inherently voltage dividers, thus there is need for one amplifier at the output.

Fig. 11.5 Example 11.4-1

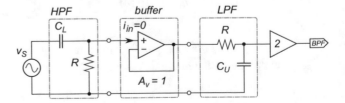

Second practical problem is that each filter is a first order network, that is to say, there is only ±20 dB/decade slope on each side of pole/zero frequency. That being the case, it is not possible to reduce amplitude by -3 dB if the frequency is changed by only $\pm BW/2 = \pm5$ kHz. The -3 dB amplitude reduction is achieved after much larger BW.

(b) Parallel RLC resonator is second order network (i.e. there are two components capable of storing energy, L and C). By choosing, for example, $L = 1\,\mu$H the calculations of BPF are as follows:

$$f_0 = \frac{1}{2\pi\sqrt{LC}} \quad \therefore \quad C \approx 25.330\,\text{nF}$$

$$Q = 100 \quad \therefore \quad R_P = 2\pi f_0 L Q \approx 628.3\,\Omega$$

Frequency simulations show difference between these two approaches to design BPF, Fig. 11.6. First order passive RC filters are not suitable for design of RF BPF circuits that require precise narrowband (i.e. high Q) BPF for each communication channel. With RC filters it is simply not possible to achieve this high Q which is needed to set $BW = 10$ kHz at $f_0 = 1$ MHz.

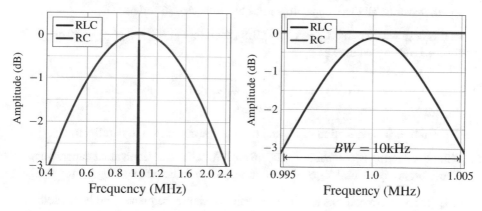

Fig. 11.6 Example 11.4-1

2. Given trimmer capacitor, by definition, two extreme resonant frequencies of ideal LC resonator are

$$f_{\min} = \frac{1}{\sqrt{LC_{max}}} = \frac{1}{2\pi\sqrt{2.533\,\text{nH} \times 120\,\text{nF}}} = 9.129\,\text{MHz}$$

$$f_{\max} = \frac{1}{\sqrt{LC_{min}}} = \frac{1}{2\pi\sqrt{2.533\,\text{nH} \times 80\,\text{nF}}} = 11.180\,\text{MHz}$$

Therefore, the tuning range is

$$\Delta f = f_{\max} - f_{\min} = (11.180 - 9.129)\,\text{MHz} = 2.051\,\text{MHz}$$

It is useful to note that the frequency tuning ratio is

$$n = \frac{f_{\max}}{f_{\min}} = \frac{11.180}{9.129} = 1.225$$

3. By definition we calculate,

$$f_0 = \frac{1}{2\pi\sqrt{LC}} \quad \therefore \quad C = \frac{1}{(2\pi f_0)^2 L} = 100\,\text{nF}$$

(a) Obviously, neither of the three given capacitors can be used alone. The solution is to search for serial/parallel combinations of capacitors to create the total $C_{\text{tot}} = 100\,\text{nF}$ capacitance. In this case, it is simply the sum of three given capacitors,

$$C_{\text{tot}} = C_1 + C_2 + C_3 = (10 + 40 + 50)\,\text{nF} = 100\,\text{nF}$$

In other words, all three capacitors must be connected in parallel.

(b) Given three capacitances, each of them greater that the required $100\,\text{nF}$, we must try their serial connections. In this case we find

$$\frac{1}{C_{\text{tot}}} = \frac{1}{C_1} + \frac{1}{C_2} + \frac{1}{C_3} = \left(\frac{1}{200} + \frac{1}{300} + \frac{1}{600}\right)\,\text{nF} = 100\,\text{nF}$$

In other words, all three capacitors must be connected in series.

(c) In this case we note that $C_2 = C_3 = 60\,\text{nF}$, therefore $(1/C_2 + 1/C_3) = 30\,\text{nF}$, which allows for the total capacitance as

$$C_{\text{tot}} = C_1 + \frac{1}{\frac{1}{C_2} + \frac{1}{C_3}} = (70 + 30)\,\text{nF} = 100\,\text{nF}$$

Thus, the serial connection of C_2 and C_3 must be connected in parallel to C_1.

(d) After connecting two capacitors $200\,\text{nF}$ each in series, the equivalent capacitance is $100\,\text{nF}$. Therefore, only C_1 and C_2 connected in series are sufficient, sparing C_3 for another project.

4. One of Q factor interpretations is the simple ratio of centre frequency and bandwidth, as

$$Q = \frac{\omega_0}{BW} \quad \therefore \quad BW = \frac{\omega_0}{Q}$$

which illustrates that, given centre frequency, BW is inversely proportional to Q factor. In other words, if the frequency is increased in order to keep the same BW it is necessary to use higher Q resonator.

5. In the case of an ideal LC resonator there are no parasitics, while realistic model shows parasitic and external capacitors in parallel, Fig. 11.7, thus we write

$$\omega_0 = \frac{1}{\sqrt{LC}} \quad \therefore$$

$$C = \frac{1}{(2\pi f_0)^2 L} = 40.523\,\text{pF}$$

$$C' = C + C_L = 45.523\,\text{pF}$$

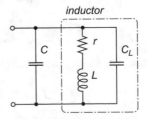

Fig. 11.7 Example 11.4-5

Realistic inductor model includes both the inductor wire resistance r and its parasitic capacitance C_L.

$$Y(\omega) = \frac{1}{r + j\omega L} + j\omega\, C' = \frac{r - j\omega L}{r^2 + (\omega\, L)^2} + j\omega\, C'$$

$$= \frac{r}{r^2 + (\omega\, L)^2} + j\left(\omega\, C' - \frac{\omega\, L}{r^2 + (\omega\, L)^2}\right) \tag{11.24}$$

at resonance (i.e. $\omega = \omega_{p0}$), the two reactances are equal $|Z_L| = |Z_C|$, which is to say that the imaginary part is $\Im(Y) = 0$, hence we write

$$\omega_{p0}\, C' = \frac{\omega_{p0}\, L}{r^2 + (\omega_{p0}L)^2} \quad \therefore \quad r^2 + (\omega_{p0}L)^2 = \frac{L}{C'} \tag{11.25}$$

which leads to the conclusion,

$$\omega_{p0} = \sqrt{\frac{1}{LC'} - \frac{r^2}{L^2}} = 23.574\,\text{MHz} \tag{11.26}$$

We note that the resonant frequency ω_{p0} of a realistic LC network that includes realistic inductance has the additional term $(r/L)^2$ due to the finite wire resistance, as well as $C' = C + C_L$, which changes the resonant frequency relative to the case of ideal LC resonator. When $(r, C_L \to 0)$, then (11.26) becomes the same as ω_0 of the ideal LC resonator, i.e. $\omega_{p0} \to \omega_0$.

We estimate realistic inductor's parameters alone by excluding C in (11.24), after noting the $(\omega\, L)^2 \gg r^2$ therefore $r^2 + (\omega\, L)^2 \approx (\omega\, L)^2$, we write

$$Y_L(\omega) = \frac{1}{r + j\omega L} + j\omega\, C_L = \frac{r - j\omega L}{r^2 + (\omega\, L)^2} + j\omega\, C_L$$

$$= \frac{r}{r^2 + (\omega\, L)^2} + j\left(\omega\, C_L - \frac{\omega\, L}{r^2 + (\omega\, L)^2}\right)$$

$$\approx \frac{r}{(\omega\, L)^2} + j\left(\omega\, C_L - \frac{1}{\omega\, L}\right)$$

$$= \frac{r}{(\omega\, L)^2} + \frac{1}{j\omega}\left(\frac{1 - \omega^2\, L\, C_L}{L}\right) = \frac{1}{R_{\text{eff}}} + \frac{1}{j\omega\, L_{\text{eff}}} \tag{11.27}$$

which gives us the effective inductance as

$$L_{\text{eff}} = \frac{L}{1 - \omega^2\, L\, C_L} = \frac{1\,\mu\text{H}}{1 - (2\pi \times 25\,\text{MHz})^2 \times 1\,\mu\text{H} \times 5\,\text{pF}} = 1.141\,\mu\text{H}$$

and the effective parallel resistance as

$$R_{\text{eff}} = \frac{(\omega\, L)^2}{r} = \frac{(2\pi \times 25\,\text{MHz} \times 1\,\mu\text{H})^2}{5\,\Omega} = 4.935\,\text{k}\Omega$$

Model of a parallel RLC resonator gives dynamic resistance as

$$R_D = Q\omega_0 L \quad \therefore$$

$$Q_{\text{eff}} = \frac{R_{\text{eff}}}{\omega_0\, L_{\text{eff}}} = \frac{4.935\,\text{k}\Omega}{2\pi \times 23.574\,\text{MHz} \times 1.141\,\mu\text{H}} = 29.2$$

Exercise 11.5, page 252

1. Resonant frequency is

$$f_0 = \frac{1}{2\pi\sqrt{LC}} = 9.189\,\text{kHz}$$

We calculate Q_S factor of this serial RLC network as

$$Q_S = \frac{\omega_0 L}{R} = \frac{2\pi \times 9.189\,\text{kHz} \times 3\,\text{mH}}{30\,\Omega} = 5.77$$

This Q factor is considered low, most approximations used to simplify RLC resonator analysis assume $Q > 10$ when serial and parallel RLC networks are interchangeable.
At given frequency $f = 10\,\text{kHz}$, impedance of serial RLC network is

$$Z(\omega) = R + j\omega L + \frac{1}{j\omega C}\frac{j}{j} = R + j\left(\omega L - \frac{1}{\omega C}\right)$$

$$\therefore$$

$$|Z(\omega)| = \sqrt{R^2 + \left(\omega L - \frac{1}{\omega C}\right)^2} \approx 42\,\Omega$$

At the resonance however, $\omega = \omega_0 = 1/\sqrt{LC}$, the imaginary part $\Im(Z(\omega_0)) = 0$, thus $|Z(\omega_0)| = R = 30\,\Omega$.

2. Given data and after the inspection of the frequency response curve, by definitions we find

$$BW = f_2 - f_1 = 10\,\text{kHz}$$

$$Q = \frac{f_0}{BW} = 45.5$$

$$f_0 = \frac{1}{2\pi\sqrt{LC}} \quad \therefore \quad L = \frac{1}{(2\pi\,f_0)^2 C} = 122.35\,\mu\text{H}$$

$$Q = \frac{\omega_0 L}{R} \quad \therefore \quad R = \frac{\omega_0 L}{Q} = 7.6877\,\Omega$$

3. By definitions,

$$f_0 = \frac{1}{2\pi f_0\sqrt{LC}} = 15.9\,\text{kHz}$$

$$Q = \frac{\omega_0 L}{r} = 100$$

$$Q = \frac{f_0}{BW} \quad \therefore \quad BW = \frac{f_0}{Q} = 159\,\text{Hz}$$

$$R_D = \frac{L}{rC} == \omega_0 L Q = \frac{Q}{\omega_0 C} = Q^2 r = 10\,\text{k}\Omega$$

Matching Networks

12

In this chapter, we study a simple basic methodology for interfacing two stages in the signal processing chain which is commonly used in the design of RF electronic systems, with the main criterion being maximum power transfer between the stages. This approach is justified by the argument that wireless RF signals that have arrived at the system input terminals (e.g. at the antenna) are very weak, thus subsequent power loss would have broad consequences for the overall system performance. This objective is achieved by using "power matching" techniques.

12.1   Important to Know

1. Q factor matching

$$Q_S = \frac{X_S}{R_0} \quad \text{and,} \quad Q_P = \frac{R_L}{X_P} \tag{12.1}$$

$$Q_S = Q_P = Q = \sqrt{\frac{R_L}{R_0} - 1} \quad (R_L > R_0) \tag{12.2}$$

$$Q_S = Q_P = Q = \sqrt{\frac{R_0}{R_L} - 1} \quad (R_0 > R_L) \tag{12.3}$$

© Springer Nature Switzerland AG 2021
R. Sobot, *Wireless Communication Electronics by Example*,
https://doi.org/10.1007/978-3-030-59498-5_12

12.2   Exercises

12.1 **   Matching networks

1. Using the Q matching technique, design a single-section LC network to match a source resistance $R_0 = 5\,\Omega$ to a resistive load $R_L = 50\,\Omega$ at $f = 10\,\text{MHz}$. Maintain DC connection between the source and the load, Fig. 12.1.

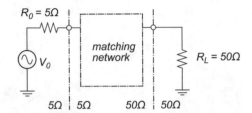

Fig. 12.1 Example 12.1-1

2. Source whose resistance is $50\,\Omega$ is to be connected to the load that consists of $50\,\Omega$ resistor and $100\,\text{pF}$ capacitor connected in series. Design matching network for $10\,\text{MHz}$ signal, Fig. 12.2.

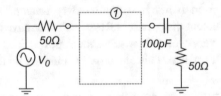

Fig. 12.2 Example 12.1-2

3. Using the Q matching technique, design a single-section LC network to match a source resistance R_0 to a resistive load R_L. Maintain DC connection between the source and the load, Fig. 12.3.
Data: $R_0 = 50\,\Omega$, $R_L = 5\,\Omega$, $f = 10\,\text{MHz}$.

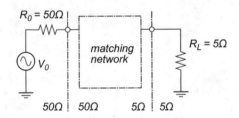

Fig. 12.3 Example 12.1-3

4. Using the Q matching technique, design a single-section LC network to match an inductive RL source to a capacitive RC load, Fig. 12.4. Source to load connection is DC.
Data:
$R_0 = 5\,\Omega$, $L_S = 138.732\,\text{nH}$, $R_L = 50\,\Omega$, $C_L = 454.910\,\text{pF}$ $f = 10\,\text{MHz}$.

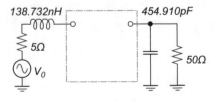

Fig. 12.4 Example 12.1-4

5. Using the Q matching technique, design a single-section LC matching network between real source R_0 and capacitive RC load R_L, CL, Fig. 12.5. The source and load are to be DC connected.
Data: $R_0 = 5\,\Omega$, $R_L = 50\,\Omega$, $C_L = 1.055\,\text{nF}$, $f = 10\,\text{MHz}$.

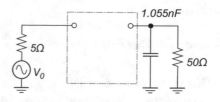

Fig. 12.5 Example 12.1-5

6. Using the Q matching technique, design a single-section LC matching network between real source R_0 and capacitive RC load R_L, CL, Fig. 12.6. The source and load are to be DC connected.
Data: $R_0 = 50\,\Omega$, $R_L = 50\,\Omega$, $C_L = 100\,\text{pF}$, $f = 10\,\text{MHz}$.

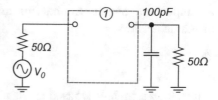

Fig. 12.6 Example 12.1-6

12.2 ** Two-stage matching networks

1. Design two-stage LC matching network while maintaining DC connection between source R_S and load R_L resistances. The bandwidth of this matching network should be increased relative to the single-stage LC matching network solution. **Data:** $R_S = 5\,\Omega$, $R_L = 50\,\Omega$, $f = 10\,\text{MHz}$.

2. Design two-stage LC matching network while maintaining DC connection between source R_S and load R_L resistances. The bandwidth of this matching network should be decreased relative to the single-stage LC matching network solution. Choose two "ghost" resistances R_{INT} and compare the resulting bandwidths. **Data:** $R_S = 5\,\Omega$, $R_L = 50\,\Omega$, $f = 10\,\text{MHz}$.

12.3 ** Serial/parallel transformations

1. Using Q matching technique, find equivalent parallel network to serial connection of $R_S = 5\,\Omega$ and $L_S = 238.732\,\text{nH}$ at $f = 100\,\text{MHz}$.

2. Given matching network, Fig. 12.7, find reflection coefficient Γ and mismatch loss ML at the interface between the serial and parallel parts of the matching network, i.e. at the node one.

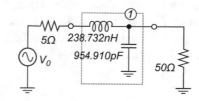

Fig. 12.7 Example 12.3-2

3. Using results from Example 12.1-1, assuming symmetrical frequency response of the calculated matching network estimate its -3dB bandwidth.

Solutions

Exercise 12.1, page 266

1. This is the case when $R_0 < R_L$. Therefore, serial reactance is needed to increase the source side impedance, and parallel reactance to reduce the load side impedance, see Fig. 12.8. By doing so, DC connection is created between the input and output terminals of this LPF matching network.

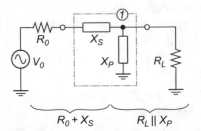

Fig. 12.8 Example 12.1-1

Q matching technique is based on identity

$$Q_S = Q_P = \sqrt{\frac{R_L}{R_S} - 1} = \sqrt{\frac{50}{5} - 1} = 3$$

First, we calculate the serial component as

$$X_S = Q_S\, R_0 = 3 \times 5\,\Omega = 15\,\Omega \quad \therefore \quad L = \frac{X_S}{\omega} = \frac{15\,\Omega}{2\pi \times 10\,\mathrm{MHz}} = 238.732\,\mathrm{nH}$$

Then, we calculate the parallel component as

$$X_P = \frac{R_L}{Q_P} = \frac{50\,\Omega}{3} = 16.667\,\Omega \quad \therefore$$

$$C = \frac{1}{\omega X_P} = \frac{1}{2\pi \times 10\,\mathrm{MHz} \times 16.667\,\Omega} = 954.910\,\mathrm{pF}$$

Let us verify the result. Looking into the source side relative to node one, there is serial connection of R_0 and inductor X_S, Fig. 12.9. Therefore, the total serial impedance between the input side terminal and node one is

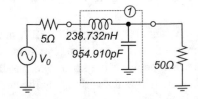

Fig. 12.9 Example 12.1-1

$$|Z_0| = \sqrt{R_0^2 + X_S^2} = \sqrt{(5^2 + 15^2)}\,\Omega = 15.811\,\Omega$$

Similarly, looking into the load side terminal into node one, there is parallel connection of R_L and X_P. Therefore, the parallel impedance at the load side is

$$|Z_L| = \cfrac{1}{\sqrt{\cfrac{1}{R_L^2} + \cfrac{1}{X_P^2}}} = \cfrac{1}{\sqrt{\cfrac{1}{50^2} + \cfrac{1}{16.667^2}}} \, \Omega = 15.811 \, \Omega$$

In conclusion, the source side impedance increased and the load side impedance decreased, with the apparent matching of the two sides at 15.811 Ω relative to node one in the middle of matching network.

It is worth mentioning that design of matching networks results in non-standard very precise component values. As a consequence, the use of "trimer capacitors" is standard practice, in addition to manufacturing of customized inductors.

2. This is a special case of matching because excluding the 100 pF capacitor, the source, and load 50 Ω resistances are already matched. That being the case, the idea is then to create short connection between the source and load. Recall that, at resonance, resistance of serial LC branch equals to 0 Ω, which is to say that the addition of appropriate inductance in series with this capacitor creates short connection and therefore "removes" the capacitor.

First, we calculate the capacitive (i.e. negative) reactance as

$$X_{S_C} = -j\frac{1}{\omega C} = -j\frac{1}{2\pi \times 10\,\text{MHz} \times 100\,\text{pF}} = -j159.155 \, \Omega$$

In order to nullify the existing negative reactance we must add inductive (i.e. positive) serial reactance $X_{S_L} = +j159.155 \, \Omega$ towards the source side. In other words, at $f = 10\,\text{MHz}$, we need an inductor as

$$X_{S_L} = j\omega L \quad \therefore \quad L = \frac{|X_{S_L}|}{\omega} = \frac{+159.155 \, \Omega}{2\pi \times 10\,\text{MHz}} = 2.533 \, \mu\text{H}$$

At $f = 10\,\text{MHz}$ this serial L, C branch resonates and therefore presents the total zero resistance between the source and load. This elegant one additional component solution illustrates the "resonating out" technique for matching network design in the case of serial RC load, Fig. 12.10.

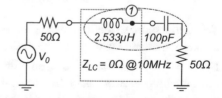

Fig. 12.10 Example 12.1-2

To conclude, we note that in this case of serial RC complex load there is not much choice left; the original connection was already blocking DC signals, thus adding inductor in series does not change this situation.

3. This case is when the source resistance is greater than the load resistance, $R_0 > R_L$. Therefore, on the input side we add reactance in parallel to the source resistance R_0 and on the output side reactance in series to the load resistance R_L. In order to keep DC connection, obviously the serial component must be inductor, therefore the parallel component must be capacitor.

By using Q matching technique we calculate

$$Q_S = Q_P = Q = \sqrt{\frac{R_0}{R_L} - 1} = \sqrt{\frac{50}{5} - 1} = 3$$

First, parallel reactance to be added on the input side should be,

$$X_P = \frac{R_0}{Q_P} = \frac{50\,\Omega}{3} = 16.667\,\Omega \quad \therefore \quad C = \frac{1}{2\pi \times 10\,\text{MHz} \times 16.667\,\Omega} = 954.910\,\text{pF}$$

Then, the serial component is calculated as

$$X_S = Q_S\,R_L = 3 \times 5\,\Omega = 15\,\Omega \quad \therefore \quad L = \frac{15\,\Omega}{2\pi \times 10\,\text{MHz}} = 238.732\,\text{nH}$$

Let us verify the result. Looking into the source side relative to node ①, there is parallel connection of R_0 and capacitor X_P, Fig. 12.11. Therefore, the total serial impedance between the input side terminal and node one is

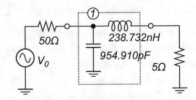

Fig. 12.11 Example 12.1-3

$$|Z_0| = \frac{1}{\sqrt{\frac{1}{R_0^2} + \frac{1}{X_P^2}}} = \frac{1}{\sqrt{\frac{1}{50^2} + \frac{1}{16.667^2}}}\,\Omega = 15.811\,\Omega$$

Looking into the load side node, there is serial connection of R_L and X_S. Therefore, the serial impedance at the load side is

$$|Z_L| = \sqrt{R_L^2 + X_S^2} = \sqrt{5^2 + 15^2}\,\Omega = 15.811\,\Omega$$

Thus, the source side impedance increased and the load side impedance decreased, with the apparent matching resistance 15.811 Ω at node one.

4. In the case of either source or load with complex impedances, general strategy is to first resolve the matching network only for the real parts of the two impedances. In Example 12.1-1, we already designed matching network for the case of real $R_0 = 5\,\Omega$ source and real $R_L = 50\,\Omega$ load at 10 MHz. Thus, we reuse the results and treat those calculations as the first phase of this example. We calculated that, in order to match real $R_0 = 5\,\Omega$ to real $R_L = 50\,\Omega$, we need $X_S' = 238.732\,\text{nH}$ inductor and $X_P' = 954.910\,\text{pF}$ capacitor.

1. *The input side calculations:* the source impedance in this example already has $L_S = 138.732\,\text{nH}$ inductance. The key point is that $X_S' > L_S$. That is to say, $X_S' = 238.732\,\text{nH}$ inductor can be imagined as the serial connection of $138.732\,\text{nH} + 100\,\text{nH}$ inductors. That is to say, L_S plus the additional 100 nH inductor. By doing this addition we "absorb" the existing $L_S = 138.732\,\text{nH}$ source inductance into the total value of inductance that is required to match real 5 Ω source to real 50 Ω load.

2. *The output side calculations:* the load impedance in this example already has $C_L = 454.910\,\text{pF}$ capacitance. The key point is that $X'_P > C_L$. That is to say, $X'_P = 954.910\,\text{pF}$ capacitor can be imagined as parallel connection of $454.91\,\text{pF} + 500\,\text{pF}$. That is to say, C_L plus the additional $500\,\text{pF}$ capacitor in parallel. By doing this addition we "absorb" the existing $C_L = 454.910\,\text{pF}$ load capacitance into the total value of capacitance that is required to match real $5\,\Omega$ source to real $50\,\Omega$ load.

Therefore, as the consequence of the "absorbing" technique used in this case, the required matching network consists of one $X_S = 100\,\text{nH}$ inductor and one $X_P = 500\,\text{pF}$ capacitor, Fig. 12.12.

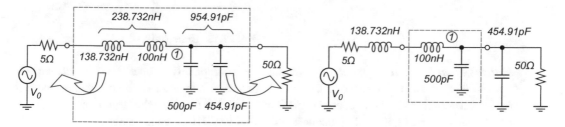

Fig. 12.12 Example 12.1-4

5. In case of either source or load complex impedance, general strategy is to first resolve the matching network only for the real parts of the two impedances. In Example 12.1-1, we already designed matching network for the case of real $R_0 = 5\,\Omega$ source and real $R_L = 50\,\Omega$ load at $10\,\text{MHz}$. Thus, we reuse those calculations as the first design phase of this example: in order to match real $R_0 = 5\,\Omega$ to real $R_L = 50\,\Omega$, we need $X'_S = 238.732\,\text{nH}$ inductor and $X'_P = 954.910\,\text{pF}$ capacitor.

The load impedance in this example already has $C_L = 1.055\,\text{nF}$ capacitance. The key point is that $C_L > X'_S$. Therefore, we cannot absorb it into the total required capacitance as in Example 12.1-4. Instead, in general, there are two possible ways to approach this kind of problems: either to "resonate out" the total C_L value; or, to "partially resonate out" only the "excess" portion of C_L. In both cases, the main idea is to exploit the fact that LC resonator's resistance equals to infinity at the resonant frequency. Therefore, at the given resonant frequency, the addition of inductor L creates ideal resonator with C_L that, in effect, replaces the newly created LC resonator with infinite resistance. Consequently, this infinite LC resistance is in parallel with the load's real resistance, therefore, does not affect the overall calculations.

Method 1 *Resonating out the total capacitance value*:

1. *The "resonating out" calculations:* in this approach, at the given frequency, the addition of inductor L whose inductance creates ideal resonator with C_L, thus infinite resistance. As a result, this problem is reduced to the problem of matching real $5\,\Omega$ to real $50\,\Omega$, which is already solved in Example 12.1-1. First, we calculate inductance that is needed to resonate out C_L as

$$L = \frac{1}{(2\pi \, f_0)^2 \, C} = \frac{1}{(2\pi \times 10\,\text{MHz})^2 \times 1.055\,\text{nF}} = 240.098\,\text{nH} \approx 240.1\,\text{nH}$$

In this step, the addition of external inductor $L \approx 240.1\,\text{nH}$ in parallel with load capacitor $C_L = 1.055\,\text{nF}$ creates ideal LC resonator so that $Z_{LC} = \infty$ at $f_0 = 10\,\text{MHz}$. With this step we "resonated out" the full value of load capacitance, Fig. 12.13.

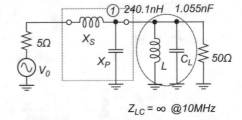

Fig. 12.13 Example 12.1-5

2. *Matching real to real resistor calculation:* once C_L is completely resonated out, its reactance is not visible anymore to the rest of the circuit. In this case, it means that the original problem is reduced to matching $5\,\Omega$ real source to $50\,\Omega$ real load. We reuse results from Example 12.1-1, i.e. we add $X_S = 238.732\,\text{nH}$ inductance and $X_P = 954.910\,\text{pF}$ capacitance.

In total, three components are used to create this matching network: two inductors and one capacitor.

Method 2 *Resonating out the partial capacitance value:*

1. *The partial "resonating out" calculations:* main idea in this technique is to imagine that the total $1.055\,\text{nF}$ capacitance consists of two capacitors connected in parallel, that is, $954.910\,\text{pF} + 100\,\text{pF}$. That being the case, we can resonate out only the $100\,\text{pF}$ "portion" of C_L and "keep" the $954.910\,\text{pF}$ "portion" for the use in the matching circuit, Fig. 12.14.

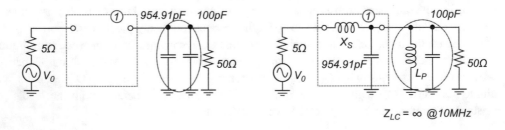

Fig. 12.14 Example 12.1-5

In order to create infinite impedance with the $100\,\text{pF}$ portion of the loading capacitance, we calculate

$$L_P = \frac{1}{(2\pi \, f_0)^2 \, C} = \frac{1}{(2\pi \times 10\,\text{MHz})^2 \times 100\,\text{pF}} = 2.533\,\mu\text{H}$$

After adding $L_P = 2.533\,\mu\text{H}$ in parallel to C_L the rest of circuit perceives as if $C_L = 954.910\,\text{pF}$.

2. *Matching network calculations:* we proceed with design of matching network assuming that $X_P = 954.910\,\text{pF}$ capacitor is already in place. Based on the already done calculations in Example 12.1-1, all that is left to do is to add serial inductance $X_S = 238.732\,\text{nH}$.

Therefore this solution requires only two new components, inductors $X_S = 238.732\,\text{nH}$ and $L_P = 2.533\,\mu\text{H}$. Portion of the original loading capacitance is resonated out with X_P, while the leftover is reused in combination with X_S inductance to match $5\,\Omega$ real source to $50\,\Omega$ real load.

Overall, both solutions are valid, while the differences are strictly practical. The decision between the three components for the first solution, or two components for the second solution, is now matter of the practicality, size, price.

6. Because real $R_0 = R_L$, that is to say, excluding C_L capacitance the source and load are already matched. Simplest solution is to completely resonate out C_L. Given data, we calculate the required inductance as

$$L = \frac{1}{(2\pi f_0)^2 C} = \frac{1}{(2\pi \times 10\,\text{MHz})^2 \times 100\,\text{pF}} = 2.533\,\mu\text{H}$$

Therefore, the required matching network consists only of $X_P = 2.533\,\mu\text{H}$ inductor in parallel to $100\,\text{pF}$ capacitor, which at $f_0 = 10\,\text{MHz}$ creates infinite Z_{LC} impedance, Fig. 12.15. We note that with this solution, DC connection between source and load is simply set by the short connection.

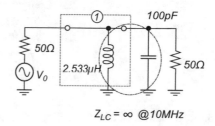

Fig. 12.15 Example 12.1-6

Exercise 12.2, page 267

1. In order to *increase* bandwidth of two-stage matching network relative to single-stage solution the intermediate resistance R_{INT} value is set using geometrical mean R_S and R_L resistances, i.e.

$$R_{INT} = \sqrt{R_S R_L}$$

With the help of R_{INT} the design of one two-stage network is reduced to design of two single-stage LC matching networks—the first one is designed to match source resistance R_S to R_{INT}, and the second one to match R_{INT} to the load R_L, Fig. 12.16. To emphasize, resistance R_{INT} is *not a real resistor* component that is added in the network. Instead, it is only a number being used to set the intermediate resistance at node between the two matching network stages.

Fig. 12.16
Example 12.2-1

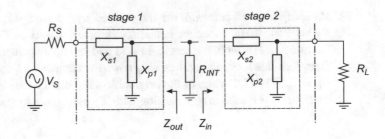

Given data, the matching network is designed as follows:

1. Calculation of the intermediate resistance:

$$R_{\text{INT}} = \sqrt{R_S \, R_L} = \sqrt{5\,\Omega \times 50\,\Omega} = 15.811\,\Omega$$

2. Calculation of Q factor at R_{INT} looking left into the first stage, and looking right into the second stage:

$$Q = \sqrt{\frac{R_{\text{INT}}}{R_S} - 1} = \sqrt{\frac{15.811\,\Omega}{5\,\Omega} - 1} = 1.470$$

$$Q = \sqrt{\frac{R_L}{R_{\text{INT}}} - 1} = \sqrt{\frac{50\,\Omega}{15.811\,\Omega} - 1} = 1.470$$

3. Calculation of first stage X_{S1} and X_{P1} components: matching $5\,\Omega$ to $15.811\,\Omega$ results in

$$X_{S1} = Q\,R_S = 1.47 \times 5\,\Omega = 7.352\,\Omega \qquad \therefore \qquad L_{S1} = \frac{X_{S1}}{2\pi\,f} = 117\,\text{nH}$$

$$X_{P1} = \frac{R_{\text{INT}}}{Q} = \frac{15.811\,\Omega}{1.47} = 10.753\,\Omega \qquad \therefore \qquad C_{P1} = \frac{1}{2\pi\,f\,X_{P1}} = 1.480\,\text{nF}$$

where series inductor and parallel capacitor maintain DC connection in the first stage.

4. Calculation of second stage X_{S2} and X_{P2} components: matching $15.811\,\Omega$ to $50\,\Omega$ results in

$$X_{S2} = Q\,R_{\text{INT}} = 1.47 \times 15.811\,\Omega = 23.250\,\Omega \qquad \therefore \qquad L_{S2} = 370\,\text{nH}$$

$$X_{P2} = \frac{R_L}{Q} = \frac{50\,\Omega}{1.47} = 34\,\Omega \qquad \therefore \qquad C_{P1} = 468\,\text{pF}$$

The complete schema of this two-stage matching network is shown in Fig. 12.17.

Fig. 12.17
Example 12.2-1

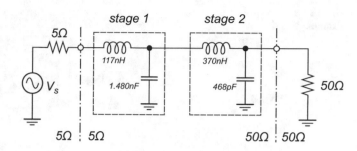

2. When the goal is to decrease bandwidth of two-stage matching network relative to single-stage solution, in general, then the intermediate resistance R_{INT} should be set to a value greater than the greater of two resistances (R_S, R_L), i.e. $R_{\text{INT}} > \max(R_S, R_L)$. Obviously, there is infinity of possible choices.

To illustrate how the bandwidth may be controlled, in this example we compare two possible bandwidths, first, when $R_{\text{INT}} = 250\,\Omega$, then $R_{\text{INT}} = 985\,\Omega$.

Case 1 $R_{\text{INT}} > \max(R_S, R_L) = 250\,\Omega$, i.e. the ghost resistance is chosen as a convenient number greater than $50\,\Omega$, for example, $250\,\Omega$.

1. Note that the following two calculated Q values, one looking left into the source side and one looking right into the load side, are not equal. However, the overall bandwidth is limited by the higher Q (i.e. narrower bandwidth), which is the one dominating the design.

$$Q_1 = \sqrt{\frac{R_{INT}}{R_S} - 1} = \sqrt{\frac{250}{5} - 1} = 7.0$$

$$Q_2 = \sqrt{\frac{R_{INT}}{R_L} - 1} = \sqrt{\frac{250}{50} - 1} = 2.0$$

thus, $Q_1 = 7$ is the one limiting the total bandwidth.

2. First stage: Matching $5\,\Omega$ to $250\,\Omega$ results in

$$X_{s1} = Q_1\,R_S = 7 \times 5\,\Omega = 35\,\Omega \qquad X_{p1} = \frac{R_{INT}}{Q_1} = \frac{250\,\Omega}{7} = 35.714\,\Omega$$

3. Second stage: Matching $250\,\Omega$ to $50\,\Omega$ results in (see Fig. 12.18)

$$X_{s2} = Q_2\,R_L = 2 \times 50\,\Omega = 100\,\Omega \qquad X_{p2} = \frac{R_{INT}}{Q_2} = \frac{250\,\Omega}{2} = 125\,\Omega$$

Fig. 12.18
Example 12.2-2

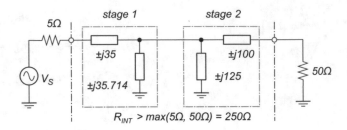

4. Considering solution, for example, that keeps DC connection between source and load, serial impedances are converted to inductances and parallel impedances into capacitances (after number rounding),

$$L_{s1} = \frac{X_{s1}}{2\pi f} = \frac{35\,\Omega}{2\pi\,10\,\text{MHz}} = 557\,\text{nH}$$

$$C_{p1} = \frac{1}{2\pi f X_{p1}} = \frac{1}{2\pi\,10\,\text{MHz} \times 35.714\,\Omega} = 445.6\,\text{pF}$$

$$L_{s2} = \frac{X_{s2}}{2\pi f} = \frac{100\,\Omega}{2\pi\,10\,\text{MHz}} = 1.592\,\mu\text{H}$$

$$C_{p2} = \frac{1}{2\pi f X_{p2}} = \frac{1}{2\pi\,10\,\text{MHz} \times 125\,\Omega} = 127.3\,\text{pF}$$

5. Thus, the complete two-stage matching network schematic, Fig. 12.19, may be reduced to T–type network after two parallel capacitors are combined into a single $C = 572.9\,\text{pF}$ component.

Fig. 12.19
Example 12.2-2

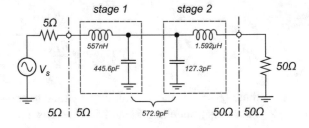

Case 2 $R_{\text{INT}} > \max(R_S, R_L) = 985\,\Omega$, i.e. the ghost resistance is chosen as a convenient number greater than $50\,\Omega$, for example, $985\,\Omega$.

1. We repeat the same calculations, however, using larger "ghost" resistor, therefore we find the new values as follows:

$$Q_1 = \sqrt{\frac{R_{INT}}{R_S} - 1} = \sqrt{\frac{985}{5} - 1} = 14.0$$

$$Q_2 = \sqrt{\frac{R_{INT}}{R_L} - 1} = \sqrt{\frac{985}{50} - 1} = 4.3$$

thus, $Q_1 = 14$ is the one limiting the total bandwidth.

2. First stage: Matching 5 Ω to 250 Ω results in

$$X_{s1} = Q_1 R_S = 14 \times 5\,\Omega = 70\,\Omega \qquad X_{p1} = \frac{R_{INT}}{Q_1} = \frac{250\,\Omega}{14} = 70.4\,\Omega$$

3. Second stage: Matching 250 Ω to 50 Ω results in

$$X_{s2} = Q_2 R_L = 4.3 \times 50\,\Omega = 216.2\,\Omega \qquad X_{p2} = \frac{R_{INT}}{Q_2} = \frac{985\,\Omega}{4.3} = 227.8\,\Omega$$

4. Again, by keeping DC connection between source and load, serial impedances are converted to inductances and parallel impedances into capacitances (after light rounding of numbers),

$$L_{s1} = \frac{X_{s1}}{2\pi f} = \frac{70\,\Omega}{2\pi\,10\,\text{MHz}} = 1.114\,\mu\text{H}$$

$$C_{p1} = \frac{1}{2\pi f\,X_{p1}} = \frac{1}{2\pi\,10\,\text{MHz} \times 70.4\,\Omega} = 226.2\,\text{pF}$$

$$L_{s2} = \frac{X_{s2}}{2\pi f} = \frac{216.2\,\Omega}{2\pi\,10\,\text{MHz}} = 3.441\,\mu\text{H}$$

$$C_{p2} = \frac{1}{2\pi f\,X_{p2}} = \frac{1}{2\pi\,10\,\text{MHz} \times 227.8\,\Omega} = 69.87\,\text{pF}$$

where the resulting circuit does not change, except for the new component values. Again, two parallel capacitors in the middle are combined into a single $C = 296\,\text{pF}$ component.

Numerical AC simulation of the two matching networks in this example illustrates that the overall bandwidth is controlled, Fig. 12.20. Higher Q limits bandwidth, in this case, to 1.1 MHz, while lower Q permits 2.5 MHz, see figure below. In both cases, the network's bandwidth is reduced relative to 7.5 MHz in Example 12.1-1.

Fig. 12.20
Example 12.2-2

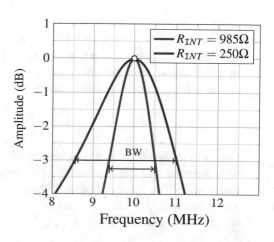

Exercise 12.3, page 267

1. Series inductor impedance at 10 MHz is calculated as

$$X_S = \omega\, L_S = 2\pi \times 10\,\text{MHz} \times 238.732\,\text{nH} = 15\,\Omega$$

which is followed by calculation of serial Q as

$$Q_S = \frac{X_S}{R_s} = 3$$

Therefore, serial to parallel conversion, see Fig. 12.21, formulas give

$$R_P = R_S(1 + Q^2) = 50\,\Omega$$

$$X_P = X_S\left(1 + \frac{1}{Q^2}\right) = 16.667\,\Omega$$

$$\therefore$$

$$L_P = 265.258\,\text{nH}$$

Fig. 12.21
Example 12.3-1

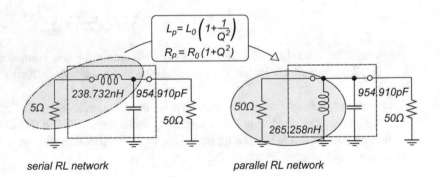

serial RL network   parallel RL network

2. One of the consequences of having non-equal source and load impedances is that part of the incoming signal energy is reflected back at the interface point and never fully reaches the load. In this example we illustrate quantitative measure of signal reflection at the interface of source and load impedances. By definition mismatch coefficient Γ is calculated as:

$$\Gamma = \frac{Z_S - Z_P}{Z_S + Z_P}$$

One of possible ways (not necessary most efficient) to compare the source side impedance Z_S with the load side impedance Z_L impedances is to do serial to parallel transformations relative to the node one. Then, it becomes obvious that the two impedances are actually matched at the given frequency.

Given source and load resistances, Q factor is

$$Q_S = Q_P = \sqrt{\frac{R_L}{R_S} - 1} = \sqrt{\frac{50}{5} - 1} = 3$$

Reactance of inductor and capacitor components are

$$X_S = j\omega L = 2\pi \times 10\,\text{MHz} \times 238.732\,\text{nH} = 15\,\Omega$$

$$X_P = \frac{1}{j\omega C} = \frac{1}{2\pi \times 10\,\text{MHz} \times 954.91\,\text{pF}} = 16.667\,\Omega$$

By looking into the source side impedance to the left we write by inspection

$$Z_S = R_S + jX_S = (5 + j\,15)\,\Omega \quad \therefore \quad |Z_S| = \sqrt{5^2 + 15^2} = 15.811\,\Omega$$

By looking into the load side impedance to the right of the node one, there is parallel RC network, thus we convert it into its equivalent serial network as

$$R_P = R_S\left(1 + Q^2\right) \quad \therefore \quad R_S = \frac{R_P}{\left(1 + Q^2\right)} = \frac{50\,\Omega}{\left(1 + 3^2\right)} = 5\,\Omega$$

$$X_P = X_S\left(1 + \frac{1}{Q^2}\right) \quad \therefore \quad X_S = \frac{X_P}{\left(1 + \frac{1}{Q^2}\right)} = \frac{16.667\,\Omega}{\left(1 + \frac{1}{3^2}\right)} = 15\,\Omega$$

therefore, impedance Z_S of the equivalent RC series network at the load side is

$$Z_S(Z_P) = R_S - jX_S = (5 - j\,15)\,\Omega \quad \therefore \quad |Z_S(Z_P)| = \sqrt{5^2 + 15^2} = 15.811\,\Omega$$

where the negative reactance is because of the capacitor ($Z_C = -j/\omega C$). In other words, source and load impedances are equal $|Z_S| = |Z_S(Z_P)| = 15.811\,\Omega$. Therefore, at matching frequency of 10 MHz, we write

$$\Gamma = \frac{Z_S - Z_P}{Z_S + Z_P} = \frac{(15.811 - 15.811)\,\Omega}{(15.811 + 15.811)\,\Omega} = 0 = \infty\,\text{dB}$$

which, by definition, gives the mismatch loss ML as

$$ML = \frac{1}{1 - \Gamma^2} = 1 = 0\,\text{dB}$$

In other words, mismatch loss $ML = 0\,\text{dB}$ obtained in this case indicates perfect loss-less matching of $5\,\Omega$ source to $50\,\Omega$ load that is valid only if inductor $L_S = 238.732\,\text{nH}$ and capacitor $C_P = 954.9\,\text{pF}$ components are used at 10 MHz. At all other frequencies, the calculated series/parallel impedances produce different impedance values.

3. As same as any other passive network that contains RLC components, matching network is a *narrowband* circuit whose bandwidth is determined by the RLC component values. In the first approximation we assume that the frequency response curve of the matching network is symmetrical, i.e. equivalent to an ideal LC resonator, which in reality is not the case. In order to find the network's impedance, first step is to transform the series part of the network (i.e. at the source side) into its equivalent parallel network (we reuse the calculated $Q = 3$), as

$$R_P = R_S(1 + Q^2) = 50\,\Omega \quad \text{and} \quad L_P = L_S\left(1 + \frac{1}{Q^2}\right) = 265.258\,\text{nH}$$

After this series to parallel transformation, Fig. 12.22, the equivalent matching network looks as

Fig. 12.22
Example 12.3-3

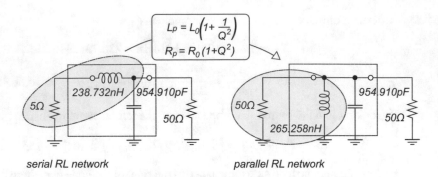

serial RL network   parallel RL network

In order to confirm that the equivalent parallel LC resonator indeed resonates at $f_0 = 10\,\text{MHz}$, we recalculate

$$f_0 = \frac{1}{2\pi\sqrt{LC}} = \frac{1}{2\pi\sqrt{265.258\,\text{nH} \times 954.9\,\text{pF}}} \approx 10\,\text{MHz} \qquad (12.4)$$

which, consequently means that the LC resonator's dynamic impedance R_D is infinite, hence the network resistance is determined only with the two real resistors $R_P = 50\,\Omega$ and $R_L = 50\,\Omega$ in parallel. In other words, the equivalent resonator's resistance $R = 50\,\Omega \| 50\,\Omega = 25\,\Omega$ at the resonant frequency. Resonator bandwidth can now be estimated by using either inductor or capacitor impedance (they are equal at the resonant frequency) versus the equivalent parallel resistor value to calculate Q factor, and therefore the bandwidth. At $f_0 = 10\,\text{MHz}$,

$$Q = \frac{R}{X_C} = \frac{25}{16.667} \approx 1.5$$

$$\therefore$$

$$BW_{3\,\text{dB}} = \frac{f_0}{Q} = \frac{10\,\text{MHz}}{1.5} = 6.667\,\text{MHz}$$

In this approximated estimate it is assumed that the matching network is *symmetrical*, which is *not* the case, consequently the implemented matching network has $BW_{3\,\text{dB}}$ a bit wider that estimated in this example. Simulation confirms validity of the estimated approach (see Fig. 12.23).

Fig. 12.23
Example 12.3-3

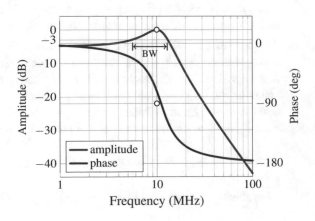

RF and IF Amplifiers

13

After a weak radio frequency (RF) signal has arrived at the antenna, it is channeled to the input terminals of the RF amplifier through a passive matching network. The matching network is made to enable maximum power transfer of the receiving signal by equalizing the antenna impedance with the RF amplifier input impedance. After that, it is the job of the RF amplifier to increase the power of the received signal and prepare it for further processing. In the first part of this chapter, we review the basic principles of linear baseband amplifiers and common circuit topologies. In the second part of the chapter, we introduce RF and IF amplifiers. Aside from their operating frequency, for all practical purposes, there is not much difference between the schematic diagrams of RF and IF amplifiers.

13.1 Important to Know

1. Insertion loss

$$Z_C = \frac{R_{ct}\, R_D}{R_{ct} + R_D} = R_{ct}\, \frac{R_D}{R_{ct} + R_D} = R_{ct} \times IL \tag{13.1}$$

$$IL_{\mathrm{dB}} = 20 \log \frac{R_D}{R_{ct} + R_D} = 20 \log \frac{1}{1 + \dfrac{R_{ct}}{R_D}} \quad [\mathrm{dB}] \tag{13.2}$$

© Springer Nature Switzerland AG 2021
R. Sobot, *Wireless Communication Electronics by Example*,
https://doi.org/10.1007/978-3-030-59498-5_13

13.2   Exercises

13.1 *   BJT RF amplifiers

1. Given a simple CE amplifier, Fig. 13.1, estimate its voltage gain and then develop its equivalent CE RF amplifier to be used for amplification of RF signal at f_{RF}.
   Data: $I_{C0} = 1.3\,\text{mA}$, $V_T = 25\,\text{mV}$, $R_C = 5\,\text{k}\Omega$, $f_{RF} = 1\,\text{MHz}$, $\beta \gg 1$, $r_o \to \infty$.

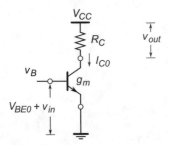

Fig. 13.1 Example 13.1-1

2. Given that simple CE amplifier in Fig. 13.2 (right) has transfer characteristics as in Fig. 13.2 (left).

   (a) Propose R_C and C_E so that CE amplifier is suitable for amplification of RF signal at f_{RF}.

   (b) Design LC resonator compatible with f_{RF} and determine Q factor and BW of CE RF amplifier.

   Data: $I_{C0} = 1.3\,\text{mA}$, $V_T = 25\,\text{mV}$, $R_E = 500\,\Omega$, $f_{RF} = 10\,\text{MHz}$, $\beta \gg 1$, $r_o \to \infty$.

Fig. 13.2 Example 13.1-2

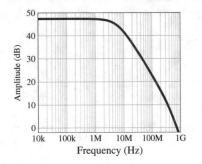

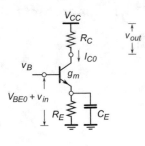

13.2 *   FET RF amplifiers

1. Given a simple CS amplifier, Fig. 13.3, estimate its voltage gain and then develop its equivalent CS RF amplifier to be used for amplification of RF signal at f_{RF}.
   Data: $I_{D0} = 1.3\,\text{mA}$, $R_D = 5\,\text{k}\Omega$, $V_{OV} = V_{GS} - V_t = 1.22\,\text{V}$, $f_{RF} = 10\,\text{MHz}$, $r_o \to \infty$.

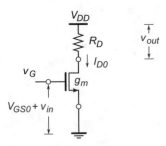

Fig. 13.3 Example 13.2-1

2. Given that simple CS amplifier in Fig. 13.4 (right) has transfer characteristics as in Fig. 13.4 (left).

(a) Propose R_D and C_S so that CS amplifier is suitable for amplification of RF signal at f_{RF}.

(b) Design LC resonator compatible with f_{RF} and determine Q factor and BW of CS RF amplifier.

Data: $I_{D0} = 1.3\,\text{mA}$, $V_{OV} = 1.22\,\text{V}$, $f_{RF} = 10\,\text{MHz}$, $R_S = 500\,\Omega$. $r_o \to \infty$.

Fig. 13.4 Example 13.2-2

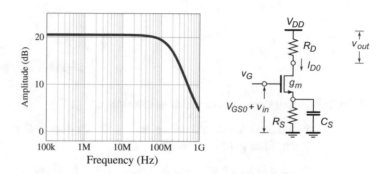

13.3 * JFET RF amplifiers

1. Given JFET transistor characteristics used in Example 7.6-2, choose biasing point and then calculate

(a) g_m, (b) $R_D(max)$, (c) $A_V(max)$.

Data: $V_{DD} = 15\,\text{V}$, $V_{DS}(min) = 2\,\text{V}$, $R_S = 833\,\Omega$.

2. We reuse JFET transistor and calculations from Example 5.1-1. Given the family of frequency characteristics where R_D is parameter, see Fig. 13.5, and assuming f_{RF} signal and C_S capacitor:

(a) Estimate values R_{D1} to R_{D4} and then choose R_D. Briefly explain your reasoning.

(b) Calculate C_S and briefly explain your reasoning.

Data: $f_{RF} = 10\,\text{MHz}$.

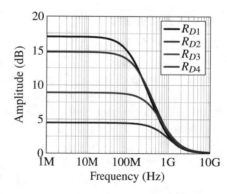

Fig. 13.5 Example 13.3-2

3. Following up the design of JFET cascoded transistors in Example 7.6-2, by simulation show the influence of C_G capacitor that is providing small signal ground $\varnothing$ to gate of J_2 in Fig. 13.6.

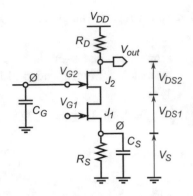

Fig. 13.6 Example 13.3-3

4. After reusing the results of previous examples:

(a)   the real load R_D should be replaced with non-ideal LC resonator to design cascoded JFET RF amplifier to be used for f_{RF} signal,

(b)   calculate Q factor, BW, and inductor's wire resistance r,

(c)   calculate load resistance $R_L(min)$ for which the output voltage is reduced by half,

(d)   calculate Q factor and BW if $R_L = R_L(min)$,

(e)   show the complete schematic diagram of this cascoded JFET RF amplifier.

Data: $f_{\mathrm{RF}} = 10\,\mathrm{MHz}, R_D = 1\,\mathrm{k\Omega}, r_o \to \infty$.

Solutions

Exercise 13.1, page 282

1. Wideband CE amplifier is converted into narrowband RF amplifier by replacing resistive load R_C by LC resonator, Fig. 13.7. If $R_C = R_D$, where R_D is the dynamic resistance of realistic LC tank at the resonant frequency, then voltage gain A_V of wideband and narrowband amplifiers are the same. This is because, we recall, the impedance of ideal LC tank at the resonant frequency is infinite, which leaves only R_D that is modelled as if connected in parallel to the ideal LC components.

Fig. 13.7 Example 13.1-1

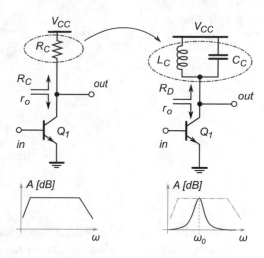

Given data, we calculate

$$g_m \stackrel{\text{def}}{=} \frac{1}{r_e} = \frac{I_{C0}}{V_T} = \frac{1.3\,\text{mA}}{25\,\text{mV}} = 52\,\text{mS} \quad \therefore \quad |A_V| = g_m R_C = 260 \approx 48\,\text{dB}$$

Assuming that the given RF frequency is within the transistor's frequency range, we set up LC resonator to replace R_C. The resonant frequency $f_0 = f_{\text{RF}}$ is

$$f_0 = \frac{1}{2\pi\sqrt{LC}} \quad \therefore \quad LC = \frac{1}{(2\pi f_0)^2} = \frac{1}{4\times\pi^2} \times 10^{-12}\,[\text{HF}] \tag{13.3}$$

Obviously, there is infinity of possible LC pairs that can produce the product (13.3). The choice of each of the individual values of L and C is driven by practical component limitations. In this exercise, for example, we rewrite (13.3)

$$LC = \frac{1}{100 \times 2\pi}\,\text{mH} \times \frac{1}{10 \times 2\pi}\,\mu\text{F}$$

and choose $L = 1/(100 \times 2\pi)\,\text{mH} \approx 1.59\,\mu\text{H}$ and $C = 1/(10 \times 2\pi)\,\mu\text{F} \approx 15.91\,\text{nF}$.
In practice, in order to achieve the exact resonating frequency f_0 and due to non-standard component values, LC resonator must be "tuned"—usually with custom-made inductor and trimmer capacitor.
The resonant frequency f_0 is set by ideal LC components; however, realistic inductor has small wire resistance r that limits Q factor, Fig. 13.8, where the equivalent parallel dynamic resistance is calculated as $R_D = Q^2 r$. If we are to keep the voltage gain A_V of CE amplifier unchanged, $R_D = R_C$.

Fig. 13.8 Example 13.1-1

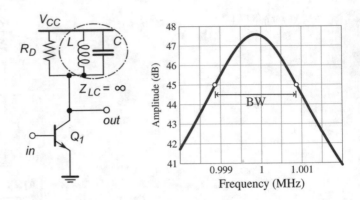

Simulation of CE RF amplifier shows that hand approximative analysis produced very close results. From the simulated frequency transfer function, we find $BW = 2\,\text{kHz}$ and voltage gain $A_V = 47.5\,\text{dB}$. In addition, we calculate

$$Q = \frac{f_0}{BW} = \frac{1\,\text{MHz}}{2\,\text{kHz}} = 500$$

and to verify again dynamic resistance that is set at the beginning of this problem, from the simulation results we calculate $R_D = \omega_0 Q L = 5\,\text{k}\Omega$. This parallel resistance is equivalent to the inductor's wire resistance $r = R_D/Q^2 = 20\,\text{m}\Omega$. It is important to note that we assumed $r_o \to \infty$, however that assumption is not exactly correct; the simulations show results that account for the total collector resistance.

2. By inspection of graph in Fig. 13.8 (left), we conclude that as developed in Example 13.1-1 CE amplifier has BW suitable for $f_{\text{RF}} = 1\,\text{MHz}$ signal. However, its voltage gain is high, thus there is possibility to trade in gain for BW and make it suitable for $f_{\text{RF}} = 10\,\text{MHz}$ signal.

(a) Using graphical method, while allowing for the non-linear section of the frequency transfer function, for example, we could reduce gain to $A_V = 30\,\text{dB}$. A horizontal line drawn at 30 dB level extends BW to beyond 10 MHz, as illustrated in Fig. 13.9. In order to reduce the gain, we recalculate R_C as follows. Given biasing current and room temperature, we calculate

$$g_m \overset{\text{def}}{=} \frac{1}{r_e} = \frac{I_{C0}}{V_T} = \frac{1.3\,\text{mA}}{25\,\text{mV}} = 52\,\text{mS}$$

Therefore,

$$|A_V| = 30\,\text{dB} = 31.6 = g_m\,R_C \quad \therefore \quad R_C = \frac{|A_V|}{g_m} \approx 610\,\Omega$$

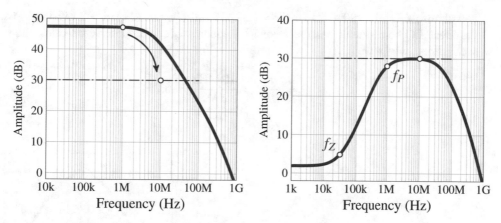

Fig. 13.9 Example 13.1-2

The addition of R_E, C_E creates pole/zero pair in the transfer function, assuming $\beta \gg 1$, as

$$f_Z = \frac{1}{2\pi R_E C_E} \quad \therefore \quad f_P = \frac{1}{2\pi (r_e || R_E) C_E} \approx \frac{1}{2\pi r_e C_E}$$

where $r_e || R_E = 19.2\,\Omega || 500\,\Omega = 18.5\,\Omega \approx r_e$. Setting up the pole frequency at least to one decade below $f_{RF} = 10\,MHz$ signal results in

$$f_P = \frac{1}{2\pi (r_e || R_E) C_E} = 0.1\,f_{RF} \quad \therefore \quad C_E \geq \frac{1}{2\pi r_e 0.1\,f_{RF}} = 8.3\,nF \approx 10\,nF$$

which results in the zero frequency as

$$f_Z = \frac{1}{2\pi R_E C_E} \approx 32\,kHz$$

The modified BW is shown in Fig. 13.9 (right), and obviously greater values of C_E would move the pole/zero pair to lower frequencies. However, in RF amplifier design, in order to reduce the noise entering we do not want to permit the BW wider than necessary.

(b) Using same technique as in Example 13.1-1, we create LC resonator to transform wideband CE amplifier into narrowband RF amplifier centred at $f_{RF} = 10\,MHz$ carrier. Then, the resonant frequency $f_0 = f_{RF}$ is

$$f_0 = \frac{1}{2\pi \sqrt{LC}} \quad \therefore \quad LC = \frac{1}{(2\pi f_0)^2} = \frac{1}{4 \times \pi^2} \times 10^{-14}\,[HF] \tag{13.4}$$

Obviously, there is infinity of possible LC pairs that can produce the product (13.4). In this exercise, for example, we rewrite (13.4)

$$LC = \frac{1}{100 \times 2\pi}\,mH \times \frac{1}{2\pi}\,nF$$

and choose $L = 1/(100 \times 2\pi)\,\text{mH} \approx 1.59\,\mu\text{H}$ and $C = 1/(2\pi)\,\text{nF} \approx 159.15\,\text{pF}$, where the ideal LC components set the resonant frequency f_0. We confirm the calculations by simulations, where we find

$$Q = \frac{f_0}{BW} \approx \frac{10\,\text{MHz}}{1.6\,\text{MHz}} = 6.25$$

Fig. 13.10
Example 13.1-2

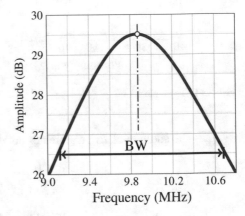

Knowing Q, we calculate

$$R_D = Q^2 r \quad \therefore \quad r = \frac{R_D}{Q^2} = \frac{610\,\Omega}{6.25^2} = 15.6\,\Omega$$

and we estimate the shifted resonant frequency due to non-negligible serial resistance r of realistic inductor, as

$$f_0 = \frac{1}{2\pi}\sqrt{\frac{1}{LC} - \frac{r^2}{L^2}} = 9.877\,\text{MHz}$$

which is in good agreement with the simulation results, Fig. 13.10.

Exercise 13.2, page 282

1. Given data, we calculate

$$g_m = \frac{2I_D}{V_{\text{OV}}} = \frac{2 \times 1.3\,\text{mA}}{1.22\,\text{V}} = 2.13\,\text{mS} \quad \therefore \quad |A_V| = g_m R_D = 10.65 = 20.5\,\text{dB}$$

For LC resonator design, we reuse results of the previous examples and use $L = 1.59\,\mu\text{H}$ and $C = 159.15\,\text{pF}$.

2. Wideband NMOS CS amplifier given in Example 5.1-1 is transformed into an RF amplifier as follows.

(a) task By inspection of the frequency transfer function in Fig. 13.4 (left), we conclude that this is a HF NMOS transistor where f_{RF} is found within its bandwidth. Thus, we can keep $R_D = 5\,k\Omega$ and $A_V = 20.5\,dB$.

In order to calculate C_S, similar to BJT case, we need to derive transfer function at source node. Knowing that $i_G = 0$, $i_D = i_S$, and $r_S = 1/g_m$, by inspection of source network in Fig. 13.4 (right), source current is calculated as

$$i_D = i_S = \frac{v_G}{Z_S} = \frac{v_G}{r_S + R_S||Z_{C_S}} = \frac{v_G}{\frac{1}{g_m} + R_S||Z_{C_S}} = v_G \frac{g_m}{1 + g_m R_S||Z_{C_S}}$$

where impedance of the parallel connection $Z = R_S||Z_{C_S}$ is

$$\frac{1}{Z} = \frac{1}{R_S} + j\omega C_S = \frac{1 + j\omega C_S R_S}{R_S} \quad \therefore \quad Z = \frac{R_S}{1 + j\omega C_S R_S}$$

Therefore, the output current i_D to the input voltage v_G function is

$$\frac{i_D}{v_G} = \frac{g_m}{1 + g_m R_S||Z_{C_S}} = \frac{g_m}{1 + g_m \left(\frac{R_S}{1 + j\omega C_S R_S}\right)} = g_m \frac{1 + j\omega C_S R_S}{1 + g_m R_S + j\omega C_S R_S}$$

$$= \frac{g_m}{1 + g_m R_S} \frac{1 + j\frac{\omega}{1/C_S R_S}}{1 + j\frac{\omega C_S R_S}{1 + g_m R_S}} \tag{13.5}$$

By inspection of (13.5), we write expressions for zero and pole of transfer function at source node as

$$\omega_Z = \frac{1}{C_S R_S} \quad \text{and} \quad \omega_P = \frac{1 + g_m R_S}{C_S R_S} = \frac{1/R_S + g_m}{C_S}$$

Given data, by setting the pole frequency f_P at one decade or more below f_0, we calculate

$$0.1 f_0 = \frac{1/R_S + g_m}{2\pi C_S} \quad \therefore \quad C_S = \frac{1/R_S + g_m}{2\pi \times 0.1 f_0} = \frac{1/500\,\Omega + 2.13\,mS}{2\pi \times 0.1 \times 10\,MHz} \approx 660\,pF$$

Pole/zero pair is tied together through R_S, C_S values, and, consequently, zero frequency f_Z is pushed to

$$f_Z = \frac{1}{C_S R_S} = \frac{1}{2\pi\,660\,pF \times 500\,\Omega} \approx 480\,kHz$$

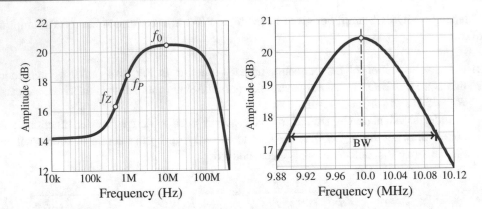

Fig. 13.11 Example 13.2-2

(b) We reuse earlier calculations $L = 1/(100 \times 2\pi)\,\mathrm{mH} \approx 1.59\,\mu\mathrm{H}$ and $C = 1/(2\pi)\mathrm{nF} \approx 159.15\,\mathrm{pF}$. Simulated transfer functions illustrate band shaping due to pole/zero pair (wideband) in Fig. 13.11 (left), and after using LC resonator (narrowband) in Fig. 13.11 (right). By inspection, we find $BW = 200\,\mathrm{kHz}$, and therefore $Q = 50$. Then, $r = R_D/Q^2 = 2\,\Omega$.

Exercise 13.3, page 283

1. By inspection of Fig. 7.8 (left), we find $I_{\mathrm{DSS}} = 12\,\mathrm{mA}$, $V_\mathrm{P} = -3\,\mathrm{V}$. Good choice of JFET biasing current is $I_{\mathrm{D0}} = I_{\mathrm{DSS}}/2 = 6\,\mathrm{mA}$.

(a) Instead of graphical method used earlier, for completeness, from definition of JFET drain current, we calculate

$$I_D = I_{\mathrm{DSS}} \left[1 - \frac{V_{\mathrm{GS}}}{V_\mathrm{P}} \right]^2 \quad \therefore \quad V_{\mathrm{GS}} = \left[1 - \sqrt{\frac{I_D}{I_{\mathrm{DSS}}}} \right] V_\mathrm{P} = -878\,\mathrm{mV}$$

(b) By definition, g_m of JFET transistor is

$$g_m \overset{\mathrm{def}}{=} \left. \frac{dI_D}{dV_{\mathrm{GS}}} \right|_{V_{Ds}=\mathrm{const.}} = \frac{2\,I_{DSS}}{|V_P|} \left(1 - \frac{V_{GS}}{V_P} \right) = 5.66\,\mathrm{mS}$$

(c) Given drain current, the range of R_D is limited by power supply voltage and $V_{\mathrm{DS}}\,(min)$ that must be kept so that the transistor does not leave the constant current mode. Therefore, by inspection of Fig. 13.12 (left), we write

Fig. 13.12
Example 13.3-1

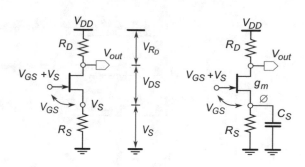

$$V_S = R_S I_S = R_S I_D = 5\,\text{V}$$

Therefore,

$$V_S + V_{\text{DS}}\,(min) + V_{R_D}\,(max) = V_{\text{DD}} \quad \therefore \quad V_S + V_{\text{DS}}\,(min) + R_D(max)\,I_D = V_{\text{DD}}$$

$$\therefore$$

$$R_D(max) = \frac{V_{\text{DD}} - V_S - V_{\text{DS}}\,(min)}{I_D} = 1.33\,\text{k}\Omega$$

(d) Knowing $R_D(max)$ and R_S and assuming $R_S \gg 1/g_m$, we estimate voltage gain as

$$|A_V| = \frac{R_D}{R_S} = \frac{1.33\,\text{k}\Omega}{833\,\Omega} = 1.6 = 4\,\text{dB}$$

which, obviously, is not too much. We recall that bypassing R_S with C_S, see Fig. 13.12 (right), effectively removes R_S from the signal path leaving only $1/g_m$ connected to the signal ground $\varnothing$, and therefore we write

$$|A_V|(max) = g_m\,R_D(max) = 5.66\,\text{mS} \times 1.33\,\text{k}\Omega = 7.53 = 17.5\,\text{dB}$$

which is therefore the upper voltage gain limit that is the consequence of the biasing current and power supply voltage choices.

2. By reusing the results from Example 5.1-1 and inspecting the given frequency characteristics, we proceed as follows:

(a) Given frequency characteristics indicate that JFET is HF transistor whose unity gain is in GHz range. That being the case, we could try to trade in bandwidth for more gain until the BW is closer to $f_{\text{RF}} = 10\,\text{MHz}$. However, the upper gain limit set by $R_D = 1.33\,\text{kHz}$ is $|A_V|(max) = 17.5\,\text{dB}$. Therefore, we cannot reduce BW by further increasing the gain. Parametric characteristics show approximately:

$$\underline{|A_V| = 17\,\text{dB}} : \quad \therefore \quad R_{D1} = \frac{|A_V|}{g_m} = \frac{7.1}{5.66\,\text{mS}} = 1.25\,\text{k}\Omega$$

$$|A_V| = 15\,\text{dB} : \quad \therefore \quad R_{D2} = \frac{|A_V|}{g_m} = \frac{5.6}{5.66\,\text{mS}} = 1.0\,\text{k}\Omega$$

$$|A_V| = 9\,\text{dB} : \quad \therefore \quad R_{D3} = \frac{|A_V|}{g_m} = \frac{2.8}{5.66\,\text{mS}} = 500\,\Omega$$

$$|A_V| = 4.5\,\text{dB} : \quad \therefore \quad R_{D4} = \frac{|A_V|}{g_m} = \frac{1.7}{5.66\,\text{mS}} = 300\,\Omega$$

We can choose relatively high gain, not necessarily the maximum. In addition, we recall that in RF amplifier the real resistance R_D will be replaced by dynamic resistance R_D of non-ideal LC resonator. Dynamic resistance is the consequence of inductor's wire resistance r that by itself determines Q factor, therefore the resonator's BW. Depending on the final RF amplifier specification, there is space to trade in between gain and BW.

Without the additional BW constrains, we choose, for example, $R_D = 1\,\text{k}\Omega$ to achieve high but not maximum gain.

(b) Capacitor C_S together with R_S creates pole/zero pair. In order to limit noise entering RF amplifier, we limit the amplifier's BW. Given R_S, by setting the pole frequency f_P at one decade or more below f_0, since this is FET, we calculate

$$0.1 f_0 = \frac{1/R_S + g_m}{2\pi\,C_S} \quad \therefore \quad C_S = \frac{1/R_S + g_m}{2\pi \times 0.1 f_0} = \frac{1/833\,\Omega + 5.66\,\text{mS}}{2\pi \times 0.1 \times 10\,\text{MHz}} \geq 1.1\,\text{nF}$$

Pole/zero pair is tied together through R_S, C_S values, and, consequently, zero frequency f_Z is pushed to

$$f_Z = \frac{1}{C_S R_S} = \frac{1}{2\pi\,1.1\,\text{nF} \times 833\,\Omega} \approx 175\,\text{kHz}$$

Simulations illustrate the control of BW by means of C_S capacitance, Fig. 13.13. The HF pole close to 100 MHz is due to JFET's parasitic capacitances.

Fig. 13.13
Example 13.3-2

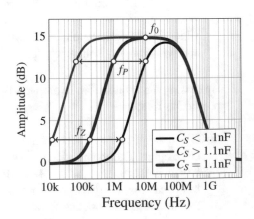

3. In Example 7.6-2, we calculated DC setup for cascoded JFET transistors. In DC calculation, we assume that DC voltages are stable and ideal. However, from the perspective of a small signal, the circuit node associated with J_2 gate does have its own time constant τ due to the equivalent resistance and parasitic capacitances connected to the gate node. Consequently, the parasitic pole reduces HF gain, Fig. 13.14. The addition of a relatively large C_G moves this pole to LF range and restores the unattenuated gain level. In order to do hand analysis, we would need to know JFET's parasitic capacitances.

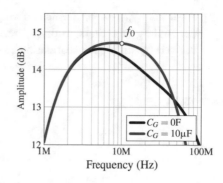

Fig. 13.14 Example 15.2-3

4. Following the previous examples, we proceed as follows:

(a) Ideal LC resonator determines the resonant frequency, thus we reuse values for $f_0 = f_{RF} = 10\,\text{MHz}$ as $L \approx 1.59\,\mu\text{H}$ and $C \approx 159\,\text{pF}$. Assuming $r_o \to \infty$, non-ideal LC resonator whose dynamic resistance equals R_D is modelled as parallel RLC network, Fig. 13.15 (left). In practice, in order to achieve non-standard LC values, a standard inductor value L is connected in parallel with standard capacitor C and trimmer capacitor C_{trim} so that it is possible to tune LC resonator to the exact frequency.

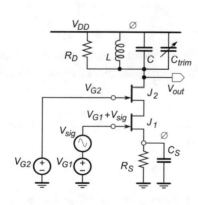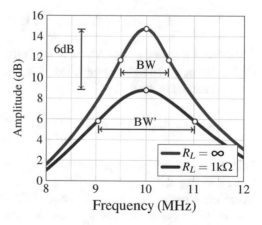

Fig. 13.15 Example 13.3-4

(b) By inspection of Fig. 13.15 (right), we find $BW = 1.0\,\text{MHz}$, and therefore

$$Q = \frac{f_0}{BW} = 10 \quad \text{and} \quad r = \frac{R_D}{Q^2} = \frac{1\,\text{k}\Omega}{10^2} = 10\,\Omega$$

where r is the inductor's wire resistance.

(c) Assuming $r_o \to \infty$ (which is very good approximation for cascoded transistors), large decoupling capacitors, and load resistor R_L, the small signal equivalent circuit network is derived by simple transformations, Fig. 13.16.

Fig. 13.16
Example 13.3-4

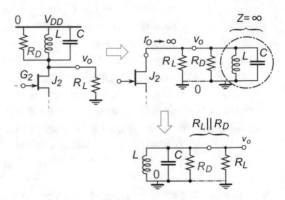

By inspection of Fig. 13.16, we find that $R_D||R_L$. Therefore, if after adding R_L the output voltage dropped by half, i.e. by 6 dB, that is to say that the reduced voltage gain $|A_V|'$ relative to unloaded amplifier gain $|A_V|$ is

$$\frac{|A_V|'}{|A_V|} = \frac{g_m\, R_D||R_L(min)}{g_m\, R_D} = \frac{1}{2} \quad \therefore \quad R_D||R_L(min) = \frac{1}{2} R_D$$

$$\therefore$$

$$R_L(min) = R_D = 1\,\text{k}\Omega$$

(d) At resonant frequency, now $R'_D = R_D/2$ is the *effective* dynamic resistance that "absorbed" load resistance. The inductance value itself and the resonant frequency do not change; thus, for RLC resonator we find

$$R'_D = Q'\omega_0 L \quad \therefore \quad Q' = \frac{R'_D}{\omega_0 L} = \frac{R_D}{2 \times 2\pi f_0\, L} = 5 \quad \therefore \quad BW' = \frac{f_0}{Q'} = 2\,\text{MHz}$$

This gain and BW change due to the addition of load resistance R_L is illustrated in Fig. 13.15 (right).

(e) Complete schematic diagram of cascoded CS JFET RF amplifier developed in this example, see Fig. 13.17 (left), includes signal generator resistance R_{sig}, decoupling capacitors C_1, C_2 as well as the two voltage divider references needed to provide biasing voltages V_{G1}, V_{G2}. Transient simulation included $R_L = 10\,\text{k}\Omega$, and estimated voltage gain is therefore

$$|A_V| = g_m R'_D = g_m R_D||R_L = 5.66\,\text{mS} \times (1\,\text{k}\Omega||10\,\text{k}\Omega) \approx 5\,\text{V/v} = 14\,\text{dB}$$

Circuits that include components that are capable of storing energy, i.e. LC components, need some time to reach steady-state level, Fig. 13.17 (right).

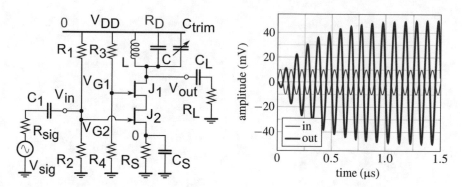

Fig. 13.17 Example 13.3-4

Sinusoidal Oscillators

Communication transceivers require oscillators that generate pure electrical sinusoidal signals (i.e. stable time reference signals) for further use in modulators, mixers, and other circuits. Although oscillators may be designed to deliver other waveforms as well, e.g. square, triangle, and sawtooth waveforms, if intended for applications in wireless radio communications, the sinusoidal and square waveforms are the most important ones. A good sinusoidal oscillator is expected to deliver either a voltage or a current signal that is stable both in amplitude and frequency. Because a variety of oscillator structures are available that are suitable for generation of periodic waveforms, circuit designers make the choice mostly based on their personal preference for one particular type of oscillator.

14.1 Important to Know

1. Closed-loop gain

$$A_c = \frac{v_o}{v_i} = \frac{A}{1 - \beta A} \tag{14.1}$$

2. Barkhausen Stability Criterions

$$|A\,\beta| \geq 1 \ \text{ and, } \ |\Delta\phi| = n \times 360° \quad (n = 0, 1, 2, \ldots) \tag{14.2}$$

3. Tapped L, Centre-Grounded Feedback Network

$$\omega_0^2 = \frac{1}{(L_1 + L_2)\,C} \tag{14.3}$$

$$\beta = -\frac{L_1}{L_2} \tag{14.4}$$

$$R_{\text{eff}} = R_{\text{eff}1} || R_{\text{eff}2} = R_L \left(\frac{L_2}{L_1}\right)^2 || \frac{Q\omega_0 L_2^2}{L_1 + L_2} \tag{14.5}$$

© Springer Nature Switzerland AG 2021
R. Sobot, *Wireless Communication Electronics by Example*,
https://doi.org/10.1007/978-3-030-59498-5_14

4. Tapped C, Centre-grounded Feedback Network

$$\omega_0^2 = \frac{C_1 + C_2}{LC_1C_2} \tag{14.6}$$

$$\beta = -\frac{C_2}{C_1} \tag{14.7}$$

$$R_{\text{eff}} = R_L \left(\frac{C_1}{C_2}\right)^2 \;||\; Q\omega_0 L \left(\frac{C_1}{C_1 + C_2}\right)^2 \tag{14.8}$$

5. Tapped L, Bottom-grounded Feedback Network

$$\omega_0^2 = \frac{1}{C(L_1 + L_2)} \tag{14.9}$$

$$\beta = \frac{L_1}{L_1 + L_2} \tag{14.10}$$

$$R_{\text{eff}} = R_L \left(\frac{L_1 + L_2}{L_1}\right)^2 \;||\; Q\omega_0(L_1 + L_2) \tag{14.11}$$

6. Tapped C, Bottom-grounded Feedback Network

$$\omega_0^2 = \frac{C_1 + C_2}{LC_1C_2} \tag{14.12}$$

$$\beta = \frac{C_2}{C_1 + C_2} \tag{14.13}$$

$$R_{\text{eff}} = R_L \left(\frac{C_1 + C_2}{C_2}\right)^2 \;||\; Q\omega_0 L \tag{14.14}$$

14.2   Exercises

14.1 *   Basic principle

1. Given the average propagation delay through a single inverter gate is estimated as t_d. How many inverter gates are needed to design a ring oscillator working at frequency f.
   Data: $t_d = 18\,\text{ns}$, $f = 1\,\text{MHz}$.

2. Derive general conditions for oscillations in a closed-loop circuit where the forward path consists of an amplifier whose gain is A and the feedback path consists of a passive RLC circuit whose gain β.

3. Given a hypothetical simple CE amplifier, Fig. 14.1, biased at I_C whose loading resistor is R_C, estimate gain of the feedback network β if the goal is to design an oscillator. Are there any additional specifications that need to be determined?

Data: $I_C = 1\,\text{mA}$, $R_C = 1\,\text{k}\Omega$, $V_T = 25\,\text{mV}$.

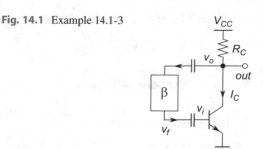

Fig. 14.1 Example 14.1-3

4. Given hypothetical oscillator in Fig. 14.2, estimate its resonant frequency ω_0. Decoupling capacitors are assumed large.

Data: $L = 2\,\mu\text{H}$, $C_1 = C_2 = 253.3\,\text{pF}$.

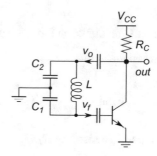

Fig. 14.2 Example 14.1-4

14.2 ** Feedback Networks

1. Given RLC resonator in Fig. 14.3 calculate its:

   (a) resonant frequency f_0,

   (b) voltage gain $\beta \stackrel{\text{def}}{=} v_{out}/v_{in}$,

   (c) dynamic resistance R_D.

   Data: $Q = 50$, $C = 76.758\,\text{pF}$, $L_1 = L_2 = 1.65\,\mu\text{H}$.

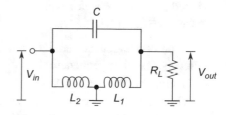

Fig. 14.3 Example 14.2-1

2. Given RLC resonator in Fig. 14.4 calculate its:

   (a) resonant frequency f_0,

   (b) voltage gain $\beta \stackrel{\text{def}}{=} v_{out}/v_{in}$,

   (c) dynamic resistance R_D.

   Data: $Q = 50$, $L = 159.15\,\text{nH}$, $C_1 = C_2 = (2 \times 159.15)\,\text{nF}$.

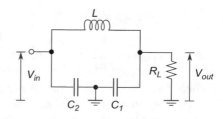

Fig. 14.4 Example 14.2-2

3. Given RLC resonator in Fig. 14.5 calculate its:

  (a) resonant frequency f_0,

  (b) voltage gain $\beta \stackrel{\text{def}}{=} v_{out}/v_{in}$,

  (c) dynamic resistance R_D.

  Data: $Q = 50$, $C = 159.15\,\text{nF}$, $L_1 = L_2 = (159.15/2)\,\text{nF}$.

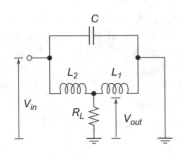

Fig. 14.5 Example 14.2-3

4. Given RLC resonator in Fig. 14.6 calculate its:

  (a) resonant frequency f_0,

  (b) voltage gain $\beta \stackrel{\text{def}}{=} v_{out}/v_{in}$,

  (c) dynamic resistance R_D.

  Data: $Q = 50$, $L = 1.5915\,\mu\text{H}$, $C_1 = C_2 = (2 \times 1.5915)\,\mu\text{F}$.

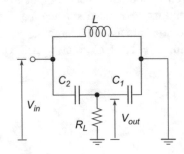

Fig. 14.6 Example 14.2-4

5. Quartz piezoelectric resonator (a.k.a. crystal) is modelled as RLC resonator with two parallel branches, see Fig. 14.7.

  (a) Given data, calculate serial f_S and parallel f_P resonant frequencies,
  (b) Calculate Q factor and BW,
  (c) While keeping same Q and r, recalculate LC values so that $f_P = 1\,\text{MHz}$, and $f_S = 998\,\text{kHz}$.

Data: $L = 1.59154943\,\text{mH}$, $r = 5\,\Omega$, $C_S = 159.793478\,\text{fF}$, $C_P = 39.8285643\,\text{pF}$.

Fig. 14.7 Example 14.2-5

14.3 **   Varicap VCO

1. Given a voltage controlled LC resonator based on varicap diode, Fig. 14.8, calculate the range of control voltages V_{ctrl} if this resonator is to be used in the commercial FM receiver. Transfer characteristics of a typical varicap diode are in Fig. 14.9.
Data: $f_{\text{FM}} = 87.5\,\text{MHz}$ to $108\,\text{MHz}$, $L = 500\,\text{nH}$, $C_p = 10\,\text{pF}$, $C_{\text{D0}} = 140\,\text{pF}$, $V_{\text{ctrl}} = (0.5 - -7\,\text{V})$, $R_S = 20\,\text{k}\Omega$

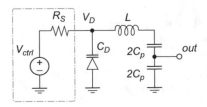

Fig. 14.8 Example 14.3-1

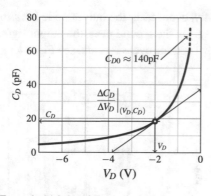
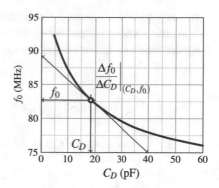

Fig. 14.9 Example 14.3-1 and Example 14.1-2

2. Voltage controlled Clapp oscillator is based on CB amplifier in the forward signal path, and voltage controlled LC resonator based on varicap diode in the feedback path, see Fig. 14.10. Assume varicap diode characteristics as in Fig. 14.9. Calculate the oscillating frequency f_0 at:

   (a) zero bias of the varicap diode,

   (b) $V_D = -7\,\text{V}$.

   Data: $L = 100\,\mu\text{H}$, $C_1 = C_2 = 300\,\text{pF}$.

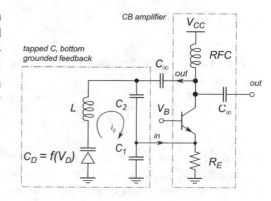

Fig. 14.10 Example 14.3-2, 3

3. Assuming LC Clapp oscillator circuit diagram in Fig. 14.10, the resonant frequency is tuned to wo by means of diode voltage V_D. The ratio of varicap to fixed capacitance in the resonator tank is n. Derive expression for and calculate the frequency deviation constant k of this oscillator.
   Data: $\omega_0 = 2\pi\,10\,\text{MHz}$, $V_0 = 6\,\text{V}$, $C_1 = C_2 = 2C_p$, $n = 0.1$.

14.4 ***   JFET oscillator

1. Design CG JFET oscillator that produces a sinusoid waveform at $f_0 = 9.545\,\text{MHz}$ and uses crystal to improve its performance.

Solutions

Exercise 14.1, page 298

1. Being the inverting amplifier, signal is inverted 180° after first inverter, then it is inverted again by the second inverter to 360° = 0°, then again to 180° by the third, and so on. In conclusion, if the number of inverters is even, i.e. $n = 2k, k = 1, 2, \ldots$, then signal's phase at the output stays as same as at the beginning of the path. Therefore, after the first signal's pass, connecting the output back to the input does not cause subsequent changes of phase along the signal path, thus there is no oscillation. However, if the number of inverters is odd, i.e. $n = 2k + 1, k = 1, 2, \ldots$, then signal's

phase at the output node is inverted. Consequently, connecting the output node back to the input of the first inverter forces another change of phase. On the second pass, the output signal changes its phase again, which completes the first period. Then the cycle repeats, see Fig. 14.11 (left).

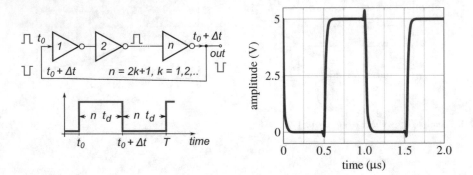

Fig. 14.11 Example 14.1-1

Period of $f = 1\,\text{MHz}$ signal equals to $T = 1/f = 1\,\mu\text{s}$. Therefore, we calculate the number of required gates as

$$T = 2n\,t_d \quad \therefore \quad n = \frac{T}{2\,t_d} = \frac{1\,\mu\text{s}}{2 \times 18\,\text{ns}} = 27.8$$

therefore, either $n = 27$ or $n = 29$ inverters must be used to create digital oscillator operating close to $1\,\text{MHz}$. Simulation of $n = 29$ shows square waveform, see Fig. 14.11 (left). Note that in simulations there is no thermal noise to start transient process, practical solution is to use `.ic` option in SPICE to set some of the internal circuit nodes to non-zero voltage level.

2. Fundamental condition for sustained circuit oscillation is the existence of a closed-loop signal path, Fig. 14.12. Once the right conditions are set, we find sustained sinusoidal waveform generated at the output node. Note that in the closed-loop circuit configuration the "input" terminal does not exist, instead there is only the "output" node.

Fig. 14.12
Example 14.1-2

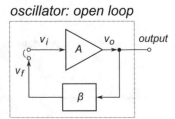

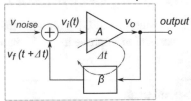

oscillator: open loop   *oscillator: clossed loop*

We start the circuit analysis with the "open-loop" configuration, Fig. 14.12 (left). Assuming existence of v_i, in the forward path it is multiplied by amplifier A and delivered to the output terminal. The amplifier is either non-inverting or inverting.[1] At the same time, there is feedback path where v_o signal is multiplied by β. The feedback network is also either non-inverting or inverting.

---

[1]This is relevant for constructive or destructive signal addition.

By inspection of Fig. 14.12 (left) we write

$$v_o = A \, v_i \tag{14.15}$$

$$v_f = \beta \, v_o = \beta \, A \, v_i \tag{14.16}$$

Therefore, a signal that follows the $(v_i \rightarrow v_o \rightarrow v_f)$ path is amplified by the "open-loop gain" A_o, i.e. ratio of the final and initial signal amplitudes, which is found from (14.16) as

$$A_o = \frac{v_f}{v_i} = A \, \beta \tag{14.17}$$

In the open-loop configuration there is nothing else to observe, as long as there is the input v_i signal it will be amplified along two paths.

However, in the closed-loop configuration where the signal follows $(v_i \rightarrow v_o \rightarrow v_f \rightarrow v_i)$ path, Fig. 14.12 (right), it is the sum of signals $v_i(t)$ and $v_f(t + \Delta t)$ that circulates along the loop and is being amplified again and again. With this understanding, by inspection of Fig. 14.12 (right) we write expression for the "closed-loop gain" A_c as

$$v_o = A(v_i + v_f) = A \, v_i + A \, \beta v_o \quad \therefore \quad A_c = \frac{v_o}{v_i} = \frac{A}{1 - \beta A} \tag{14.18}$$

Very important observation is that, as the consequence of closed-loop configuration, expression for the closed-loop gain (14.18) is *rational* function that can limit to infinity if the denominator limits to zero. Consequently, if

$$1 - \beta A = 0 \quad \therefore \quad \beta A = 1 \quad \therefore \quad A_c = \infty \tag{14.19}$$

which is to say, given right combination of A and β gains, i.e. if one is inverse of the other, there is possibility for the output signal amplitude to become infinite.

The gain condition $\beta A = 1$, however, is *not* sufficient. As the consequence of the propagation time Δt, i.e. phase, after traveling along the $A\beta$ path there are two possible outcomes:

1. Phase difference between $v_f(t + \Delta t)$ and $v_i(t)$ equals to π: in that case the two signals are inverted, therefore theirs sum equals to zero. That being the case, there is nothing to amplify and there is no waveform found at the output node.
2. Phase difference between $v_f(t + \Delta t)$ and $v_i(t)$ equals to zero, i.e. $n \times 2\pi$: in that case the two signals are aligned ("in phase") and the summing operation produces signal whose amplitude is increased. That signal is amplified again with $A\beta$, added again, amplified, etc. Therefore, after some number of "round trips", unless constrained by some other mechanism, the signal's amplitude tends to infinity.

In summary, there are *two* conditions, commonly known as "Barkhausen Stability Criterion", necessary for establishing periodic sustainable waveform that the output terminal:

$$|A \, \beta| \geq 1 \quad \text{and} \quad |\Delta\phi| = n \times 360° \tag{14.20}$$

A question to answer is: if there is no input node, where v_i is coming from? It is the internal thermal noise that provides the initial "seed" that is subsequently amplified through the loop gain.

Then, the last question is: the thermal noise frequency spectrum is flat, i.e. there are all possible frequencies present, how is it possible to produce a single sinusoid signal at the output node? It is the feedback network that serves as a narrowband filter that suppresses all tones except f_0 that is selected by the network's transfer function.

3. A CE amplifier biased at $I_C = 1\,\text{mA}$ at room temperature, i.e. $V_T \approx 25\,\text{mV}$, is therefore set to

$$g_m = \frac{1}{r_e} = \frac{I_C}{V_T} = \frac{1\,\text{mA}}{25\,\text{mV}} = 40\,\text{mS}$$

thus we estimate CE amplifier voltage gain as

$$A_V = -g_m\,R_C = -40\,\text{mS} \times 1\,\text{k}\Omega = -40$$

In accordance to (14.20), it follows that the feedback network gain should be

$$A\beta \geq 1 \quad \therefore \quad \beta \geq \frac{1}{A} \geq \frac{1}{40}$$

In practice, the βA gain is set only slightly larger than one. Additional condition for creating sustained oscillations is that the total phase shift must satisfy the $n \times 2\pi$ phase shift condition. As CE amplifier by itself is inverting, i.e. it introduces 180° phase shift, therefore, it is necessary to design the feedback network with the additional 180° shift so that the total phase shift is set to zero.

4. This circuit diagram represents oscillator whose forward signal path consists of CE amplifier, while the feedback circuit is made of L, C_1, C_2. Oscillation frequency is selected by the LC resonant frequency. By inspection of circuit diagram, we follow the resonant current i_0 inside LC loop, Fig. 14.13. From the resonant current's perspective, two capacitors C_1 and C_2 are connected in series (i.e. the same current flows through both capacitors).

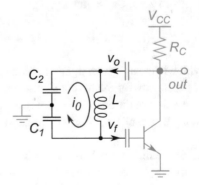

Fig. 14.13 Example 14.1-4

Given data, the equivalent capacitance C_{eq} is

$$\frac{1}{C_{\text{eq}}} = \frac{1}{C_1} + \frac{1}{C_2} \quad \therefore \quad C_{\text{eq}} = \frac{C_1 C_2}{C_1 + C_2} = \frac{253.3\,\text{pF}}{2} = 126.65\,\text{pF}$$

That is to say that the resonant frequency is

$$f_0 = \frac{1}{2\pi\sqrt{LC_{\text{eq}}}} = \frac{1}{2\pi\sqrt{2\,\mu\text{H} \times 126.65\,\text{pF}}} = 10\,\text{MHz}$$

Exercise 14.2, page 299

1. We observe that the resonant current i_0 circulates along $C \to L_1 \to L_2 \to C$ path. Therefore, we calculate as follows:

   (a) The equivalent inductance in the LC loop equals $L_{eq} = L_1 + L_2 = 3.3\,\mu H$, therefore the resonant frequency is

   $$f_0 = \frac{1}{2\pi \sqrt{(L_1 + L_2)\,C}} = \frac{1}{2\pi \sqrt{3.3\,\mu H \times 76.758\,pF}} = 10\,MHz$$

   (b) We observe that the input voltage v_{in} is found across L_2, while the output voltage v_{out} is found across L_1. By inspection, we write two equations as:

   $$\left. \begin{array}{l} v_{in} = i_0\,Z_{L_2} = i_0\,j\omega L_2 \\ v_{out} = -i_0\,Z_{L_1} = -i_0\,j\omega L_1 \end{array} \right\} \quad \therefore \quad \beta \stackrel{def}{=} \frac{v_{out}}{v_{in}} = \frac{-i_0\,j\omega L_1}{i_0\,j\omega L_2} = -\frac{L_1}{L_2}$$

   that is, the voltage gain β of a tapped L, centre-grounded feedback network is set by the inductive voltage divider.

   (c) Dynamic resistance of this LC resonator is found at the resonant frequency relative to the calculated L_{eq}, that is,

   $$R_D = Q\,\omega_0 L_{eq} = 50 \times 2\pi \times 10\,MHz \times 3.3\,\mu H = 10.3\,k\Omega$$

2. We observe that the resonant current i_0 circulates along $L \to C_1 \to C_2 \to L$ path. Therefore, we calculate as follows:

   (a) The equivalent capacitance in the LC loop equals

   $$\frac{1}{C_{eq}} = \frac{1}{C_1} + \frac{1}{C_2} = \frac{C_1 + C_2}{C_1\,C_2} \quad \therefore \quad C_{eq} = 159.15\,nF$$

   therefore the resonant frequency is

   $$f_0 = \frac{1}{2\pi \sqrt{L\,C_{eq}}} = \frac{1}{2\pi \sqrt{159.15\,nH \times 159.15\,nF}} = 1\,MHz$$

   (b) We observe that the input voltage v_{in} is found across C_2, while the output voltage v_{out} is found across C_1. By inspection, we write two equations as:

   $$\left. \begin{array}{l} v_{in} = i_0\,Z_{C_2} = i_0\,\frac{1}{j\omega C_2} \\ v_{out} = -i_0\,Z_{C_1} = -i_0\,\frac{1}{j\omega C_1} \end{array} \right\} \quad \therefore \quad \beta \stackrel{def}{=} \frac{v_{out}}{v_{in}} = \frac{-i_0\,j\omega\,C_2}{i_0\,j\omega\,C_1} = -\frac{C_2}{C_1}$$

   that is, the voltage gain β of a tapped C, centre-grounded feedback network is set by the capacitive voltage divider.

(c) Dynamic resistance of this LC resonator is found at the resonant frequency relative to the calculated L_{eq}, that is,

$$R_D = Q\,\omega_0 L = 50 \times 2\pi \times 1\,\text{MHz} \times 159.15\,\text{nH} = 50\,\Omega$$

3. We observe that the resonant current i_0 circulates along $C \to L_1 \to L_2 \to C$ path. Therefore, we calculate as follows:

(a) The equivalent inductance in the LC loop equals $L_{eq} = L_1 + L_2 = 159.15\,\text{nH}$, therefore the resonant frequency is

$$f_0 = \frac{1}{2\pi\sqrt{(L_1 + L_2)\,C}} = \frac{1}{2\pi\sqrt{159.15\,\text{nH} \times 159.15\,\text{nF}}} = 1\,\text{MHz}$$

(b) We observe that the input voltage v_{in} is found across $L_1 + L_2$, while the output voltage v_{out} is found across L_1. By inspection, we write two equations as:

$$\left.\begin{array}{l} v_{in} = i_0\,(Z_{L_1} + Z_{L_2}) = i_0\,j\omega\,(L_1 + L_2) \\ v_{out} = i_0\,Z_{L_1} = -i_0\,j\omega L_1 \end{array}\right\}$$

$$\therefore$$

$$\beta \stackrel{\text{def}}{=} \frac{v_{out}}{v_{in}} = \frac{i_0\,j\omega L_1}{i_0\,j\omega\,(L_1 + L_2)} = \frac{L_1}{L_1 + L_2}$$

that is, the voltage gain β of a tapped L, bottom-grounded feedback network is set by the inductive voltage divider.

(c) Dynamic resistance of this LC resonator is found at the resonant frequency relative to the calculated L_{eq}, that is,

$$R_D = Q\,\omega_0 L_{eq} = 50 \times 2\pi \times 1\,\text{MHz} \times 2 \times 159.15\,\text{nH} = 100\,\Omega$$

4. We observe that the resonant current i_0 circulates along $L \to C_1 \to C_2 \to L$ path. Therefore, we calculate as follows:

(a)   The equivalent capacitance C_{eq} in LC loop equals

$$\frac{1}{C_{eq}} = \frac{1}{C_1} + \frac{1}{C_2} \quad \therefore \quad C_{eq} = \frac{C_1 C_2}{C_1 + C_2} = \frac{2 \times 1.5915\,\mu F}{2} = 1.5915\,\mu F$$

That is to say that the resonant frequency is

$$f_0 = \frac{1}{2\pi \sqrt{LC_{eq}}} = \frac{1}{2\pi \sqrt{1.5915\,\mu H \times 1.5915\,\mu F}} = 100\,kHz$$

(b)   We observe that the input voltage v_{in} is found across (C_1, C_2) branch, while the output voltage v_{out} is found across C_1. By inspection, we write two equations as:

$$\left. \begin{aligned} v_{in} &= i_0\, Z_{C_{eq}} = i_0\, \frac{1}{j\omega\, \dfrac{C_1 + C_2}{C_1 C_2}} \\[2mm] v_{out} &= i_0\, Z_{C_1} = i_0\, \frac{1}{j\omega\, C_1} \end{aligned} \right\}$$

$$\therefore$$

$$\beta \stackrel{def}{=} \frac{v_{out}}{v_{in}} = \frac{i_0\, j\omega\, C_1\, C_2}{i_0\, j\omega\, C_1\, (C_1 + C_2)} = \frac{C_2}{C_1 + C_2}$$

that is, the voltage gain β of a tapped C, bottom-grounded feedback network is set by the capacitive voltage divider.

(c)   Dynamic resistance of this LC resonator is found at the resonant frequency relative to L, that is,

$$R_D = Q\,\omega_0 L = 50 \times 2\pi \times 100\,kHz \times 1.5915\,\mu H = 50\,\Omega$$

5. By inspection of Fig. 14.14 we find that in quartz piezoelectric resonator's RLC model we can follow two resonant paths:

1. serial resonance path f_S along: $A \rightarrow r \rightarrow L \rightarrow C_S \rightarrow B$, and
2. parallel resonance circular path f_P along: $\cdots \rightarrow A \rightarrow r \rightarrow L \rightarrow C_S \rightarrow B \rightarrow C_P \rightarrow A \rightarrow r \rightarrow \cdots$

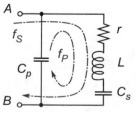

Fig. 14.14 Example 14.2-5

By definition, given data, we calculate

(a) Serial resonant frequency f_S is found as,

$$f_S = \frac{1}{2\pi\sqrt{LC_S}} = \frac{1}{2\pi\sqrt{1.59154943\,\text{mH} \times 159.793478\,\text{fF}}} = 9.980\,\text{MHz}$$

The equivalent capacitance C_{eq} in series is

$$\frac{1}{C_{eq}} = \frac{1}{C_S} + \frac{1}{C_P} = 159.154943\,\text{fF}$$

$\therefore$

$$f_P = \frac{1}{2\pi\sqrt{LC_{eq}}} = \frac{1}{2\pi\sqrt{1.59154943\,\text{mH} \times 159.154943\,\text{fFfF}}} = 10.0\,\text{MHz}$$

(b) By definition, Q factor of a series RLC network is calculated as

$$Q = \frac{\omega_0 L}{r} = \frac{2\pi \times 9.980\,\text{MHz} \times 1.59154943\,\text{mH}}{5\,\Omega} = 19,960 \approx 20,000$$

We note extremely high Q factor relative to discrete RLC resonators. It is not practically possible to manufacture large inductor (e.g. more than 1 mS) with wire resistance of only a few ohms. This resonator model shows *the equivalent* Q factor of crystal electro-mechanical resonator that is cut very precisely to produce very precise serial/parallel resonant frequencies. Serial and parallel resonant frequencies are very close to each other and are separated by only

$$\Delta f = f_P - f_S = (10.000 - 9.980)\,\text{MHz} = 20\,\text{kHz}$$

By definition, we calculate BW as

$$BW \stackrel{\text{def}}{\equiv} \frac{f_0}{Q} = \frac{10\,\text{MHz}}{20,000} = 500\text{Hz}$$

which illustrates very high frequency selectivity of RF carriers that is possible by using crystal resonators, as shown by simulations in Fig. 14.15.
We note that phase flips between $+\pi/2$ and $-\pi/2$ at the two resonant frequencies, and of course the phase equals to zero at the resonant frequencies where the resonators impedance is real.

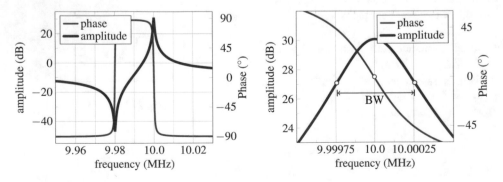

Fig. 14.15 Example 14.2-5

(c) By keeping $Q = 20,000$ and $r = 5\,\Omega$, at $f_P = 1\,\text{MHz}$, and $f_S = 998\,\text{kHz}$ we calculate

$$Q = \frac{\omega_0 L}{r} \quad \therefore \quad L = \frac{Q\,r}{2\pi\,f_S} = \frac{20,000 \times 5\,\Omega}{2\pi \times 998\,\text{kHz}} = 15.9154943\,\text{mS}$$

$$\therefore$$

$$C_S = \frac{1}{(2\pi f_S)^2\,L} = 1.59793478\,\text{pF}$$

Similarly, the parallel resonant frequency is

$$f_P = \frac{1}{2\pi\sqrt{L\,C_{eq}}} \quad \therefore \quad C_{eq} = \frac{1}{(2\pi f_P)^2\,L} = 1.59154943\,\text{pF}$$

$$\therefore$$

$$\frac{1}{C_{eq}} = \frac{1}{C_S} + \frac{1}{C_P} \quad \therefore \quad C_P = 398.285643\,\text{pF}$$

We note that because the two resonant frequencies are very close, the two serial capacitors are very different in size so that the equivalent capacitance is very close to C_S.

Exercise 14.3, page 300

1. By inspection of Fig. 14.8 we conclude that the resonant current flows through three capacitors connected in series: $2C_p$, $2C_p$, and C_D. Furthermore, the two $2C_p$ capacitors in series are equal, consequently their equivalent capacitance is half, i.e. C_p. Which is to say that the resonant frequency is determined by inductor L and C_{eq}, i.e.

$$C_{eq} = \frac{1}{L\omega_0^2} \quad \therefore \quad \frac{C_p C_D}{C_p + C_D} = \frac{1}{L\omega_0^2} \quad \therefore \quad C_D = \frac{C_p}{C_p L\omega_0^2 - 1} = \frac{C_p}{C_p L(2\pi\,f_0)^2 - 1}$$

Given data, we calculate at $f_0 = 87.5\,\text{MHz}$ capacitance $C_D = 19.559\,\text{pF}$, and at $f_0 = 108\,\text{MHz}$ capacitance $C_D = 7.678\,\text{pF}$.

To set these two capacitance values, from a typical varicap diode transfer function in Fig. 14.9 (left) we derive that

$$C_D = \frac{C_{D0}}{(1 - 2V_D)^{\frac{5}{4}}} \quad \therefore \quad V_D = \frac{1}{2}\left[1 - \left(\frac{C_{D0}}{C_D}\right)^{\frac{4}{5}}\right]$$

Thus, to set $C_D = 7.678\,\text{pF}$ the diode must be biased with $V_D = -4.601\,\text{V}$, and to set $C_D = 19.559\,\text{pF}$ the diode must be biased with $V_D = -1.914\,\text{V}$. AC simulation of circuit in Fig. 14.8 illustrates these results, see Fig. 14.16.

Fig. 14.16
Example 14.3-1

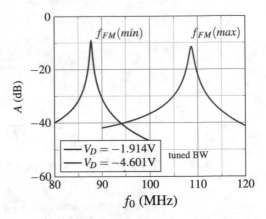

2. By inspection of Fig. 14.8 we conclude that from the resonating current i_0 perspective there are three capacitors on series: C_1, C_2, and C_D, therefore the equivalent resonator's capacitance is

$$\frac{1}{C_{\text{eq}}} = \frac{1}{C_1} + \frac{1}{C_2} + \frac{1}{C_D} \tag{14.21}$$

where, the varicap diode capacitance C_D is calculated as

$$C_D = \frac{C_{D0}}{(1 - 2V_D)^{\frac{5}{4}}} \tag{14.22}$$

From graph in Fig. 14.9 we find $C_{D0} = 140\,\text{pF}$, then

(a) Given diode voltage $V_D = 0$ we calculate

$$C_D = C_{D0} = 140\,\text{pF} \quad \therefore \quad C_{\text{eq}} = 72.4\,\text{pF} \quad \therefore \quad f_0 = \frac{1}{2\pi\,L\,C_{\text{eq}}} = 1.87\,\text{MHz}$$

(b) Given diode voltage $V_D = -7$, from (14.21) and (14.22), we calculate

$$C_D = C_{D0} = 4.75\,\text{pF} \quad \therefore \quad C_{\text{eq}} = 4.6\,\text{pF} \quad \therefore \quad f_0 = \frac{1}{2\pi\,L\,C_{\text{eq}}} = 7.4\,\text{MHz}$$

At zero bias voltage $C = 17.65\,\text{pF}$, therefore $f_0 = 3.789\,\text{MHz}$. With the biasing $V_D = -7\,\text{V}$ it follows that $C = 5\,\text{pF}$, therefore $f_0 = 7.126\,\text{MHz}$.

3. By inspection of circuit diagram in Fig. 14.10 we conclude that, from the perspective of the resonant current i_0, the three capacitors in LC loop are connected in series, thus we write

$$\omega_0 = \frac{1}{\sqrt{LC_{eq}}} = \frac{1}{\sqrt{L\,\frac{C_p C_D}{C_p + C_D}}} = \sqrt{\frac{C_p + C_D}{L\,C_p C_D}} \quad \therefore \quad L = \frac{C_p + C_D}{\omega_0^2\,C_p C_D} \tag{14.23}$$

With the help of (14.23) we find the ω_0 vs. C_D sensitivity is as follows:

$$\frac{\partial \omega_0}{\partial C_D} = \frac{\partial}{\partial C_D}\left(\sqrt{\frac{C_p + C_D}{L\, C_p\, C_D}}\right) = \frac{1}{2}\frac{1}{\sqrt{\frac{C_p+C_D}{L C_p C_D}}}\frac{L\,C_p C_D - (C_p + C_D)\,L\,C_p}{(L\,C_p C_D)^2}$$

$$= -\frac{1}{2}\frac{1}{\omega_0}\frac{1}{L\,C_D^2} = -\frac{1}{2}\frac{1}{\omega_0}\frac{1}{C_D^2}\frac{\omega_0^2\,C_p\,C_D}{C_p + C_D} == -\frac{1}{2}\frac{\omega_0}{C_D}\frac{C_p}{C_p + C_D} \tag{14.24}$$

where, it is convenient to use the ratio of varicap and fixed capacitances as $n = C_D/C_p$. Therefore, after substituting $\omega_0 = 2\pi f_0$ on both sides of (14.24) we find

$$\frac{\partial f_0}{\partial C_D} = -\frac{1}{2}\frac{f_0}{C_D}\frac{1}{1+n} \tag{14.25}$$

Assuming varicap diode model (14.22), we can analytically derive the varicap capacitance versus its biasing voltage as follows:

$$\frac{\partial C_D}{\partial V_D} = \frac{5}{2}\frac{C_{D0}}{(1-2V_D)^{\frac{9}{4}}} = \frac{5}{2}\underbrace{\frac{C_{D0}}{(1-2V_D)^{\frac{5}{4}}}}_{C_D}\frac{1}{(1-2V_D)} = \frac{5}{2}\frac{C_D}{(1-2V_D)} \tag{14.26}$$

In reality, a VCO functions as a voltage-to-frequency converter. That being the case, sensitivity of the output frequency relative to changes variations of varicap's biasing voltage V_D must be derived. Knowing (14.25) and (14.26), we write

$$k = \frac{\partial f_0}{\partial V_D}\bigg|_{V_0,C_0} = \frac{\partial f_0}{\partial C_D}\frac{\partial C_D}{\partial V_D} = \left[-\frac{1}{2}\frac{f_0}{C_D}\frac{1}{1+n}\right]\left[\frac{5}{2}\frac{C_D}{(1-2V_D)}\right]$$

$$= -\frac{5}{4}\frac{f_0}{(1+n)(1-2V_D)}$$

Given data, we calculate

$$k = -\frac{5}{4}\frac{f_0}{(1+n)(1-2V_D)} = -\frac{5}{4}\frac{10\,\text{MHz}}{(1+0.1)(1+2\times 6\,\text{V})} = -874\frac{\text{kHz}}{\text{V}}$$

We conclude that, for each volt of increase/decrease in varicap bias, there is decrease/increase in the oscillator resonant frequency by about 874 kHz, thus this circuit is an example of VCO.

Exercise 14.4, page 301

1. One possible design flow procedure, where we reuse JFET characteristics studied in the previous examples, may be as follows:

   (a) *Power supply:* is chosen at the system level, for example, $V_{DD} = 15\,\text{V}$.
   (b) *Transistor:* we reuse JFET transistor from previous examples whose DC characteristics and biasing point are $(I_{DSS}, V_P) = (12\,\text{mA}, -3\,\text{V})$, $(I_D, V_{GS}) = (6\,\text{mA}, -884\,\text{mV})$, $g_m = 1/r_s = 5.67\,\text{mS}$, $r_s = 176.3\,\Omega$, and $r_o = 77\,\text{k}\Omega$. JFET biasing point is set by R_S resistor, given (I_D, V_{GS}) we calculate

$$V_{\mathrm{GSO}} = -V_S = R_S I_{\mathrm{D0}} \quad \therefore \quad R_S = 147\,\Omega$$

(c) *Inductor:* The choice of inductor whose wire resistance is r sets on Q factor, therefore on BW. These two specifications, Q and BW, are determined at the system level. For example, we choose commercial $L = 3.3\,\mu\mathrm{H}$ inductor whose $Q = 10.2$. The resonant frequency $f_0 = 1/(2\pi\sqrt{LC}) = 9.545\,\mathrm{MHz}$ requires that $C = 84.2508824\,\mathrm{pF}$. This capacitance value is achieved by tuning trimmer C_{trim} in parallel with fixed C_0 capacitor, i.e. $C = C_{\mathrm{trim}} + C_0$. Knowing Q factor and f_0 we calculate

$$BW = \frac{f_0}{Q} = \frac{9.545\,\mathrm{MHz}}{10.2} = 936\,\mathrm{kHz}$$

Then, the equivalent parallel dynamic resistance R_D of this inductor is calculated as

$$R_D = Q\,2\pi\,f_0\,L = 10.2 \times 2\pi \times 9.545\,\mathrm{MHz} \times 3.3\,\mu\mathrm{H} = 2019\,\Omega \approx 2\,\mathrm{k}\Omega$$

(d) *LC Resonator:* any of the LC networks studied in Example 14.1-1 to Example 14.1-4 could be used. Let us choose, for example, tapped C, bottom-grounded feedback network, see Fig. 14.6.

(e) *Feedback loop calculations:* the forward path amplifier whose gain is A and the feedback path network whose gain is β must satisfy the Barkhausen gain and phase criterion. From JFET perspective, there is R_{eff} load connected to its drain. We note that $R_{\mathrm{eff}} \neq R_D$ because there is resistance R_{in} loading the resonator's network. Therefore, we find R_{in} resistance at source node, as

$$R_{\mathrm{in}} = R_S || r_s = 147\,\Omega || 176\,\Omega = 80\,\Omega$$

Fig. 14.17
Example 14.4-1

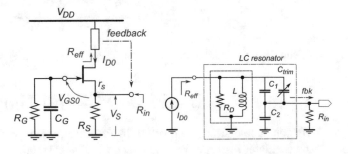

This design is iterative process, for example, we can start by assuming a relatively modest voltage gain of CG amplifier. This assumption is valid, because the LC feedback resonator is loaded by the rather low $R_{\mathrm{in}} = 80\,\Omega$, thus its effective resistance R_{eff} is reduced, which subsequently limits the overall voltage gain (Fig. 14.17).

Let us assume gain $A_V = 5$, consequently, in order to satisfy the Barkhausen gain criterion it follows that the feedback gain must be $\beta \geq 1/A_V = 0.2$, thus we calculate as follows: Parameters of tapped C, bottom-grounded feedback network are

$$\beta = \frac{C_2}{C_1 + C_2} \quad ; \quad \omega_0^2 = \frac{C_1 + C_2}{LC_1 C_2} = \frac{1}{L\,C_1\,\beta} \quad ; \quad R_{\mathrm{eff}} = \frac{R_{\mathrm{in}}}{\beta^2} \, || \, Q\omega_0 L$$

where, the effective resistance of loaded resonator R_{eff} is derived in the textbook. Therefore, we calculate

$$\omega_0^2 = \frac{C_1 + C_2}{LC_1C_2} = \frac{1}{LC_1\beta}$$

$$\therefore$$

$$C_1 = \frac{1}{L\omega_0^2\beta} = 421.25441\,\text{pF} \quad \therefore \quad C_2 = \frac{1}{\frac{1}{\beta} - 1} = 105.31360\,\text{pF}$$

$$R_{\text{eff}} = \frac{R_{\text{in}}}{\beta^2} \parallel Q\omega_0 L = 2\,\text{k}\Omega \parallel 2\,\text{k}\Omega = 1\,\text{k}\Omega$$

(f) *Open-loop simulation* technique is used to verify that the amplifier indeed perceives dynamic resistance $R_D = 1\,\text{k}\Omega$, and provides the gain as expected. This test is done in the "open-loop" configurations, see Fig. 14.18!(left). By inspection, we calculate voltage gain as

$$A_v = A_1 \times A_2 = \frac{R_{\text{in}}}{R_{\text{x}} + R_{\text{in}}} \times g_m\,(R_D \| r_o)$$

$$= \frac{80\,\Omega}{80\,\Omega + 80\,\Omega} \times 5.67\,\text{mS}\,(2\,\text{k}\Omega \| 77\,\text{k}\Omega) \approx 0.5 \times 11 = 5.5$$

where, for example, the signal generator's resistance is $R_{\text{x}} = 80\,\Omega$ and $V_x = 100\,\text{mV}$. As originally assumed, this voltage gain is not too high (it is also function of the source resistance), as confirmed by simulations, Fig. 14.18 (right). We note that due to LC elements, there is about $1.5\,\mu\text{s}$ transition period before the full signal amplitude at the output is reached. In addition, being an RF amplifier, average of the output voltage signal is at V_{DD} and the maximum voltage is above the power supply rail (Through L, there is DC connection between drain and power supply nodes.).

Fig. 14.18
Example 14.4-1

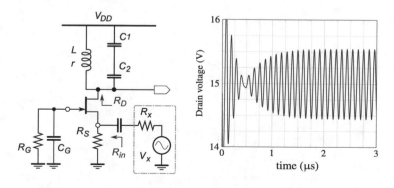

(g) *Closed loop simulations*: simply closing the loop with a wire connection between resonator's output node (between C_1 and C_2) and JFET source node completes the oscillator design. Correct functionality is verified by (longer) simulations and Fourier transformation of time domain waveform. In order to reduce phase jitter and improve the stability of the resonant frequency, discrete crystal component is inserted in the feedback path. Following

Example 14.1-5, model of quartz crystal whose *serial* resonant frequency is $f_S = 9.545$ MHz is inserted instead of the wire feedback path, Fig. 14.19 (left).

We note the quartz model whose $Q \approx 24{,}000$ causes extremely narrowband frequency response, see approximate response in Fig. 14.19 (left). As a direct consequence of this high Q factor, the oscillator needs about 25 ms to finish its transition period and reach full signal amplitude.

In the practical experiment, adding this component and observing the response on spectrum analyser is not difficult at all. However, to produce a good frequency response curve by Fourier analysis of transient data, it is necessary that the simulation is run for a long time, thus the resulting data file is accordingly large. The problem is that the resolution of the produced data file must be sufficient so that SPICE can separate the serial and parallel frequency modes of the crystal. Approximated plot of the oscillator transient domain simulations with and without crystal are shown in Fig. 14.19 (right).

As a closing note, by choosing large R_G and C_G we set very low pole frequency at the gate node, that is to say, very good small signal ground, see Example 5.1-3.

Fig. 14.19
Example 14.4-1

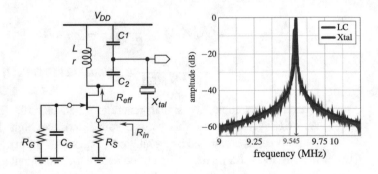

Frequency Shifting

15

In this chapter, we focus on the mathematical operation of "frequency shifting" that is fundamental to wireless communication systems. Frequency shifting (or "frequency translation") is complementary to the frequency tuning mechanism used in VCOs. However, as will be shown, it is a much broader concept with a much wider range of applications. As it turns out, mathematical multiplication of two sinusoidal waveforms with given frequencies results in waveforms that contain both higher and lower frequencies. This phenomenon is known as "frequency shifting", where the term "up-conversion" refers to the process of shifting of lower frequency tone to the upper frequency range (used in RF transmitters), while "down-conversion" refers to the frequency shifting from higher to lower frequency ranges (used in RF receivers). Hence, in a complete wireless communication system, the information-carrying signal is shifted in both directions.

15.1 Important to Know

1. Trigonometric identities

$$\cos \alpha \cos \beta = \frac{1}{2} \left(\cos(\alpha - \beta) + \cos(\alpha + \beta) \right) \tag{15.1}$$

$$\sin \alpha \sin \beta = \frac{1}{2} \left(\cos(\alpha - \beta) - \cos(\alpha + \beta) \right) \tag{15.2}$$

$$\sin \alpha \cos \beta = \frac{1}{2} \left(\sin(\alpha + \beta) + \sin(\alpha - \beta) \right) \tag{15.3}$$

$$\cos \alpha \sin \beta = \frac{1}{2} \left(\sin(\alpha + \beta) - \sin(\alpha - \beta) \right) \tag{15.4}$$

2. LC resonator's attenuation

$$A_r \triangleq \frac{|V_0|}{V_0(\omega_0)} = \frac{1}{\sqrt{1 + (\delta Q)^2}} \tag{15.5}$$

$$\delta = \frac{\omega}{\omega_0} - \frac{\omega_0}{\omega} \tag{15.6}$$

© Springer Nature Switzerland AG 2021
R. Sobot, *Wireless Communication Electronics by Example*,
https://doi.org/10.1007/978-3-030-59498-5_15

15.2 Exercises

15.1 *   Frequency shifting

1. Given two sine waveforms, f_1 and f_2, find waveform generated at the output of frequency multiplier circuit. **Data:** $f_1 = 10\,\text{MHz}$, $f_2 = 9.545\,\text{MHz}$.

2. Given two sine waveforms, f_1 and f_2 at the input of frequency multiplier:

   (a) Derive analytical model of ideal frequency multiplier that generates only one up-converted sine waveform at its output.
   (b) Create behavioural model that implements equations used in 2 and illustrate the analytical result by simulations.

   Data: $f_1 = 10\,\text{kHz}$, $f_2 = 13\,\text{kHz}$.

3. Given $s(t) = \sin(2\pi \times 10\,\text{MHz} \times t)$, find *two* other sine functions that could be used to generate a single sinusoid at $f = 1\,\text{kHz}$. Explain the process and the result.

4. A radio receiver is tuned to receive AM modulated wave transmitted at carrier frequency of f_{RF}. Local oscillator inside the receiver is set at f_{LO}.

   (a) Calculate frequencies coming out of the receiver's mixer,
   (b) Which one is IF frequency?
   (c) Calculate frequency of radio station which would be transmitting at the image frequency.
   (d) Show the frequency spectrum graph.

   Data: $f_{RF} = 980\,\text{kHz}$, $f_{LO} = 1435\,\text{kHz}$.

15.2 *   Image suppression

1. Calculate resonant frequency f_0 of a serial RLC network. Then, calculate its gain at given frequency f that is close to f_0.
   Data: $R = 1.73\,\Omega$, $L = 3\,\text{mH}$, $C = 100\,\text{nF}$. $f = 10\,\text{kHz}$.

2. RLC resonator of AM receiver is tuned to resonant frequency f_0. Calculate signal rejection in dB of unwanted signal being transmitted at frequency $f \neq f_0$.
   Data: $f_0 = 500\,\text{kHz}$ $Q = 50$, $f = 1430\,\text{kHz}$

3. An RF amplifier has RLC tank and is tuned at RF frequency f_0. Estimate attenuation of the image signal, if the image frequency is 10% higher than the RF signal. **Data:** $Q = 20$.

15.3 **   Mixers

1. Illustrate the operation of a diode mixer and explain how its non-linear function is used to implement the frequency multiplication operation.

2. Practical mixer may be created by JFET transistors.

   (a) Explain how non-linear function JFET is used to implement the frequency multiplication operation.
   (b) Design JFET mixer based on cascaded (i.e. "dual–gate") transistor that is to be used to down-convert $f_{RF} = 10\,\text{MHz}$ RF carrier to $f_{IF} = 455\,\text{kHz}$ IF frequency and the input side $BW = 10\,\text{kHz}$.

Solutions

Exercise 15.1, page 316

1. Assuming "true" sinusoids, where the initial phase is not critical for the final result, the ideal frequency shifting is achieved by trigonometric identity for product of two sine functions,

$$s(1) \times s(2) = \sin(\omega_1 t) \sin(\omega_2 t)$$

$$= \frac{1}{2} \left(\cos(\omega_1 - \omega_2) t - \cos(\omega_1 + \omega_2) t \right)$$

$$= \frac{1}{2} \left[\cos 2\pi (10\,\mathrm{MHz} - 9.545\,\mathrm{MHz}) t \right.$$

$$\left. - \cos 2\pi (10\,\mathrm{MHz} + 9.545\,\mathrm{MHz}) t \right]$$

$$= \left(\frac{1}{2} \cos 2\pi (455\,\mathrm{kHz}) t - \frac{1}{2} \cos 2\pi (19.545\,\mathrm{MHz}) t \right)$$

$$= (f_{\mathrm{DIF}} + f_{\mathrm{SUM}})$$

That is to say, multiplication of two individual sine waveforms with f_1 and f_2 generates one waveform at the output whose spectrum contains "shifted" waveforms f_{DIF} and f_{SUM}. The frequency shifting is consequence of the *non-linear* operation—that is main difference between linear operation like the addition, see Example 10.2–1 and Fig. 10.5.

2. In order to eliminate f_{DIF} part of trigonometric identity for product of two sine functions, we exploit its symmetry with the equivalent identity for product of two cos functions. Starting with two sine functions

$$s(1) = \sin \omega_1 t$$

$$s(2) = \sin \omega_2 t$$

we proceed as follows:

(a) Generate two equivalent cos functions with ω_1 and ω_2. This is achieved by *phase shifter* that adds $\pi/2$ phase to $s(1)$ and $s(2)$ as

$$s(3) = \sin \left(\omega_1 t + \frac{\pi}{2} \right) = \cos \omega_1 t$$

$$s(4) = \sin \left(\omega_2 t + \frac{\pi}{2} \right) = \cos \omega_2 t$$

We recall that "phase shifting" is as same as "time delay" relative to the sine period.

(b)   Products of sin and cos pairs are

$$s(1) \times s(2) = \sin \omega_1 t \times \sin \omega_2 t$$

$$= \frac{1}{2} \left[\cos(\omega_1 - \omega_2)t - \cos(\omega_1 + \omega_2)t \right] \tag{15.7}$$

$$s(3) \times s(4) = \cos \omega_1 t) \times \cos \omega_2 t$$

$$= \frac{1}{2} \left[\cos(\omega_1 - \omega_2)t + \cos(\omega_1 + \omega_2)t \right] \tag{15.8}$$

Difference between (15.7) and (15.8) gives

$$[s(3) \times s(4)] - [s(1) \times s(2)] = \cos(\omega_1 + \omega_2)t \tag{15.9}$$

Where (15.9) is analytical model of ideal up-converter whose block diagram and I/O
spectrum is shown in Fig. 15.1.

In summary, in order to implement up/down-converter that generates only either f_{SUM} (a.k.a.
"upper-side band"—USB) or f_{DIF} (a.k.a. "lower-side band"—LSB) waveforms, in total we need
to design five circuits: two phase-shifters (i.e. time delay), two multipliers, and one adder (with
the inverting and non-inverting inputs).

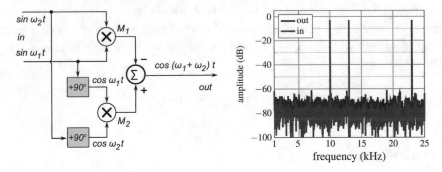

Fig. 15.1   Example 15.1–2

3. Given waveform whose frequency is $f_1 = 10\,\text{MHz}$, there are two other waveforms whose
   frequencies are separated by 1kHz: on the lower side there is 9.999 MHz and on the upper side
   there is 10.001 MHz. In both cases we use the multiplying operation to generate

$$\sin \omega_1 t \times \sin \omega_2 t = \frac{1}{2} \left[\cos((10 - 9.999)\,\text{MHz})\,t - \cos((10 + 9.999)\,\text{MHz})\,t \right]$$

$$= \frac{1}{2} \left[\cos(1\,\text{kHz})t - \cos(19.999\,\text{MHz})t \right]$$

and

$$\sin \omega_1 t \times \sin \omega_2 t = \frac{1}{2} \left[\cos((10 - 10.001)\,\text{MHz})\,t - \cos((10 + 10.001)\,\text{MHz})\,t \right]$$

$$= \frac{1}{2} \ [\cos(1\,\mathrm{kHz})t - \cos(20.001\,\mathrm{MHz})t]$$

In both cases, there is LSB at 1 kHz, while two USB frequencies are separated by 2 kHz at 19.999 MHz and 20.001 MHz. Consequently, if looking only at LSB it is not possible to know which one of the two frequencies, 9.999 MHz or 10.001 MHz, is multiplied with the 10 MHz waveform. Even worse, it may be the case that both of them found their way to enter the multiplier and generated multiple sum/difference combinations.

This example illustrates the fundamental consequence of the multiplying operation. Relative to 10 MHz waveform the other two are often referred to as one being "ghost" (or, "image") of the other. The "image" expression evokes the analogy of mirror surface (here, 10 MHz waveform) and two images, one in front of mirror (the "real" one) and one behind the mirror (the "image"). Filtering and suppressing "ghost" frequencies constitute large percentage of RF circuit design.

4. Given data, we calculate

   (a) Mixing f_{LO} and f_{RF} generates the sum and difference frequencies, i.e. $f_{\mathrm{SUM}} = 1435\,\mathrm{kHz} + 980\,\mathrm{kHz} = 2.415\,\mathrm{MHz}$ and $f_{\mathrm{DIF}} = 1435\,\mathrm{kHz} - 980\,\mathrm{kHz} = 455\,\mathrm{kHz}$.

   (b) Standard IF in AM radio communication systems is the down-converted frequency $f_{\mathrm{IF}} = 455\,\mathrm{kHz}$. Therefore, USB HF component generated by mixer is removed by filtering.

   (c) Relative to f_{RF}, aside from the f_{LO} frequency that is used to generate f_{IF}, it is possible to generate 455 kHz if

   $$f_{\mathrm{image}} = f_{\mathrm{LO}} + f_{\mathrm{IF}} = 1435\,\mathrm{kHz} + 455\,\mathrm{kHz} = 1.890\,\mathrm{MHz}$$

   That is to say, if there is another RF station transmitting at 1.890 MHz and if its signal reaches the mixer's input, then its down-converted version is found at the same $f_{\mathrm{IF}} = 455\,\mathrm{kHz}$ as the first RF station, therefore the two RF transmissions would be received and irrevocably overlapped after the receiver's mixer output.

   (d) Frequency spectrum of all involved frequencies illustrates the system level schema, Fig. 15.2.

Fig. 15.2
Example 15.1–4

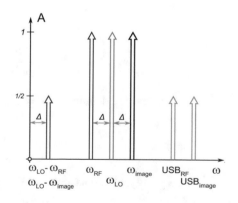

Exercise 15.2, page 316

1. By definition, RLC resonant frequency is

$$f_0 = \frac{1}{2\pi\sqrt{LC}} = 9.189\,\text{kHz}$$

We calculate Q_S factor of this serial RLC network as

$$Q_S = \frac{\omega_0 L}{R} = \frac{2\pi \times 9.189\,\text{kHz} \times 3\,\text{mH}}{1.73\,\Omega} = 100$$

At resonance, $\omega_0 = 1/\sqrt{LC}$, thus given alternative Q factor identities

$$Q_S = \frac{\omega_0 L}{R} = \frac{\omega_0}{R}\frac{1}{\omega_0^2 C} = \frac{1}{\omega_0 RC} \tag{15.10}$$

from (15.10), we write the following two relations:

$$\frac{Q_S}{\omega_0} = \frac{L}{R} \quad\text{and}\quad Q_S\omega_0 = \frac{1}{RC}$$

In general, Q factor is considered high—most approximations that are used to simplify RLC resonator analysis assume $Q > 10$ so that serial and parallel RLC networks are interchangeable, i.e. $Q = Q_S = Q_P$. That being the case, at resonance serial to parallel RLC network transformations are

$$X_S \approx X_P \quad\text{and}\quad R_P \approx QR_S$$

At non-resonant frequencies, admittance of parallel RLC network is written by inspection as

$$Y(\omega) = \frac{1}{R + j\omega L} + j\omega C = \frac{R - j\omega L}{R^2 + (\omega L)^2} + j\omega C$$

$$= \frac{R}{R^2 + (\omega L)^2} + j\left(\omega C - \frac{\omega L}{R^2 + (\omega L)^2}\right)$$

$$\approx \frac{R}{(\omega L)^2} + j\left(\omega C - \frac{\omega L}{(\omega L)^2}\right) \tag{15.11}$$

At resonance, by definition $X_L = X_C$, thus

$$\omega_0 C - \frac{\omega_0 L}{(\omega_0 L)^2} = 0 \quad \therefore \quad \omega_0 = \sqrt{\frac{1}{LC} - \frac{R^2}{L^2}}$$

Consequently, at resonance (15.11) becomes

$$Y(\omega_0) = \frac{1}{R_D} \approx \frac{R}{(\omega_0 L)^2} = \frac{R}{\frac{1}{LC}L^2} = \frac{RC}{L} \tag{15.12}$$

Ratio of admittances at frequencies ω and ω_0 gives

$$
\begin{aligned}
\frac{Y(\omega)}{Y(\omega_0)} &\approx \frac{L}{RC}\left[\frac{R}{(\omega L)^2} + j\left(\omega C - \frac{1}{\omega L}\right)\right] \\
&= \frac{1}{\omega^2 LC} + j\left(\omega\frac{L}{R} - \frac{1}{\omega}\frac{1}{RC}\right) \\
&= \frac{\omega_0^2}{\omega^2} + j\left(\omega\frac{Q}{\omega_0} - \frac{Q\omega_0}{\omega}\right) \\
&= \frac{\omega_0^2}{\omega^2} + jQ\left(\frac{\omega}{\omega_0} - \frac{\omega_0}{\omega}\right) \\
&= \frac{\omega_0^2}{\omega^2} + j\,\delta\,Q \quad \text{where,} \quad \delta = \frac{\omega}{\omega_0} - \frac{\omega_0}{\omega}
\end{aligned}
\tag{15.13}
$$

This ratio is a complex function, therefore, from (15.13) we write

$$
\left|\frac{Y(\omega)}{Y(\omega_0)}\right| = \sqrt{\left(\frac{\omega_0}{\omega}\right)^4 + (\delta\,Q)^2} \approx \sqrt{1 + (\delta\,Q)^2}
$$

$$
\therefore
$$

$$
|Y(\omega)| \approx Y(\omega_0)\sqrt{1 + (\delta\,Q)^2}
\tag{15.14}
$$

in close proximity to ω_0 where $\omega \approx \omega_0$. In general, this assumption is valid for all frequencies within the RLC bandwidth centred at ω_0. Assuming the signal current I_S, we deliver the voltage gain equation by rearranging (15.14) as follows:

$$
V(\omega) = \frac{I_S}{Y(\omega)} = \frac{I_S}{Y(\omega_0)\sqrt{1 + (\delta\,Q)^2}}
\tag{15.15}
$$

Also, at $\omega = \omega_0$ from (15.13) it follows that $\delta = 0$, thus

$$
V(\omega_0) = \frac{I_S}{Y(\omega_0)}
\tag{15.16}
$$

Voltage gain/attenuation A_V is calculated by definition as the ratio of two voltages (15.15) and (15.16), i.e.

$$
A_V \overset{\text{def}}{=} \frac{V(\omega)}{V(\omega_0)} = \frac{\dfrac{I_S}{Y(\omega_0)\sqrt{1 + (\delta\,Q)^2}}}{\dfrac{I_S}{Y(\omega_0)}} = \frac{1}{\sqrt{1 + (\delta\,Q)^2}}
\tag{15.17}
$$

Given data, we calculate

$$
Q = Q_S = Q_P = 100
$$

$$\delta = \frac{\omega}{\omega_0} - \frac{\omega_0}{\omega} = \frac{10\,\text{kHz}}{9.189\,\text{kHz}} - \frac{9.189\,\text{kHz}}{10\,\text{kHz}} = 169.358 \times 10^{-3}$$

$$\therefore \quad \text{(from (15.17))}$$

$$A_V = \frac{1}{\sqrt{1 + (\delta\,Q)^2}} \approx \frac{1}{17} = -24.4\,\text{dB}$$

The frequency profile of this RLC resonator, Fig. 15.3, illustrates how depending on Q factor, tones close to the resonant frequency are attenuated. Because Q factor of resonators is not infinite, i.e. there is finite BW that allows for the non-resonant tones, it is necessary to control the allowed power of these image signals. The amount of image suppression is usually specified by the communication standard.

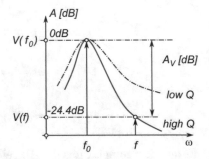

Fig. 15.3 Example 15.2–1

2. Obviously, because the front-end LC tank is tuned at f_0, the RF radio transmitter emitting at that frequency is the desired signal for this receiver. Thus, by reusing (15.13) it follows that

$$\delta\,Q = \left(\frac{\omega}{\omega_0} - \frac{\omega_0}{\omega} \right) Q = \left(\frac{1430}{500} - \frac{500}{1430} \right) 50 = 126$$

which, after substituting in (15.14) and approximating $\sqrt{1 + 126^2} \approx 126$, yields

$$A_V = 20 \log \frac{1}{126} = -42\,\text{dB}$$

Therefore, if a second radio station is transmitting at 1430 kHz, its signal is received as 126 times weaker than the signal from the desired radio station. As a note, using two tuned amplifiers would double the selectivity and further suppress the image signal down to $-84\,\text{dB}$.

3. Given $Q = 20$, a waveform close to the resonant frequency f_0, where the interfering RF signal is 10% higher, that is to say $f = 1.1\,f_0$. By using (15.13) we write

$$\delta\,Q = \left(\frac{f}{f_0} - \frac{f_0}{f} \right) Q = \left(\frac{1.1\,f_0}{f_0} - \frac{f_0}{1.1\,f_0} \right) Q \approx 0.191\,Q$$

$$\approx 3.82$$

Which after being substituted into (15.14) results

$$A_V = \frac{1}{\sqrt{1 + (\delta\,Q)^2}} = 0.253 \approx -12\,\text{dB}$$

Exercise 15.3, page 316

1. Any non-linear element, therefore diodes and transistors, can be used to implement the frequency shifting operation. For example, voltage/current relationship of a resistor is linear because $V = R \times I$, where R is simply the proportionality constant. If, for example, $I = I_1 + I_2$, then

$$V = R \times I = R \times (I_1 + I_2) = R \times I_1 + R \times I_2$$

which is still linear voltage/current relationship.

However, current/voltage relationship of FET transistor is quadratic because $I_D = f(V_{GS}{}^2)$. For example, if $V_{GS} = V_{GS1} + V_{GS2}$, then

$$I_D = f(V_{GS}{}^2) = f((V_{GS1} + V_{GS2})^2) = f(V_{GS1}{}^2 + 2V_{GS1} V_{GS2} + V_{GS1}{}^2)$$

as it is well known result from basic algebra. The main point is that among the resulting three terms there is one term that calculates the product of two input voltages V_{GS1} and V_{GS2}. The other two terms can be eliminated by some means.

Similarly, current/voltage relationship of diode and BJT transistor is exponential because $I_C \propto \exp(V_{BE})$. In this case, in order to find the term that calculates product of V_{GS1} and V_{GS2}, first we convert a non-linear function into a polynomial function by well known calculus technique of development into the Taylor series of exponential function as

$$e^x = \sum_{n=0}^{\infty} \frac{x^n}{n!} = 1 + x + \frac{x^2}{2} + \frac{x^3}{6} + \frac{x^4}{24} + \cdots \qquad (15.18)$$

where, the first term is constant (i.e. DC), the second term is linear, and the third term is quadratic—it is the non-linear term to be used. The remaining terms are to be cancelled by some means.

We start with two voltage single-tone signals, see Fig. 15.4,

$$v_1 = V_1 \cos(\omega_1 t) \qquad (15.19)$$

$$v_2 = V_2 \cos(\omega_2 t) \qquad (15.20)$$

that are first added and then passed through ideal diode whose voltage/current function is given as

$$i_D = I_S \left[\exp\left(\frac{v_D}{V_t}\right) - 1 \right] \qquad (15.21)$$

which is non-linear function.

In the following analysis, for the simplicity, we assume a small diode current and ignore the voltage drop across the loading resistor R_L, Fig. 15.4. That is, the diode voltage V_D is approximately equal to the voltage at node ①, i.e. $V_D \approx V(1)$.

Fig. 15.4
Example 15.3–1

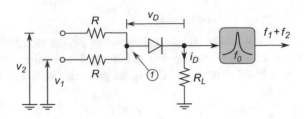

Assuming two equal resistors R that serve as a linear voltage adder, by inspection of schematic diagram Fig. 15.4, we write

$$v_D = v_1 = \frac{1}{2}(v_1 + v_2) = \frac{1}{2}[V_1 \cdot \cos(\omega_1 t) + V_2 \cdot \cos(\omega_2 t)] \qquad (15.22)$$

After Taylor development (15.18) of exponential function in (15.21) and substituting (15.22) we derive

$$i_D = I_S \left\{ \left[1 + \frac{v_D}{V_t} + \frac{1}{2}\left(\frac{v_D}{V_t}\right)^2 + \frac{1}{6}\left(\frac{v_D}{V_t}\right)^3 + \frac{1}{24}\left(\frac{v_D}{V_t}\right)^4 + \cdots \right] - 1 \right\} \qquad (15.23)$$

We focus on the first two terms:

1. The linear term:

$$\frac{v_D}{V_t} = \frac{1}{2V_t}[V_1 \cdot \cos(\omega_1 t) + V_2 \cdot \cos(\omega_2 t)] = f(\omega_1, \omega_2) \qquad (15.24)$$

   We conclude that the linear term of the series expansion has frequency spectrum that is equal to the original spectrum of the signal v_D, i.e. ω_1 and ω_2. Hence, this term is not useful.

2. The square term:

$$\frac{1}{2}\left(\frac{v_D}{V_t}\right)^2 = \frac{1}{2V_t^2}\left[\frac{1}{2}(V_1 \cdot \cos(\omega_1 t) + V_2 \cdot \cos(\omega_2 t)) \right]^2$$

$$= \frac{1}{8V_t^2}\left[V_1^2 \cos^2(\omega_1 t) + 2V_1 V_2 \cos(\omega_1 t)\cos(\omega_2 t) + V_2^2 \cos^2(\omega_2 t) \right]$$

$$= \frac{1}{8V_t^2}\left[V_1^2 \frac{1}{2}(1 + \cos(2\omega_1 t)) + \right.$$

$$V_1 V_2(\cos(|\omega_1 - \omega_2|t) + \cos((\omega_1 + \omega_2)t) +$$

$$\left. V_2^2 \frac{1}{2}(1 + \cos(2\omega_2 t)) \right] \qquad (15.25)$$

which states that part of diode's output frequency spectrum due to the second (non-linear) term contains

$$\frac{1}{2}\left(\frac{v_D}{V_t}\right)^2 = f\left[(\omega_1 - \omega_2), 2\omega_1, 2\omega_2, (\omega_1 + \omega_2)\right]$$

In other words, aside from the up and down shifted tones ($\omega_1 + \omega_2$) and ($|\omega_1 - \omega_2|$), there are additional tones ($2\omega_1$ and $2\omega_2$) present that are not present in the result of the ideal multiplication operation. They are HF tones and are to be removed by filtering.

Higher order terms due to cubic, fourth, etc. powers in Taylor series, they all generate HF tones that are to be removed by filtering.

Fig. 15.5
Example 15.3–1

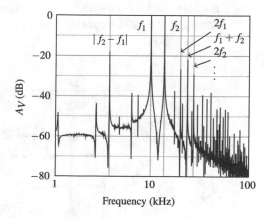

In conclusion, even though diode mixer is very inefficient (it generates many tones that are not needed), see Fig. 15.5, it is simplest circuit used in practice. What is more, at extremely high frequencies they may be the only active circuit capable of doing the mixing operation.

2. JFET is also non-linear device whose I/O transfer function is quadratic.

(a) Therefore we write,

$$I_D - I_{DSS} \left(1 - \frac{v_{GS}}{V_p}\right)^2 = I_{DSS} \left[1 - 2\frac{v_{GS}}{V_p} + \frac{v_{GS}^2}{V_p^2}\right] \tag{15.26}$$

which gives us polynomial functions without the Taylor series. Given two sine waveforms as

$$v_1 = V_1 \cos(\omega_1 t) \tag{15.27}$$

$$v_2 = V_2 \cos(\omega_2 t) \tag{15.28}$$

so that $v_{GS} = v_1 + v_2$. After substituting in (15.26) and by focusing only on the non-linear terms in (15.26), the square term is

$$I_D \sim -I_{DSS} \frac{1}{4} \frac{[V_1 \cdot \cos(\omega_1 t) + V_2 \cdot \cos(\omega_2 t)]^2}{V_p^2}$$

$$\sim -I_{DSS} \frac{V_1 V_2}{2 V_p^2} \left[\cos(|\omega_1 - \omega_2|t) + \cos((\omega_1 + \omega_2)t)\right] \tag{15.29}$$

where (15.26) focuses only at the cos product term from the previous step. Because there are no other higher-order terms, there is no strict limitation to the amplitudes of V_1 and V_2, as long as the JFET is not cut off or becomes forward biased. JFETs are commonly used in RF mixer applications because of their tolerance for high signal levels and good conversion efficiency.

(b) Dual gate JFET amplifier developed in Example 5.1–4 is, in reality, already prepared to function as mixer. Minor modification is that instead of being fixed to the small signal ground, gate of J_2 is released to be used as the second input, see Fig. 15.6.

Given data, the LC resonator must be centred at $f_{IF} = 455$ kHz, which is implemented, for example, with $L = 15\,\mu H$ and $C = 8.156920\,nF$. The capacitance such as this one is implemented by a parallel connection of fixed and trim capacitors. In addition, in order to satisfy $BW = 10$ kHz specification we must choose the inductor with $Q = 46$, or equivalently $R_D = 1950\,\Omega$.

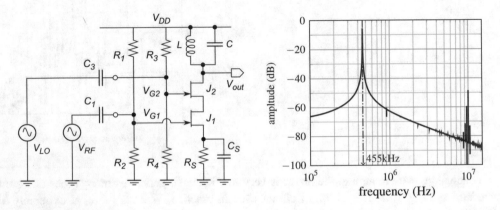

Fig. 15.6 Example 15.3–2

Modulation

16

In a broad sense, the term "modulation" implies a change in time of a certain parameter, where the "change" itself is the message being transmitted. For instance, while listening to a steady single-tone signal with constant amplitude and frequency coming out of a speaker, we merely receive the simplest message that conveys information only about the existence of the signal source and nothing else. If the source is turned off, then we cannot even say if there is a signal source out there or not. For the purpose of transmitting a more sophisticated message, the communication system must use at least the simplest modulation scheme, based on time divisions, i.e. turning on and off the signal source. By listening to short and long beeps, we can decode complicated messages letter by letter. As slow and inefficient as it is, Morse code does work and is used even today in special situations, for example, in a very low SNR environment.

16.1 Important to Know

1. AM modulation principle:

$$e(t) = C + b(t) \tag{16.1}$$

$$\begin{aligned}
c_{AM}(t) &= e(t) \sin \omega_c t \\
&= (C + B \sin \omega_b t) \sin \omega_c t \\
&= C \left(1 + \frac{B}{C} \sin \omega_b t\right) \sin \omega_c t \\
&= C (1 + m \sin \omega_b t) \sin \omega_c t \tag{16.2} \\
&= \sin \omega_c t + m \sin \omega_b t \sin \omega_c t \tag{16.3} \\
&= \sin \omega_c t + \frac{m}{2} [\cos |\omega_c - \omega_b| \, t - \cos (\omega_c + \omega_b) \, t] \tag{16.4}
\end{aligned}$$

2. AM modulation index

$$m \stackrel{\text{def}}{=} \frac{\text{the modulating signals maximum amplitude}}{\text{the carriers maximum amplitude}} = \frac{B}{C} \tag{16.5}$$

© Springer Nature Switzerland AG 2021
R. Sobot, *Wireless Communication Electronics by Example*,
https://doi.org/10.1007/978-3-030-59498-5_16

3. Average power in AM waveform

$$\langle P_C \rangle = \frac{c_{C\mathrm{rms}}^2}{R} = \frac{\left(\frac{C}{\sqrt{2}}\right)^2}{R} = \frac{C^2}{2R} \tag{16.6}$$

$$\langle P_L \rangle = \frac{c_{L\mathrm{rms}}^2}{R} = \frac{\left(\frac{mC/2}{\sqrt{2}}\right)^2}{R} = \frac{m^2}{4}\frac{C^2}{2R} = \frac{m^2}{4}\langle P_C \rangle \tag{16.7}$$

$$\langle P_U \rangle = \frac{c_{U\mathrm{rms}}^2}{R} = \frac{\left(\frac{mC/2}{\sqrt{2}}\right)^2}{R} = \frac{m^2}{4}\frac{C^2}{2R} = \frac{m^2}{4}\langle P_C \rangle = \langle P_L \rangle \tag{16.8}$$

$$\therefore$$

$$P_T = P_C\left(1 + \frac{m^2}{2}\right) \tag{16.9}$$

4. FM modulation principle

$$b(t) = B\cos\omega_b t \tag{16.10}$$

$$\Delta\omega_c = k\,b(t) \tag{16.11}$$

$$\omega(t) = \omega_c + \Delta\omega_c = \omega_c + k\,b(t) \tag{16.12}$$

$$f(t) = \frac{\omega(t)}{2\pi} = f_c + \frac{kB}{2\pi}\cos\omega_b t \tag{16.13}$$

5. FM modulation index and deviation ratio

$$m_f \stackrel{\mathrm{def}}{=} \frac{\Delta f}{f_m} = \frac{kB}{\omega_b} \tag{16.14}$$

$$\delta \stackrel{\mathrm{def}}{=} \frac{\Delta f}{f_c} = \frac{kB}{\omega_c} \tag{16.15}$$

6. FM waveform

$$\begin{aligned}
c_{FM} = {}& J_0(m_f)\,C\,\sin\omega_c t \\
& + J_1(m_f)\,C\,[\sin(\omega_c+\omega_b)\,t - \sin(\omega_c-\omega_b)\,t] \\
& + J_2(m_f)\,C\,[\sin(\omega_c+2\omega_b)\,t + \sin(\omega_c-2\omega_b)\,t] \\
& + J_3(m_f)\,C\,[\sin(\omega_c+3\omega_b)\,t - \sin(\omega_c-3\omega_b)\,t] \\
& + \cdots
\end{aligned} \tag{16.16}$$

7. Carson's rule

$$B_{FM} = 2(m_f + 1)f_b = 2(\Delta f + f_b) \tag{16.17}$$

8. Phase deviation constant (varicap diode)

$$K = \frac{d\phi}{dV_D} = \frac{d\phi}{dC} \frac{dC}{dC_D} \frac{dC_D}{dV_D} \qquad (16.18)$$

16.2 Exercises

16.1 * AM modulation

1. Basic principles of AM:

    (a) Derive equation that is behind AM modulator, then

    (b) Design block diagram of ideal AM modulator that is literal implementation of the equation.

2. Given audio signal $b(t)$ and carrier $c(t)$ as

$$b(t) = 0.5\,\text{V}\sin(2\pi \times 1\,500\,t)$$
$$c(t) = 1\,\text{V}\sin(2\pi \times 100\,000\,t)$$

    (a) What are amplitudes and frequencies of given audio and carrier signals?

    (b) Determine the modulation factor.

    (c) What frequencies are found in the frequency spectrum of this AM wave?

    (d) Using behavioural model from Example 16.1-1 simulate and show AM modulated wave.

16.2 * AM spectrum

1. The full 20 kHz audio band is transmitted by a 10 MHz RF carrier. If the USB and LSB are separated by $\Delta f = 100$ kHz, what are frequency ranges of the two sidebands?

2. In a typical AM radio system, the signal bandwidth is Δf. Estimate the type of SSB filter that is needed to suppress the LSB by A_{dB}, if the centre frequency is

    (a) $f_C = 100$ kHz, (b) $f_C = 1$ MHz

    Data: $\Delta f = \pm 100$ Hz, $A_{\text{dB}} = 80\,dB$.

3. How many AM broadcasting stations can be accommodated in a 100 kHz bandwidth if the highest frequency modulating the carriers is 5 kHz?

16.3 * AM power

1. Given that AM signal is modulated at 85% while the total RF power is 1200 W, determine the power content in each of sideband and AM carrier signals.

2. An AM signal whose carrier waveform is modulated 70% contains 1 500 W at the carrier frequency.

    (a) Determine the power content in the USB and LSB.

    (b) Calculate the power at the carrier and the power of each of the sidebands if the modulation index drops to 50%.

16.4 *   FM definitions

1. A $f_c = 107.6\,\text{MHz}$ carrier is frequency modulated by $f_m = 7\,\text{kHz}$ sine wave. The resultant FM signal has frequency deviation of $\Delta f = 50\,\text{kHz}$.

   (a)   Calculate the FM carrier frequency swing,

   (b)   Determine min and max frequencies attained by the modulated signal,

   (c)   Determine the modulation index of the FM wave.

2. FM transmitter operates with the total RF power of $P_T = 100\,\text{W}$ and modulation index of $m_f = 2.0$.

   (a)   Determine power levels contained in the first six frequency components,

   (b)   Estimate the bandwidth requirement if modulation signal is $f_m = 1.0\,\text{kHz}$.

3. Given frequency modulation index $m_f = 1.5$ and modulation signal $f_b = 10\,\text{kHz}$, find:

   (a)   The required bandwidth B_{FM} (using Carson's rule),

   (b)   Ratio of the total power relative P_T to the power in the FM unmodulated waveform,

   (c)   The highest amplitude harmonics.

16.5 **   Reactance modulator

1. The output node of circuit in Fig. 16.1 is connected with parallel L_T, C_T resonator (not shown, while decoupling capacitor is assumed large). Given data and the resonant frequency f_{out} of the resonator as found at the output node, calculate value of capacitor C and illustrate how this circuit is used in FM modulators. **Data:** $f_{out} = 3.5\,\text{MHz}$, $C_T = 83.4\,\text{nF}$, $L_T = 20\text{nH}$, $R = 10\,\Omega$, $g_m(M_1) = 10\,\text{mS}$.

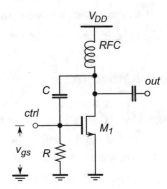

Fig. 16.1 Example 16.5–1

Solutions

Exercise 16.1, page 329

1. Given carrier $c(t)$ and a single frequency signal $b(t)$ waveforms we write

$$b(t) = B \sin \omega_b t \tag{16.19}$$

$$c(t) = C \sin \omega_c t \tag{16.20}$$

(a) The addition of signal $b(t)$ waveform to the carrier's amplitude C as

$$e(t) = C + b(t) \tag{16.21}$$

results in the carrier waveform being

$$
\begin{aligned}
c_{AM}(t) &= e(t) \sin \omega_c t \\
&= (C + B \sin \omega_b t) \sin \omega_c t \\
&= C \left(1 + \frac{B}{C} \sin \omega_b t \right) \sin \omega_c t \\
&= C (1 + m \sin \omega_b t) \sin \omega_c t \tag{16.22} \\
&= \sin \omega_c t + m \sin \omega_b t \sin \omega_c t \tag{16.23} \\
&= \sin \omega_c t + \frac{m}{2} [\cos |\omega_c - \omega_b| \, t - \cos (\omega_c + \omega_b) \, t] \tag{16.24}
\end{aligned}
$$

where the modulation index is defined as

$$m \overset{\text{def}}{=} \frac{\text{the modulating signals maximum amplitude}}{\text{the carriers maximum amplitude}} = \frac{B}{C} \tag{16.25}$$

Note that, without losing in generality, after setting $C = 1$ in (16.22) everything afterwards is consequently normalized to the carrier's amplitude.

Therefore, (16.23) is underlying equation behind AM modulator, while (16.24) shows frequency spectrum of AM carrier. We note the similarity with the mixer function, however, aside from the two frequency components f_{SUM} and f_{DIF}, in addition AM waveform contains the carrier tone $\sin \omega_c$ itself as well.

(b) Literal implementation of (16.24) requires two multiplier and one adder circuits. Assuming m, $c(t)$ and $b(t)$ as the input variables, behavioural block diagram of AM modulator is shown in Fig. 16.2 (left). Simulation example, where $m = 0.5$, $f_c = 10$, and $f_b = 1$ kHz, illustrates the frequency spectrum of $c_{AM}(t)$ at the AM modulator output, Fig. 16.2 (right).

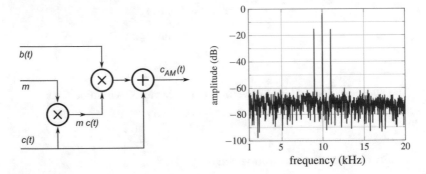

Fig. 16.2 Example 16.1–1

2. By definitions of AM signals, we write

(a) By inspection of $b(t)$ and $c(t)$ we write: audio signal amplitude is $B = 0.5$ V and frequency is 1.5 kHz, the carrier signal's amplitude is $C = 1$ V and frequency 100 kHz.
(b) By definition, AM index is $m = B/C = 0.5$.
(c) AM modulated signal spectrum contains three tones: the carrier itself 100 kHz, the sum $f_{SUM} = 101.5$ kHz and difference $f_{DIF} = 98.5$ kHz. We note that the sum and difference tones are multiplied by half, i.e. -6 dB relative to its carrier.
(d) Carrier frequency is much higher than the signal, thus time domain plot shows its envelope with $m = 0.5 = 50\%$ index, Fig. 16.3. That is to say, the envelope amplitude is ± 0.5 V whose average is $C = 1$ V and its frequency is 1.5 kHz (because the envelope is where signal is embedded into the carrier's amplitude).

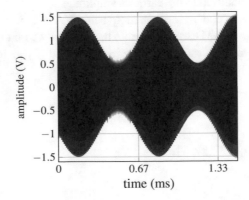

Fig. 16.3 Example 16.1–2

Exercise 16.2, page 329

1. In DSB-FC modulation scheme, after multiplication, each frequency in the signal's spectrum is shifted both to USB and LSB, Fig. 16.4. Because USB and LSB are symmetrically centred around the carrier frequency, given that the separation between USB and LSB is 100 kHz, it follows that USB starts at

$$f_{min}(USB) = 10\,\text{MHz} + (100/2)\,\text{kHz}$$

$$= 10.050\,\text{MHz}$$

Therefore, the audio band ends at

$$f_{max}(USB) = f_{min}(USB) + 20\,\text{kHz}$$

$$= 10.070\,\text{MHz}$$

Similarly, LSB occupies space as

$$f_{max}(LSB) = 10\,\text{MHz} - (100/2)\,\text{kHz}$$

$$= 9.950\,\text{MHz}$$

$$f_{min}(USB) = f_{max}(LSB) - 20\,\text{kHz}$$

$$= 9.030\,\text{MHz}$$

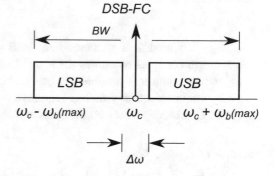

Fig. 16.4 Example 16.2–1

2. By direct implementation of textbook equation for this type of filter, Fig. 16.5, we write,

(a)   given $f_C = 100\,\text{kHz}$,

$$Q = \frac{f_C}{\Delta f} \frac{\sqrt{10^{\left(\frac{A_{dB}}{20}\right)}}}{4}$$

$$= \frac{100\,\text{kHz}}{200\text{Hz}} \frac{\sqrt{10^{\left(\frac{80}{20}\right)}}}{4} = 12\,500$$

i.e. we need crystal filter or better.

(b)   given $f_C = 1\,\text{MHz}$,

$$Q = \frac{f_C}{\Delta f} \frac{\sqrt{10^{\left(\frac{A_{dB}}{20}\right)}}}{4}$$

$$= \frac{1\,\text{MHz}}{200\text{Hz}} \frac{\sqrt{10^{\left(\frac{80}{20}\right)}}}{4} = 125\,000$$

i.e. we need several SAW filters in cascade.

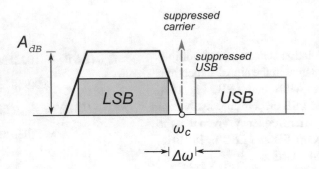

Fig. 16.5 Example 16.2–2

3. Using DSB modulation scheme both USB and LSB bands are transmitted, consequently AM waveform occupies two times the signal frequency, i.e. $2 \times 5\,\text{kHz} = 10\,\text{kHz}$. Therefore, in the given bandwidth it is possible to create maximum of $100/10 = 10$ channels. However, in reality every communication standard requires some "guard-band" between each two neighbouring channels thus, depending on the specific standard, smaller number of channels is available.

Exercise 16.3, page 329

1. Given carrier $c(t)$ and a single frequency signal $b(t)$ waveforms we write

$$b(t) = B \sin \omega_b t \tag{16.26}$$

$$c(t) = C \sin \omega_c t \tag{16.27}$$

and AM waveform,

$$c_{AM}(t) = \sin \omega_c t + \frac{m}{2} \cos \omega_L t - \frac{m}{2} \cos \omega_U t = c_C + c_L - c_U \tag{16.28}$$

Hence, the instantaneous power of the AM wave across load resistor R is

$$p_{AM} \stackrel{\text{def}}{=} \frac{c_{AM}^2}{R} = \frac{(c_C + c_L - c_U)^2}{R} = \frac{c_C^2}{R} + \frac{c_L^2}{R} + \frac{c_U^2}{R} + \frac{2}{R}\left(c_C c_L - c_L c_U - c_C c_U\right) \tag{16.29}$$

With this notification, we write expressions for average power of each component as

$$\langle P_C \rangle = \frac{c_{C\text{rms}}^2}{R} = \frac{\left(\dfrac{C}{\sqrt{2}}\right)^2}{R} = \frac{C^2}{2R} \tag{16.30}$$

$$\langle P_L \rangle = \frac{c_{L\text{rms}}^2}{R} = \frac{\left(\dfrac{m\,C/2}{\sqrt{2}}\right)^2}{R} = \frac{m^2}{4}\frac{C^2}{2R} = \frac{m^2}{4}\langle P_C \rangle \tag{16.31}$$

$$\langle P_U \rangle = \frac{c_{U\text{rms}}^2}{R} = \frac{\left(\dfrac{m\,C/2}{\sqrt{2}}\right)^2}{R} = \frac{m^2}{4}\frac{C^2}{2R} = \frac{m^2}{4}\langle P_C \rangle = \langle P_L \rangle \tag{16.32}$$

Hence, the total average power P_T of an AM modulated waveform is therefore

$$\langle P_T \rangle = \langle P_C \rangle + \frac{m^2}{4}\langle P_C \rangle + \frac{m^2}{4}\langle P_C \rangle = \langle P_C \rangle \left(1 + \frac{m^2}{2} \right) \qquad (16.33)$$

In order to simplify the syntax, we recall that (16.33) refers to the average power simply write

$$P_T = P_C \left(1 + \frac{m^2}{2} \right) \qquad (16.34)$$

Given data,

$$P_C = \frac{P_T}{1 + \dfrac{m^2}{2}} = \frac{1200\,\text{W}}{1 + \dfrac{0.85^2}{2}} = 881.5\,\text{W}$$

which is to say that combined power in two sidebands $P_{\text{SB}} = P_{\text{USB}} + P_{\text{LSB}}$ is

$$P_{\text{SB}} = P_T - P_C = 318.5\,\text{W} \quad \therefore \quad P_{\text{USB}} = P_{\text{LSB}} = \frac{P_{\text{SB}}}{2} = 159.25\,\text{W}$$

2. By reusing derivations in Example 16.1–1, where

$$P_T = P_C \left(1 + \frac{m^2}{2} \right) \quad \text{and} \quad P_{\text{SB}} = P_T - P_C$$

$$P_{\text{SB}} = P_C \left(1 + \frac{m^2}{2} \right) - P_C = \frac{P_C\, m^2}{2}$$

Therefore, because USB and LSB are symmetric and equal in power,

$$P_{\text{USB}} = P_{\text{LSB}} = \frac{P_{\text{SB}}}{2} = \frac{m^2 P_C}{4}$$

(a) Given $m = 0.7$ we find

$$P_T = P_C \left(1 + \frac{m^2}{2} \right) = 1\,500\,\text{W} \left(1 + \frac{0.7^2}{2} \right) = 1867.5\,\text{W}$$

And,

$$P_{\text{USB}} = P_{\text{LSB}} = \frac{0.7^2 \times 1\,500\,\text{W}}{4} = 183.75\,\text{W}$$

(b)   Given $m = 0.5$, while keeping the total power $P_T = 1867.5$ W, we calculate

$$P_C = \frac{P_T}{1 + \dfrac{m^2}{2}} = \frac{1867.5 \text{ W}}{1 + \dfrac{0.5^2}{2}} = 1.660 \text{ kW}$$

$$P_{\text{USB}} = P_{\text{LSB}} = \frac{m^2 P_C}{4} = \frac{0.5^2 \times 1.660 \text{ kW}}{4} = 103.75 \text{ W}$$

Exercise 16.4, page 330

1. Given $\Delta f = 50$ kHz, that is to say that the carrier swing is ± 50 kHz $= 100$ kHz. Therefore, the carrier frequency changes between $f_c(min) = 107.55$ MHz and $f_c(max) = 107.65$ MHz. By definition, frequency modulation index is

$$m_f = \frac{\Delta f}{f_m} = \frac{50 \text{ kHz}}{7 \text{ kHz}} = 7.143$$

2. In accordance to Bessel functions for $m_f = 2.0$, we write

$$P_0 = 100 \text{ W} \times 0.224^2 = 5.0176 \text{ W}$$

$$P_1 = 100 \text{ W} \times 2 \times 0.577^2 = 66.5858 \text{ W}$$

$$P_2 = 100 \text{ W} \times 2 \times 0.353^2 = 24.9218 \text{ W}$$

$$P_3 = 100 \text{ W} \times 2 \times 0.129^2 = 3.3282 \text{ W}$$

$$P_4 = 100 \text{ W} \times 2 \times 0.034^2 = 0.2312 \text{ W}$$

$$P_5 = 100 \text{ W} \times 2 \times 0.007^2 = 0.0098 \text{ W}$$

$$P_6 = 100 \text{ W} \times 2 \times 0.001^2 = 0.0002 \text{ W}$$

which adds up to 100 W (when neglecting all remaining harmonics). Given $f_m = 1$ kHz and $m_f = 2.0$, by using Carson's rule we estimate

$$B_{\text{FM}} = 2(m_f + 1)f_m = 6 \text{ kHz}$$

3. By using Carson's rule we calculate $B_{\text{FM}} = 2(m_f + 1)f_b = 50$ kHz. The sum of Bessel's functions for $m_f = 1.5$ results in

$$\frac{P_T}{P_c} = J_0^2 + 2(J_1^2 + J_2^2 + \cdots + J_5^2)$$

$$= 0.512^2 + 2(0.558^2 + 0.232^2 + 0.061^2 + 0.012^2 + 0.002^2)$$

$$\approx 1.000$$

for the first five functions. That is to say, the total power is constant, just redistributed among the harmonics. When $m_f = 1.5$ the first sideband harmonic has highest amplitude $J_1 = 0.558$ relative to the unmodulated signal.

Exercise 16.5, page 330

1. This circuit is known as reactance modulator, it behaves as voltage controlled capacitance between *out* ground nodes, Fig. 16.6.

   Effective output impedance Z_o, as seen by looking into the output node, is found by definition. That is, by the ratio of the output voltage v_o to the output current i_o, as

   $$i_C = \frac{v_o}{R + Z_C} = \frac{v_o}{R - j\dfrac{1}{\omega C}} \tag{16.35}$$

   $$\therefore$$

   $$v_{gs} = R\, i_C = \frac{R\, v_o}{R - j\dfrac{1}{\omega C}} \tag{16.36}$$

   which leads into expression for M_1 drain current i_d as

   $$i_d = g_m v_{gs} = g_m \frac{R\, v_o}{R - j\dfrac{1}{\omega C}} \tag{16.37}$$

   Assuming: current i_C through the capacitor C branch is much smaller than the M_1 drain current i_d, i.e. $i_d \gg i_C$ or $i_d + i_C \approx i_d$; and the capacitor C impedance X_C is much greater than resistance R, that is, $R - X_C \approx -X_C$.

   With these approximation we write

   $$i_o = i_C + i_d \approx i_d = g_m \frac{R\, v_o}{R - j\dfrac{1}{\omega C}} \approx g_m \frac{R\, v_o}{-j\dfrac{1}{\omega C}} = \frac{v_o}{-j\dfrac{1}{\omega\, g_m\, R\, C}}$$

   $$\therefore$$

   $$Z_o \equiv \frac{v_o}{i_o} = -j\frac{1}{\omega\,(g_m\, R\, C)} = -j\frac{1}{\omega\, C_{RM}} \tag{16.38}$$

   Therefore, because $C_{RM} = g_m R C$ is in parallel with C_L, we write

   $$f_{out} = \frac{1}{2\pi\sqrt{L_T(C_T + C_{RM})}} \quad \therefore \quad C_{RM} = 20\,\text{nF}$$

   Which gives

   $$C_{RM} = g_m R C \quad \therefore \quad C = \frac{C_{RM}}{g_m R} = \frac{20\,\text{nF}}{10\,\text{mS} \times 10\,\Omega} = 200\,\text{nF}$$

Fig. 16.6
Example 16.5–1

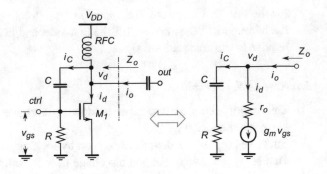

AM and FM Signal Demodulation $\qquad$ 17

When a modulated signal arrives at the receiving antenna, the embedded information must somehow be extracted by the receiver and separated from the HF carrier signal. This information recovery process is known as "demodulation" or "detection". It is based on an underlying mechanism similar to the one used in mixers, where a non-linear element is used to multiply two waves and accomplish the frequency shifting. However, the demodulation process is centred around the carrier frequency ω_0 and the signal spectrum is shifted downward to the baseband and returned to its original position in the frequency domain. Both modulation and demodulation involve a frequency shifting process; both processes shift the frequency spectrum by a distance ω_0 on the frequency axis; and both processes require a non-linear circuit to accomplish the task. Although very similar, the two processes are different in very subtle but important details. In the modulating process the carrier wave is generated by the LO circuit, and then combined with the baseband signal inside the mixer. In the demodulating process, however, the carrier signal is already contained in the incoming modulated signal and it can be recovered at the receiving point.

17.1 Important to Know

1. AM demodulation principle:

$$c_{AM}(t) = C\,(1 + m \cos \omega_b t)\, \cos \omega_c t$$

$$i(t) = a_2\, c_{AM}^2(t)$$

$$a_2\, C^2\, [1 + m \cos \omega_b t]^2\ \cos^2 \omega_c t$$

$$= a_2\, C^2\, [1 + 2m \cos \omega_b t + m^2 \cos^2 \omega_b t] \left[\frac{1}{2} + \frac{1}{2} \cos 2\omega_c t \right]$$

$$\approx a_2\, m\, C^2 \left[\cos \omega_b t + \frac{m}{4} \underset{\xcancel{}}{\cos 2\omega_b t}^{\,LPF} \right] \tag{17.1}$$

2. Distortion factor estimates:

$$\frac{1}{RC} \geqslant \omega_b\, \frac{m}{\sqrt{1 - m^2}} \quad \text{or,} \quad \frac{1}{RC} \geqslant m\, \omega_b \tag{17.2}$$

© Springer Nature Switzerland AG 2021
R. Sobot, *Wireless Communication Electronics by Example*,
https://doi.org/10.1007/978-3-030-59498-5_17

17.2   Exercises

17.1 *   Demodulation principle

1. Given AM modulated waveform

$$c_{AM}(t) = C(1 + m\cos\omega_b t)\cos\omega_c t$$

where, $(0 \leq m \leq 1)$ is AM modulation index, ω_b is the message frequency, and ω_c is the carrier frequency of this AM waveform. Derive the theoretical expression and explain the method that should be used to recover the received message signal $\cos\omega_b t$. We note that in order to make hand analysis doable, we analyse only a single-tone message (e.g. a single continuous note C). Signals that are synthesized out of many single tones (e.g. voice) are analysed by simulations—one can imagine the process where same calculations are repeated for each frequency in the message's spectrum.

2. As a follow up to Example 17.1–1, create AM demodulator behavioural model.

   (a) Illustrate by simulation its frequency spectrum,

   (b) Based on the frequency spectrum graph, estimate the order of LPF if the objective is to keep the $2\omega_b$ tone at $-40\,\mathrm{dB}$ or less,

   (c) Replace LPF from 2 with LC resonator. Try to estimate Q factor that is required to achieve same suppression of the $2\omega_b$ tone as in 2.

Data: $\omega_b = 2\pi \times 1\,\mathrm{kHz}$, $\omega_c = 2\pi \times 10\,\mathrm{kHz}$, $m = 1.0$, $C = 1\,\mathrm{V}$, $a = \sqrt{2}$.

17.2 *   Envelope detector

1. Given diode envelope detector, Fig. 17.1, where the diode resistance is r_D,

   (a) Determine the value of the loading resistor   (b) Determine the effective input resistance of
       R if the desired detection efficiency is η,   this circuit.

Data: $r_D = 100\,\Omega$, $\eta = 80\%$.

Fig. 17.1
Example 17.2–1, 2

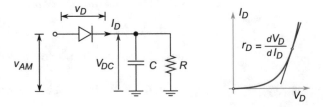

2. Assuming diode envelope circuit diagram in Fig. 17.1 that is intended to recover message signal $b(t)$, calculate boundaries of R relative to AM modulation factor m, then use simulations to illustrate conditions for "clipping". **Data:** $f_{IF} = 50\,\mathrm{kHz}$, $f_b = 1\,\mathrm{kHz}$, $C = 15.9154943\,\mathrm{nF}$.

3. Given AM diode detector and the diode I_D vs. V_D characteristics we assume AM RF waveform is already down-converted from f_{RF} to f_{IF} and is applied to node① of the envelope detector, see Fig. 17.2,

(a) Sketch frequency spectrum of all signals found in this circuit;
(b) Sketch AM waveform shapes at node①to node⑤;
(c) Sketch the equivalent circuit at the signal frequency f_S. Calculate the voltage gain between node③and node①;
(d) Sketch the equivalent circuit at f_{IF}. Calculate the voltage gain between node③and node①; item[(e)] Comment on the results.

Data: $f_{IF} = 665\,\text{kHz}$, $f_S = 5\,\text{kHz}$, $C_1 = 220\,\text{pF}$, $C_2 = 5\,\text{nF}$, $R_1 = 470\,\Omega$, $R_2 = 4.7\,\text{k}\Omega$, $R_L = 50\,\text{k}\Omega$

Fig. 17.2
Example 17.2–3

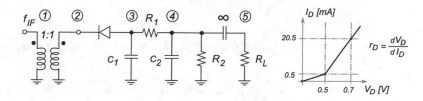

17.3 *   FM demodulator

1. In order to illustrate FM signal demodulation process, given a hypothetical VCO voltage to frequency characteristics, see Fig. 17.3,

   (a) Create FM waveform source where $f_b = 1\,\text{kHz}$,
   (b) Design RLC network suitable for this application,
   (c) Design behavioural model of quadrature demodulator and demonstrate by simulations the recovery of $b(t)$ signal.

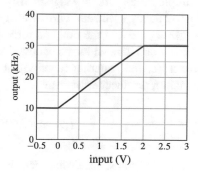

Fig. 17.3 Example 17.3–1

Solutions

Exercise 17.1, page 340

1. A simple method to de-embed transmitted message from AM waveform would be to apply the square law function to the received waveform so that

$$i(t) = a\, c_{AM}{}^2(t) \tag{17.3}$$

where, a is the proportionality constant, and $i(t)$ the squaring operation result. Then, the output signal contains the following terms:

$$i(t) = a\,C^2\,[1 + m\cos\omega_b t]^2\,\cos^2\omega_c t$$

$$= a\,C^2\,[1 + 2m\cos\omega_b t + m^2\cos^2\omega_b t]\left[\frac{1}{2} + \frac{1}{2}\cos 2\omega_c t\right]$$

$$= a\,C^2\left[\left(\frac{1}{2} + \frac{m^2}{4}\right)_{DC} + m\,\cos\omega_b t \qquad \text{(DC and message terms)}\right.$$

$$+\frac{m^2}{4}\cos 2\omega_b t \qquad\qquad \text{(the message term, found at } 2\omega_b)$$

$$+\left(\frac{1}{2} + \frac{m^2}{2}\right)\cancel{\cos 2\omega_c t}^{\;LPF}$$

$$+\frac{m}{2}\cancel{\cos(2\omega_c + \omega_b)t}^{\;LPF} \qquad +\frac{m}{2}\cancel{\cos(2\omega_c - \omega_b)t}^{\;LPF}$$

$$\left. +\frac{m^2}{8}\cancel{\cos(2\omega_c + 2\omega_b)t}^{\;LPF} \qquad +\frac{m^2}{8}\cancel{\cos(2\omega_c - 2\omega_b)t}^{\;LPF}\right] \qquad (17.4)$$

Obviously, squaring the incoming AM waveform produces term with the message itself; however, there are additional unwanted terms. In practice, the squaring function may be implemented as a simple diode. The message is also found at its double frequency which is still rather close in the frequency spectrum. Further away in the frequency spectrum, close to the carrier frequency $\omega_c \gg \omega_b$ we find the sum and difference terms, which are easily removed by LPF designed to keep ω_b and $2\omega_b$ within its bandwidth.

Therefore, the squaring function followed by LPF, after DC term is blocked by decoupling capacitor, produces waveform

$$i(t) \approx a\,m\,C^2\left[\cos\omega_b t + \frac{m}{4}\cancel{\cos 2\omega_b t}^{\;LPF}\right]$$

Again, LPF whose bandwidth permits only ω_b term is used to remove $2\omega_b$ term. This theoretical analysis is summarized in behavioural block diagram of AM demodulator as shown in Fig. 17.4.

Fig. 17.4
Example 17.1–1

2. Simulation of behavioural model in Fig. 17.4 illustrates time domain and frequency spectrum at the outputs of square function and LPF, Fig. 17.5 (left).

(a) After passing through the squaring circuit, envelope of AM waveform is obviously in form of $\cos\omega_b t$ signal. After passing through first order LPF whose $\omega_0 = 2\pi \times 1\,\text{kHz}$, the signal is already recovered with small but still visible HF harmonics. That is to say, even first order LPF is already effective to remove the HF components of the frequency spectrum.

(b) Frequency spectrum plot, Fig. 17.5 (right), shows 1 kHz signal, then 2 kHz tone then the group of HF tones around 20 kHz that are caused by multiple sum and difference terms, as predicted in (17.4). If we are to suppress the 2 kHz tone to at least -40 dB, linear approximated LPF shown in Fig. 17.5 (right) implies that the filter slope should be around -40 dB/oct, which is equivalent to approximately -130 dB/dec. This is rather high order filter, each -20 dB/dec requires first order filter, therefore it should be at least sixth order LPF.

Fig. 17.5
Example 17.1–2

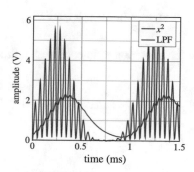 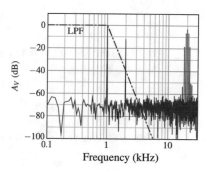

(c) Given that the resonator should be tuned at $f_0 = 1$ and $f = 2$ kHz it should provide suppression of -40 dB, we could estimate Q factor using

$$A_V = -40\,\text{dB} = \frac{1}{100}$$

$$A_V = \frac{1}{\sqrt{1 + (\delta\,Q)^2}} \quad \text{where,} \quad \delta = \frac{\omega}{\omega_0} - \frac{\omega_0}{\omega} = \frac{2}{1} - \frac{1}{2} = 1.5$$

Given data, we calculate $Q \approx 66$. However, this formula is derived as an approximation assuming $f_0 \approx f$. We can compare BPF transfer function derived by AC simulation (which actually uses linearized circuit model) with FFT response of transient simulation (which uses circuit as is, i.e. non-linear), see Fig. 17.6.

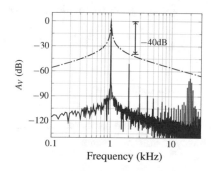

Fig. 17.6 Example 17.1–2

Exercise 17.2, page 340

1. As derived in the textbook,

(a) The following two equations

$$\eta = \frac{V_{DC}}{V_{AM}} = \cos\theta \tag{17.5}$$

and

$$\eta = \frac{R}{\pi r_D} (\sin\theta - \theta \cos\theta) \tag{17.6}$$

provide mechanism to calculate the detection efficiency η as a function of $(R/r_D, \theta)$. Given data, we calculate

$$\eta = \cos\theta \quad \therefore \quad \theta = \arccos(0.8) = 0.675\,\text{rad} \tag{17.7}$$

then, given $r_D = 100\,\Omega$, from (17.6) we write

$$R = \frac{\pi\,\eta}{(\sin\theta - \theta\cos\theta)}\, r_D = 2.57\,\text{k}\Omega \tag{17.8}$$

(b) Further derivations give

$$\frac{R_{\text{eff}}}{R} = \frac{R_{\text{eff}}}{r_D}\frac{r_D}{R} = \frac{\tan\theta - \theta}{\theta - \sin\theta\cos\theta} \tag{17.9}$$

for the ratio of the input effective resistance R_{eff} and resistor R as a function of θ. In the ideal case, detection efficiency is high, i.e. $\eta \to 1$, which implies very low $\theta \to 0$. By applying the small angle approximation, we conclude that

$$\frac{R_{\text{eff}}}{R} \approx \frac{1}{2\eta} \approx \frac{1}{2}$$

Given data, we approximate $R_{\text{eff}} \approx R/2 = 1.285\,\text{k}\Omega$.

2. As derived in the textbook, one boundary condition for the time constant RC where "clipping" is likely to start is given as

$$\frac{1}{RC} \approx \omega_b \frac{m}{\sqrt{1 - m^2}} \tag{17.10}$$

as well as its linearized version

$$\frac{1}{RC} \approx m\,\omega_b \tag{17.11}$$

where both equations give guideline on how to optimize time constant to minimize clipping of the recovered signal.

Given f_b and C we can show dependence of R versus the modulation index m, from (17.10) and (17.11) we write

$$R \approx \frac{\sqrt{1 - m^2}}{C \omega_b m} \quad \text{and,} \quad R \approx \frac{1}{C m \omega_b}$$

Given data, these two limiting functions are shown in Fig. 17.7, where both (17.10) and (17.11) limit to $R \to \infty$ when $m \to 0$. Naturally, if $m = 0$ there is no AM modulation, thus the envelope detector would measure and keep forever the value of carrier's maximum amplitude.

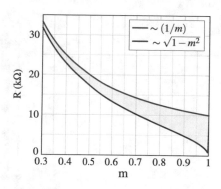

Fig. 17.7 Example 17.2–2

However, as $m \to 1$, the disagreement between the two estimates becomes larger and larger. In order to illustrate the choice of R (thus, the time constant), we compare the two boundaries by simulations. We note that the final version of circuit is derived by compromise among amplitude jitter, clipping, and m factor (Fig. 17.8).

Fig. 17.8
Example 17.2–2

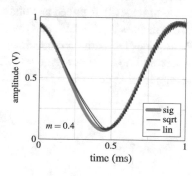 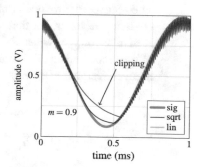

3. AM carrier is bound by two envelopes (positive and negative) that are embodiments of the same message. In Fig. 17.2 diode is forward biased if the input signal amplitude is *lower* than the capacitor voltage, thus only the *negative* envelope is detected.

(a) Given that f_{IF} is AM waveform modulated by f_S, that means there are three tones found in its frequency spectrum: $f_{\text{IF}} - f_S$, f_{IF}, and $f_{\text{IF}} + f_S$. Thus, we systematically write all f_{SUM} and f_{DIF} results generated by diode's non-linear characteristics, see Fig. 17.9. Note, the signal at 5 kHz is isolated by filtering.

$$\text{Sum frequencies:} \quad (660 + 665)\,\text{kHz} = 1325\,\text{kHz}$$
$$(660 + 670)\,\text{kHz} = 1330\,\text{kHz}$$
$$(665 + 670)\,\text{kHz} = 1335\,\text{kHz}$$

$$\text{Difference frequencies:} \quad (670 - 665)\,\text{kHz} = 5\,\text{kHz}$$
$$(665 - 660)\,\text{kHz} = 5\,\text{kHz}$$
$$(670 - 660)\,\text{kHz} = 10\,\text{kHz}$$

Fig. 17.9
Example 17.2–3

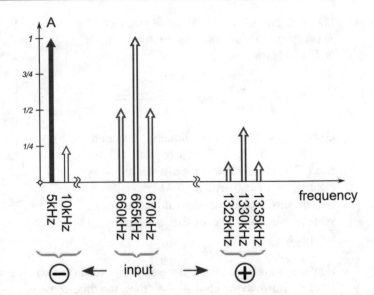

(b) In this case, due to its orientation, diode detector generates voltage that follows the negative envelope. Thus, the signal is progressively recovered at each subsequent node, Fig. 17.10. The recovered envelope amplitude at node ③ is traced as voltage across C_1; however, it still contains HF tones. For that reason, there is R_1, C_2 LPF that removes this noise, as at node ④. Resistive voltage divider R_1, R_2 provides DC path for diode detector, and also scales amplitude of the recovered signal. Finally, decoupling capacitor removes DC component and elevates the recovered signal to the zero average level at node ⑤.

Fig. 17.10
Example 17.2–3

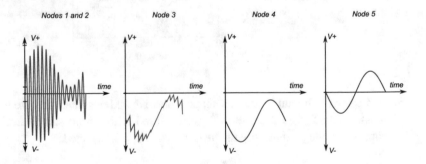

(c) At $f_S = 5\,\text{kHz}$: impedances of capacitors and diode resistance are as follows:

$$Z_{C1} = \frac{1}{2\pi \times 5\,\text{kHz} \times 220\,\text{pF}} = 144.68\,\text{k}\Omega \approx 145\,\text{k}\Omega$$

$$Z_{C2} = \frac{1}{2\pi \times 5\,\text{kHz} \times 22\,\text{pF}} = 1.4468\,\text{M}\Omega \approx 1.45\,\text{M}\Omega$$

$$R_D = \frac{\Delta V}{\Delta I} = \frac{0.7\text{V}}{7\,\text{mA}} = 100\,\Omega$$

where diode resistance R_D is found from the diode transfer characteristics graph, Fig. 17.2 (right), and the transformer is modelled as an ideal voltage source element. Hence, the peak detector is modelled with the equivalent voltage divider, where first resistance consists of $R_D = 100\,\Omega$, and the second resistance R is

$$R = Z_{C1}||(R_1 + (Z_{C2}||R_2||R_L)) = 4.6\,\text{k}\Omega \qquad (17.12)$$

That means that the voltage gain at node③ is calculated as the following ratio:

$$A \stackrel{\text{def}}{=} \frac{V(3)}{V_{in}} = \frac{R}{R_D + R} = 0.978 \qquad (17.13)$$

i.e. the $f_S = 5\,\text{kHz}$ signal is attenuated by approximately 2% relative to its input side amplitude, Fig. 17.11.

Fig. 17.11
Example 17.2–3

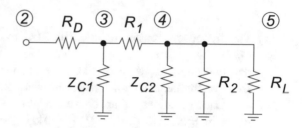

(d) At $f_{\text{IF}} = 665\,\text{kHz}$: impedances of capacitors are as follows:

$$Z_{C1} = \frac{1}{2\pi \times 665\,\text{kHz} \times 220\,\text{pF}} = 1087.86\,\Omega \approx 1.1\,\text{k}\Omega$$

$$Z_{C2} = \frac{1}{2\pi \times 665\,\text{kHz} \times 22\,\text{pF}} = 10.878\,\text{k}\Omega \approx 11\,\text{k}\Omega$$

while the diode resistance is not function of frequency, i.e. diode resistance is still $R_D = 100\,\Omega$. The equivalent circuit network is as same as at $f_S = 5\,\text{kHz}$; however, this time $R = 840\,\Omega$. Subsequently, $A = 0.894$, i.e. the 665 kHz carrier tone is attenuated approximately 10% relative to its input side amplitude.

(e) By choosing the circuit component values, designer has control over how much the carrier tone is attenuated relative to the envelope signal, as well as control over the internal timing constants that are important to avoid "clipping". Finally, LPF characteristic are important to supress HF components of the spectrum.

Exercise 17.3, page 341

1. After textbook derivations, given I/O characteristics in Fig. 17.3, we proceed as follows. In order to keep simulations manageable and to produce visually illustrating graphs, quadrature demodulator is illustrated at relatively low frequencies and low carrier to signal frequency ratio.

(a) From the graph, nominal non-modulated frequency is in the middle of the available range, i.e. 20 kHz. With that choice, we create FM source where the carrier frequency is $(20 \pm 10)\,\text{kHz}$ when the input voltage range is between 0 and 2 V. The input node of VCO is controlled by audio signal $b(t)$, in this case a single sine function whose frequency is $f_b = 1\,\text{kHz}$, DC voltage 1 V, and amplitude ± 1 V, i.e. $b(t) = 1\,\text{V} + \sin(2\pi\,f_b\,t)$, see Fig. 17.12

Fig. 17.12
Example 17.3–1

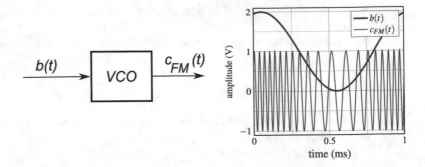

(b) The frequency transfer function of RLC resonator is set to permit the 10–30 kHz range ($f_0 = 20 \pm 10\,\text{kHz}$) only on one side of its resonant frequency. By doing so, as the instantaneous frequency of FM wave changes so does its attenuation as enforced by the slope of frequency characteristics in close proximity of the resonant frequency. Therefore, frequency variations Δf are converted into amplitude variations ΔA. We note that in example given in Fig. 17.13 (right), $Q \approx 20\,\text{kHz}/40\,\text{kHz} = 0.5$, which is set by R_p. From the perspective of node②, $C_0 \| C_p$ therefore $C_{\text{eq}} = C_0 + C_p$. Then, the resonant frequency $f_0 = 1/(2\pi\sqrt{L\,C_{\text{eq}}})$. Finally, C_0, C_p path creates capacitive voltage divider, thus the ratio between the two capacitors determines amplitude at node②.

Fig. 17.13
Example 17.3–1

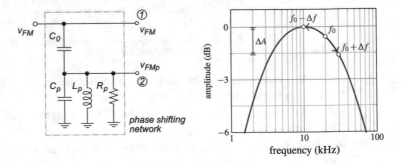

(c) The principle of quadrature decoder is based on the idea to multiply FM waveform (node①) with its 90° delayed version (node②), see Fig. 17.14. Capacitor C_0 serves as the phase shifting element, which is illustrated in the time domain plot of these two waveforms. We note that FMp is not only shifted by 90° but also amplitude modulated due to the RLC resonator's transfer function.

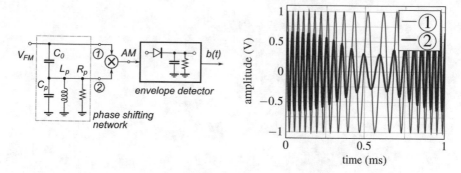

Fig. 17.14 Example 17.3–1

Given the two waveforms delivered by RLC phase shifting network, in the next step multiplier generates AM waveform, and the problem of FM demodulation is reduced to the problem of AM demodulation and design of envelope detector (see Example 17.1–1), as illustrated in Fig. 17.15 both in time and frequency domains.

Fig. 17.15
Example 17.3–1

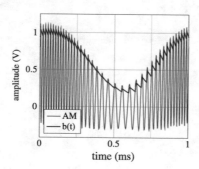 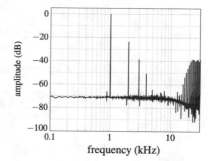

RF Receivers

18

In a general sense, radio receiver is an electronic system that is expected to detect the existence of a single, very specific EM wave in the overcrowded air space, to separate it from the rest of the frequency spectrum, and to extract the message. Hence, the literal implementation of the receiver function, which is known as a TRF receiver, consists only of a receiving antenna, an RF amplifier, and an audio amplifier. On the other hand, advanced radio receiver versions include one or more mixers and VCO blocks, which are meant to perform either a single-step frequency down-conversion (also known as a "heterodyne receiver") or multiple step frequency down-conversions (also known as a "super-heterodyne receiver") in order to shift the HF wave down to the baseband.

18.1 Important to Know

1. Harmonic distortion:

$$b(t) = B \, \cos \omega t$$

$$y(t) = a_1 B \, \cos \omega t + a_2 B^2 \, \cos^2 \omega t + a_3 B^3 \, \cos^3 \omega t + \cdots$$

$$= b_0 + b_1 \, \cos \omega t + b_2 \cos 2\omega t + b_3 \cos 3\omega t + \cdots \qquad (18.1)$$

$$D_2 = \frac{b_2}{b_1} \quad D_3 = \frac{b_3}{b_1} \quad D_4 = \frac{b_4}{b_1} \quad \cdots$$

$$THD = \sqrt{D_2^2 + D_3^2 + D_4^2 \cdots} \qquad (18.2)$$

2. Gain compression:

$$B(-1\,\mathrm{dB}) = \sqrt{0.145 \left| \frac{a_1}{a_3} \right|} \qquad (18.3)$$

$$S_{in}(-1\,\mathrm{dB}) = 20 \log \left[B(-1\,\mathrm{dB}) \right] \quad [\mathrm{dB}] \qquad (18.4)$$

© Springer Nature Switzerland AG 2021
R. Sobot, *Wireless Communication Electronics by Example*,
https://doi.org/10.1007/978-3-030-59498-5_18

3. Intermodulation:

$$x(t) = B_1 \cos \omega_a t + B_2 \cos \omega_b t$$

$$
\begin{aligned}
y(t) =& a_1 \left(B_1 \cos \omega_a t + B_2 \cos \omega_b t \right) \\
&+ a_2 \left(B_1 \cos \omega_a t + B_2 \cos \omega_b t \right)^2 \\
&+ a_3 \left(B_1 \cos \omega_a t + B_2 \cos \omega_b t \right)^3 + \cdots
\end{aligned}
\tag{18.5}
$$

$$
\begin{aligned}
=& \frac{a_2(B_1^2 + B_2^2)}{2} && (DC\ term) \\
&+ \left(a_1 B_1 + \frac{3}{4} a_3 B_1^3 + \frac{3}{2} a_3 B_1 B_2^2 \right) \cos \omega_a t && (fundamental\ terms) \\
&+ \left(a_1 B_2 + \frac{3}{4} a_3 B_2^3 + \frac{3}{2} a_3 B_2 B_1^2 \right) \cos \omega_b t \\
&+ \frac{a_2}{2} \left(B_1^2 \cos 2\omega_a t + B_2^2 \cos 2\omega_b t \right) && (second\ order\ terms) \\
&+ a_2 B_1 B_2 \left[\cos(\omega_a + \omega_b)t + \cos|\omega_a - \omega_b|t \right] \\
&+ \frac{a_3}{4} \left(B_1^3 \cos 3\omega_a t + B_2^3 \cos 3\omega_b t \right) && (third\ order\ terms) \\
&+ \frac{3a_3}{4} \big\{ B_1^2 B_2 \left[\cos(2\omega_a + \omega_b)t + \cos(2\omega_a - \omega_b)t \right] + \\
&\qquad\quad B_1 B_2^2 \left[\cos(2\omega_b + \omega_a)t + \cos(2\omega_b - \omega_a)t \right] \big\}
\end{aligned}
\tag{18.6}
$$

4. Third order intercept point (IIP3):

$$B(\text{IIP3}) = \sqrt{\frac{4}{3} \left| \frac{a_1}{a_3} \right|} \tag{18.7}$$

$$\therefore$$

$$B(-1\,\text{dB}) \approx \sqrt{\frac{4}{3} \left| \frac{a_1}{a_3} \right|} \, 0.11 = IIP3 - 9.6\,\text{dB} \tag{18.8}$$

5. Cross-modulation:

$$x(t) = B_1 \cos \omega_a t + B_2 \cos \omega_b t; \qquad B_2 \gg B_1$$

$$\therefore$$

$$y(t) \approx \left(a_1 B_1 + \frac{3}{4} a_3 B_1^3 + \frac{3}{2} a_3 B_1 B_2^2 \right) \cos \omega_a t + \cdots$$

$$(B_2 \gg B_1)$$

$$\approx \left(1 + \frac{3}{2} \frac{a_3}{a_1} B_2^2 \right) a_1 B_1 \cos \omega_a t + \cdots \tag{18.9}$$

6. Noise power, sensitivity, and dynamic range:

$$P_n = -174 \, \text{dBm} + 10 \log \Delta f + NF \quad [\text{dBm}] \tag{18.10}$$

$$S_n = P_n + \text{SNR}_{desired} \quad [\text{dBm}] \tag{18.11}$$

$$DR = 1 \, \text{dB}_{point} - S_n \quad [\text{dB}] \tag{18.12}$$

18.2 Exercises

18.1 AM receiver

1. Sketch a block diagram of a hypothetical heterodyne AM receiver, and sketch approximate
waveform shapes at output of each stage. The
transmitted message $b(t)$ waveform is shown
in Fig. 18.1.

Fig. 18.1 Example 18.1-1

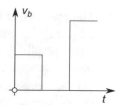

2. A double conversion AM receiver architecture is based on two IF frequencies, f_{IF1} and f_{IF2},
and the receiver is tuned to a f_{RF} signal. Sketch block diagram of possible AM receiver and find
frequencies of the local oscillators as well as the image frequencies.
Data: $f_{\text{IF1}} = 10.7 \, \text{MHz}$, $f_{\text{IF2}} = 455 \, \text{kHz}$, $f_{\text{RF}} = 20 \, \text{MHz}$.

3. An AM receiver is designed to receive RF signals in the f_{min} to f_{max} frequency range with the
required bandwidth of BW at f_0. The resonant frequency of RF amplifier is set by LC resonator.

   (a) Calculate bandwidth at f_{max} and the required capacitance C.

   (b) Calculate bandwidth at f_{min} and the required capacitance C.

   (c) Comment on the results.

   Data: $f_{\text{min}} = 500 \, \text{kHz}$, $f_{\text{max}} = 1600 \, \text{kHz}$, $BW = 10 \, \text{kHz}$, $f_0 = 1050 \, \text{kHz}$, $L = 1 \, \mu\text{H}$.

4. An AM receiver is designed to receive RF
signals in the f_{min} to f_{max} frequency range,
where the incoming RF signals are shifted to
intermediate frequency f_{IF}. There is a knob
that simultaneously tunes resonating capacitors
in both RF and LO oscillator sections, see
Fig. 18.2.

   Data: $f_{\text{min}} = 500 \, \text{kHz}$, $f_{\text{max}} = 1600 \, \text{kHz}$,
   $f_{\text{IF}} = 465 \, \text{kHz}$.

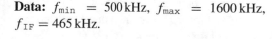

   (a) Calculate the tuning ratio $C_{\text{RF}}(max)/C_{\text{RF}}(min)$.

   (b) Calculate the tuning ratio $C_{\text{LO}}(max)/C_{\text{LO}}(min)$.

   (c) Recommend the resonating frequency f_0
   for the local oscillator.

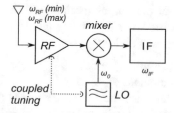

Fig. 18.2 Example 18.1-4

5. A receiver whose IF frequency is f_{IF} is tuned to an RF signal with f_{RF}. No other transmission frequency is allowed within the RF band Δ_{RF}. However, for the sake of argument, let us imagine the existence of a nearby non-linear transmitter whose emitting frequency spectrum consists of its both first and second harmonics.

Aside from the obvious $f_{\mathrm{RF}} \pm \Delta_{\mathrm{RF}}$ frequency range, what other frequency range(s) should also be prohibited for the external transmitter?

Data: $f_{\mathrm{IF}} = 455\,\mathrm{kHz}$, $f_{\mathrm{RF}} = 950\,\mathrm{kHz}$, $\Delta_{\mathrm{RF}} = \pm 10\,\mathrm{kHz}$.

6. AM transmitters operate in the $f_{\mathrm{RF}}(min)$ to $f_{\mathrm{RF}}(max)$ range with Δf bandwidth while using f_{IF} frequency. Estimate the range of the local oscillator frequencies f_{LO} and suggest bandpass filter(s) suitable for use with AM medium wave receiver?

Data: $f_{\mathrm{RF}}(min) = 540\,\mathrm{kHz}$, $f_{\mathrm{RF}}(max) = 1610\,\mathrm{kHz}$, $\Delta f = 10\,\mathrm{kHz}$, $f_{\mathrm{IF}} = 455\,\mathrm{kHz}$.

18.2 *   FM transmission

1. The local VCO of FM transmitter is controlled by audio signal to generate waveform whose frequency is centred at f_0, Fig. 18.3. The oscillator shifts its frequency by Δf_0 when the audio signal amplitude is V_{in}. Each FM station is assigned a frequency channel that is B wide. At the point of antenna, find:

(a) the carrier rest frequency f_c;
(b) the carrier frequency deviation Δf_c;
(c) the FM modulation factor m_f; and
(d) peak-to-peak voltage of the message signal V_{in} needed to achieve 100% modulation.

Fig. 18.3 Example 18.2-1

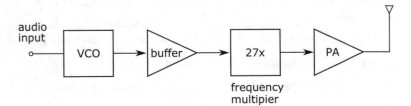

Data: $f_0 = 3.5\,\mathrm{MHz}$, $\Delta f_0 = \pm 1.6\,\mathrm{kHz}$ $V_{\mathrm{in}} = 3.6\,\mathrm{V_{pp}}$, $B = 150\,\mathrm{kHz}$.

2. A hypothetical FM signal whose centre frequency is f_0 and its frequency deviation is Δf_0 is passed through two frequency multipliers, Fig. 18.4. The two newly created waveforms are then added together by a summing block. Signal coming out of the summing block is passed through LC resonator whose centre frequency and bandwidth are aligned with the lower-side band (LSB) part of the output spectrum. At the output of the LC resonator, power of the LSB waveform is measured as P_{out} at the edges of its bandwidth.

Assuming an ideal system:

(a) calculate $f_1 \pm \Delta f_1$;
(b) calculate $f_2 \pm \Delta f_2$;
(c) calculate the LC resonator's Q factor;
(d) sketch detailed plot of the output power spectrum. For the vertical axis, use [dBm] units, and for the horizontal axis, use [Hz] units. Clearly show power levels of the relevant tones.

Data: $f_0 = 200\,\text{kHz}$, $\Delta f_0 = \pm 200\,\text{Hz}$, $P_{out} = 1\,\text{mW}$.

Fig. 18.4 Example 18.2-2

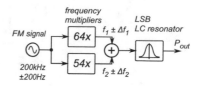

1. Given the input–output power transfer characteristics of an amplifier as given in Fig. 18.5.
   Estimate:

   (a) the power gain,
   (b) 1 dB compression point, and
   (c) the third order intercept point.

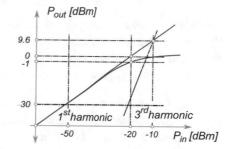

Fig. 18.5 Example 18.3-1

18.3 ** Non-linear effects

2. Cosine waveform current is measured at the output of a G_m non-inverting amplifier at three values
   of the input voltage: biasing V_b, maximum V_{max}, and minimum V_{min}, as

   $$I_{max}(V_{max}) = 1\,\text{mA}, \quad I_b(V_b) = 0.01\,\text{mA}, \quad I_{min}(V_{min}) = -0.95\,\text{mA},$$

   where V_b is the biasing voltage at the midpoint between the maximum and minimum input voltage
   amplitudes. Based on the available data, estimate the THD of this amplifier.

3. Determine the sensitivity S of RF receiver at room temperature.
   Data: NF $= 5\,\text{dB}$, $BW = 1\,\text{MHz}$, $\text{SNR}_{\text{desired}} = 10\,\text{dB}$.

4. Given that the block diagram of a hypothetical three-stage RF amplifier is in Fig. 18.6, where
   power level P_{in} of the RF source signal corresponds to the 1 dB compression point and bandwidth
   of the LC resonator B is set by the design of RF amplifier's LC resonator. Assuming that the
   total input side noise is generated only by the equivalent input resistance while each gain stage
   generates its own noise, and the load resistor R_L is ideal, estimate:

   (a) power gain of the second stage only A_2 in [dB];
   (b) noise figure NF of this amplifier;
   (c) noise voltage v_n at the output; and
   (d) dynamic range DR of this amplifier.

   Data: $R_L = 100\,\Omega$, $T = 26.85\,°\text{C}$, $A_1 = 10\,\text{dB}$, $NF_2 = 6.02\,\text{dB}$, $NF_3 = 9.0309\,\text{dB}$.

Fig. 18.6 Example 18.3-4

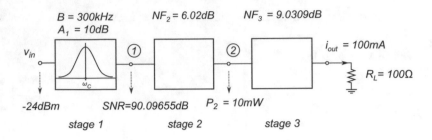

$B = 300kHz$ $NF_2 = 6.02dB$ $NF_3 = 9.0309dB$
$A_1 = 10dB$

$i_{out} = 100mA$

$R_L = 100\Omega$

$-24dBm$ $SNR=90.09655dB$ $P_2 = 10mW$

stage 1   stage 2   stage 3

5. Given the data and assuming that non-linearity of cell phone receiver is described by the following $y(b)$ function:

$$y(b) = 2b(t) - 0.267b(t)^3,$$

where $b(t)$ is the incoming signal received by cell phone. In the following scenario, there are two incoming signals: (1) the desired incoming $b_1(t) = 0.1\,\mathrm{V}\ \sin\omega_1 t$; and (2) at the same time when $b_1(t)$ signal is being received, in close proximity there is another cell phone transmitting signal $b_2(t) = A_2\ \sin\omega_2 t$, consequently $(A_2 \gg 0.1\,\mathrm{V})$. Given the described situation, calculate:

(a) the 1 dB compression point in [dB];
(b) the input intercept point IIP3 in [dB];
(c) receiver sensitivity S in [dBm];
(d) maximum dynamic range DR in [dB], and effective dynamic range DR in [dB] (assuming 2/3 of the maximum DR);
(e) amplitude of the second signal A_2 in [V] that would cause the "signal blocking" effect. In that case, when both b_1 and b_2 signals are present, we assume

$$y(b_1, b_2) \approx \left(a_1 + \frac{3}{2}\, a_3\, A_2^2\right) A_1 \cos\omega_1 t.$$

Data: $T = 16.85\,^{\circ}\mathrm{C}$, NF $= 20\,\mathrm{dB}$ $B = 1\,\mathrm{MHz}$, SNR $= 0\,\mathrm{dB}$.

Solutions

Exercise 18.1, page 353

1. The incoming RF signal is recovered in multiple stages, step by step. A typical hypothetical AM receiver my look as follows, see Fig. 18.7.

(a) The space around us is filled with RF EM waves v_{RF} from various transmitting sources that emit at the same time, as a consequence v_{RF} may look like a random signal whose frequency spectrum looks as if representing white noise.
(b) Radio antenna followed by the matching network, due to resonant narrow band BP transfer characteristics, is the first stage of signal recovering that serves as "entrance door" into receiver. Thus, the filtering effect generates a HF signal v_{AM} that is ready for down-conversion.
(c) The filtered v_{AM} waveform is mixed (i.e. multiplied) with the local voltage controlled oscillator's tone v_{VCO} to produce down-converted waveform v_{AM-}.
(d) It is the role of IF amplifier to filter and select LSB waveform v_{IF}.
(e) The filtered LSB waveform v_{IF} is suitable to enter envelope detector that extracts the transmitted message v_{pk}.

(f) The extracted envelope waveform v_{pk} is filtered with LPF or BPF to remove HF noise and to generate high fidelity version of the transmitted message $vpkLP$.

(g) The last stage consists of audio amplifier that generates the recovered message v_b so that it can be played by speakers.

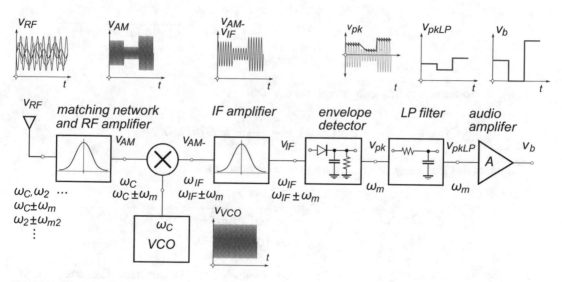

Fig. 18.7 Example 18.1-1

2. Summary of this superheterodyne architecture is shown in Fig. 18.8.

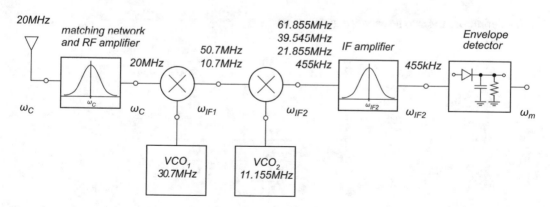

Fig. 18.8 Example 18.1-2

3. By definition, given the data, LC resonator's Q factor is calculated as follows:

$$Q = \frac{f_0}{BW} = \frac{1050\,\text{kHz}}{10\,\text{kHz}} = 105$$

Therefore, assuming constant Q factor, we calculate:

(a) At the upper end of the frequency range, we find

$$BW(f_{\max}) = \frac{1.6\,\text{MHz}}{105} = 15.238\,\text{kHz}$$

and, at this frequency, it is necessary to tune capacitor to

$$f_{\max} = \frac{1}{2\pi\sqrt{LC_{\min}}} \quad \therefore \quad C_{\min} = \frac{1}{(2\pi\,f_{\max})^2\,L} = \frac{1}{(2\pi \times 1.6\,\text{MHz})^2\,1\,\mu\text{H}}$$

$$= 9.895\,\text{nF}$$

(b) At the lower end of frequency range, we calculate

$$BW(f_{\min}) = \frac{500\,\text{kHz}}{105} = 4.762\,\text{kHz}$$

and, at this frequency, it is necessary to tune capacitor to

$$f_{\min} = \frac{1}{2\pi\sqrt{LC_{\max}}} \quad \therefore \quad C_{\max} = \frac{1}{(2\pi\,f_{\min})^2\,L} = \frac{1}{(2\pi \times 500\,\text{kHz})^2\,1\,\mu\text{H}}$$

$$= 101.321\,\text{nF}$$

(c) Assuming constant Q factor and inductor, bandwidth is not constant. Consequently, this receiver should be used to process signals whose bandwidth is $BW(b(t)) \leq 4.762\,\text{kHz}$. However, in order to avoid the inter-channel interference, the channel spacing Δf between various AM radio stations must be set to $\Delta f \geq BW(f_{\max}) = 15.238\,\text{kHz}$ assuming that "guard band" must also be added between the channels.

4. By definition, the two resonant frequencies are

$$f_{\max} = \frac{1}{2\pi\sqrt{LC_{\min}}} \quad \text{and} \quad f_{\min} = \frac{1}{2\pi\sqrt{LC_{\max}}}$$

and, therefore,

$$\frac{f_{\max}}{f_{\min}} = \frac{\dfrac{1}{2\pi\sqrt{LC_{\min}}}}{\dfrac{1}{2\pi\sqrt{LC_{\max}}}} = \sqrt{\frac{C_{\max}}{C_{\min}}}$$

(a) Given the data, the ratio of maximum and minimum RF frequencies leads to

$$\sqrt{\frac{C_{RF}(max)}{C_{RF}(min)}} = \frac{f_{RF}(max)}{f_{RF}(min)} = \frac{1.6\,\text{MHz}}{500\,\text{kHz}} = 3.2 \quad \therefore \quad \frac{C_{RF}(max)}{C_{RF}(min)} = 10.24$$

(b) However, there are two possibilities at the mixer's output:

a. Case $f_{\mathrm{LO}} > f_{\mathrm{RF}}$: in this case, we find

$$f_{\mathrm{LO}}\,(min) = f_{\mathrm{RF}}\,(min) + f_{\mathrm{IF}} = 500\,\mathrm{kHz} + 465\,\mathrm{kHz} = 965\,\mathrm{kHz}$$

$$f_{\mathrm{LO}}\,(max) = f_{\mathrm{RF}}\,(max) + f_{\mathrm{IF}} = 1600\,\mathrm{kHz} + 465\,\mathrm{kHz} = 2065\,\mathrm{kHz}$$

Again, given the data, we calculate

$$\sqrt{\frac{C_{\mathrm{LO}}\,(max)}{C_{\mathrm{LO}}\,(min)}} = \frac{f_{\mathrm{LO}}\,(max)}{f_{\mathrm{LO}}\,(min)} = \frac{2065\,\mathrm{kHz}}{965\,\mathrm{kHz}} = 2.14$$

$$\therefore$$

$$\frac{C_{\mathrm{LO}}\,(max)}{C_{\mathrm{LO}}\,(min)} = 4.58$$

b. Case $f_{\mathrm{LO}} < f_{\mathrm{RF}}$: in this case, we find

$$f_{\mathrm{LO}}\,(min) = f_{\mathrm{RF}}\,(min) - f_{\mathrm{IF}} = 500\,\mathrm{kHz} - 465\,\mathrm{kHz} = 35\,\mathrm{kHz}$$

$$f_{\mathrm{LO}}\,(max) = f_{\mathrm{RF}}\,(max) - f_{\mathrm{IF}} = 1600\,\mathrm{kHz} - 465\,\mathrm{kHz} = 1135\,\mathrm{kHz}$$

Then, given the data, we calculate

$$\sqrt{\frac{C_{\mathrm{LO}}\,(max)}{C_{\mathrm{LO}}\,(min)}} = \frac{f_{\mathrm{LO}}\,(max)}{f_{\mathrm{LO}}\,(min)} = \frac{1135\,\mathrm{kHz}}{35\,\mathrm{kHz}} = 32.43$$

$$\therefore$$

$$\frac{C_{\mathrm{LO}}\,(max)}{C_{\mathrm{LO}}\,(min)} = 1052$$

(c) The above results provide guidelines for choosing the receiver architecture and the tuning range of capacitors.

Obviously, it is not reasonable to choose tune-able capacitor whose tuning ratio is greater than one thousand, thus it is more logical to choose $f_{\mathrm{LO}} > f_{\mathrm{RF}}$ that can be realized with the tuning ratio of less than five, Fig. 18.9. Therefore, we choose the tuning range of C_{LO} from 965 to 2065 kHz.

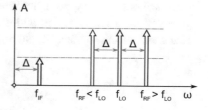

Fig. 18.9 Example 18.1-4

5. Receiver's RF front end serves as an entrance door that allows only tones whose frequencies are within the LC resonator's bandwidth to enter the mixer stage. We also keep in mind that the local oscillator frequency f_{LO} is not visible to the outside world, i.e. only frequencies that pass through the LC bandpass filter are multiplied by the oscillator's frequency. We also note that, while frequency shifting merely translates bandwidth along the frequency axis without

changing its width (because the shifting is the result of addition/subtraction of two frequencies), plain frequency multiplication (i.e. multiplication by a constant) affects the bandwidth as well. Depending upon relationship between f_{RF} and f_{Tx} (because $f_{RF} \neq f_{Tx-I}$), there are two possibilities that we should consider:

1. *Case $f_{Tx} > f_{RF}$*: If the local oscillator's frequency f_{LO} is lower than the interfering Tx signal, then both f_{Tx-I} and its second harmonic $f_{Tx-II} = 2 \times f_{Tx-I}$ are far away from the receiver's input RF frequency band f_{RF}, Fig. 18.10. Therefore, both frequencies are rejected by the input LC bandpass filter.

Fig. 18.10
Example 18.1-5

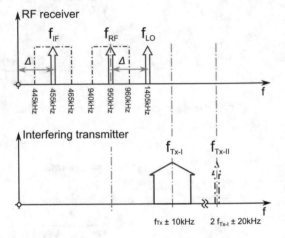

2. *Case $f_{Tx} < f_{RF}$*: If the interfering signals are to reach the mixer, their frequencies must be aligned with the LC bandpass filter's frequency range $f_{RF} = 950 \pm 200$ kHz. The first harmonic is already prohibited from that range, thus we take a look at what happens if the second harmonic is aligned with $f_{Tx-II} = f_{RF}$ band, Fig. 18.11.

In that case, the first harmonic must be further down the frequency axis, i.e. $f_{Tx-I} = \frac{1}{2} f_{Tx-I} = 475 \pm 5$ kHz. This frequency alignment results in the transmitter's second harmonic f_{Tx-II} entering the receiver along with the intended RF signal f_{RF} and reaches the mixer's input terminals. Output terminals of the mixer then generate f_{IF} waveform that is modulated by both the intended RF signal and the unintended Tx signal, thus the received message is ruined and cannot be recovered. Therefore, in this case, we conclude that transmitting stations should not be allowed to operate in the $f_{Tx} = 475 \pm 5$ kHz range.

Fig. 18.11
Example 18.1-5

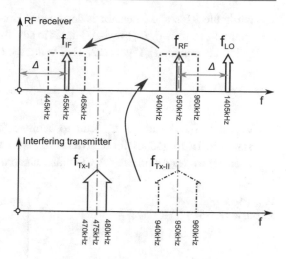

6. A bundle of frequencies close to each other can be visualized as a multi-wire cable, where each wire is reserved for one broadcasting channel. In this example, each channel occupies Δf-wide frequency space (similar to wire diameter), and thus within the medium wave AM band it is possible to have the total of n channels, i.e.

$$n = \frac{f_{RF}(max) - f_{RF}(min)}{\Delta f} = \frac{(1610 - 540)\,\text{kHz}}{10\,\text{kHz}} = 107$$

if the assumptions are that there is no need for "guard bands" (i.e. frequency spacing in between the neighbouring channels to account for manufacturing tolerances) and that all channels are available for communication (i.e. for the moment, we ignore the image frequencies inside the AM band). In order to shift all channels down to f_{IF}, the local VCO must be able to generate frequencies

$$f_{LO}(min) = f_{IF} + f_{RF}(min) = 455\,\text{kHz} + 540\,\text{kHz} = 995\,\text{kHz}$$

$$f_{LO}(max) = f_{IF} + f_{RF}(max) = 455\,\text{kHz} + 1610\,\text{kHz} = 1965\,\text{kHz}$$

$$\therefore$$

$$\frac{f_{LO}(max)}{f_{LO}(min)} = 1.975$$

which is, being less than two, considered an easy tuning ratio to design. Let us now estimate bandpass filter's Q factor that can provide $BW = \Delta f$ for each channel.

$$Q_{max} = \frac{f_{RF}(max)}{BW} = \frac{1610\,\text{kHz}}{10\,\text{kHz}} = 161$$

$$Q_{min} = \frac{f_{RF}(min)}{BW} = \frac{540\,\text{kHz}}{10\,\text{kHz}} = 54$$

Obviously, holding a constant $BW = 10\,\text{kHz}$ over the whole AM band is not a trivial requirement because Q factor is set by properties of RLC components in the RF resonator. One way to deal with the problem would be to provide the fixed $BW = 10\,\text{kHz}$ bandwidth filtering at the IF stage, while the RF stage resonator is designed to provide minimal bandwidth that allows all individual channels to pass (i.e. $Q = 54$). That is to say, the bandwidth is finally adjusted at the IF frequency (which is constant), Fig. 18.12. That being the case, bandpass filter is designed at IF frequency as

$$Q = \frac{f_{IF}}{BW} = \frac{455\,\text{kHz}}{10\,\text{kHz}} = 45.5$$

which is fixed and not difficult to achieve. This example illustrates one advantage of using standard IF frequency for a given band, which simplifies the overall receiver design because it is easier to design tuneable VCO than tuneable fixed bandwidth bandpass filter.

Fig. 18.12
Example 18.1-6

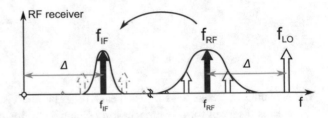

Exercise 18.2, page 354

1. As oppose to frequency shifter (which requires a mixer), plain frequency multiplication does not produce the sums and difference frequencies. Therefore, by inspection of block diagram in Fig. 18.3, we calculate as follows:

(a) The output carrier frequency is $f_c = 27 \times f_0 = 94.5\,\text{MHz}$.
(b) The output carrier deviation is $\Delta f_c = 27 \times \Delta f_0 = \pm 43.2\,\text{kHz}$, which is equivalent to finding a difference

$$\Delta f_{cout} = f_{cmax} - f_{cmin} = 94.5432\,\text{MHz} - 94.4568\,\text{MHz} = 86.4\,\text{kHz}$$

(c) Frequency modulation is found as

$$m_f = \frac{\Delta f_c}{B/2} \times 100\% = \frac{43.2\,\text{kHz}}{75\,\text{kHz}} \times 100\% = 57.6\%$$

where we keep in mind that frequency modulation occupies half bandwidth for the positive frequency deviation and half bandwidth for the negative frequency deviation.

(d) By simple proportion, we find

$$\frac{100\%}{57.6\%} = \frac{V_{max}}{V_{pp}} \qquad \therefore \qquad V_{max} = 6.25\,V_{pp}$$

2. Electronic circuits are, in reality, just implementation of the system-level mathematical functions. Hypothetical system in Fig. 18.4 takes a range of tones and applies mathematical operations to produce the output waveform whose power is measured.

(a) The input FM signal occupies the frequency range of 200 kHz $\pm$ 200 Hz, that is from 199.8 to 200.2 kHz range. Therefore,

$$f_1(max) = 64 \times 200.2\,\text{kHz} = 12.8128\,\text{MHz}$$
$$f_1(min) = 64 \times 199.8\,\text{kHz} = 12.7872\,\text{MHz}$$

$$\therefore$$

$$f_1 = 12.8\,\text{MHz} \pm 12.8\,\text{kHz}$$

(b) Similarly,

$$f_2(max) = 54 \times 200.2\,\text{kHz} = 10.8108\,\text{MHz}$$
$$f_2(min) = 54 \times 199.8\,\text{kHz} = 10.7892\,\text{MHz}$$

$$\therefore$$

$$f_2 = 10.8\,\text{MHz} \pm 10.8\,\text{kHz}$$

(c) It is important to note that linear addition of two waveforms does not introduce new tones in the output waveform (frequency shifting is done by multiplication of two waveforms, which is non-linear operation). Thus, the output of the summing block contains only two frequency bands, $f_1 \pm \Delta f_1$ and $f_2 \pm \Delta f_2$, which is to say that LSB is centred around $f_2 = 10.8\,\text{MHz}$ with the bandwidth of $B = \Delta f_2 = 21.6\,\text{kHz}$. Thus the LSB bandwidth occupies from 10.7892 to 10.8108 MHz, and it requires

$$Q = \frac{f_2}{B} = \frac{10.8\,\text{MHz}}{21.6\,\text{kHz}} = 500$$

(d) Waveform bandwidth B is defined at one-half of the maximum power level at the centre frequency, which is by definition $-3.01\,\text{dB}$, and in this example, it is centred on LSB band at $f_2 = 10.8\,\text{MHz}$. In other words, if the measured power at the edge of LSB waveform bandwidth is 1 mW, which is by definition equivalent to 0 dBm, then the maximum power is by definition doubled, i.e. $P_{f0} = 2\,\text{mW}$, which is equivalent to 3.01 dBm.

The upper-side band (USB) in this example occupies frequency space from $f_1(max) = 12.8128\,\text{MHz}$ to $f_1(min) = 12.7872\,\text{MHz}$. In order to find attenuation of these two tones, we apply the attenuation formula for tones that are not centred at the resonant frequency f_0 and for the given Q as

$$y_1 = \frac{f_1(max)}{f_2} - \frac{f_2}{f_1(max)} = \frac{12.8128\,\text{MHz}}{10.8\,\text{MHz}} - \frac{10.8\,\text{MHz}}{12.8128\,\text{MHz}} = 0.343$$

$$y_2 = \frac{f_1(min)}{f_2} - \frac{f_2}{f_1(min)} = \frac{12.7872\,\text{MHz}}{10.8\,\text{MHz}} - \frac{10.8\,\text{MHz}}{12.7872\,\text{MHz}} = 0.339$$

which we use to calculate the attenuation relative to the maximum level as

$$A_r(12.8128\,\text{MHz}) = \frac{1}{\sqrt{1 + (Q \times y_1)^2}} = \frac{1}{171.735} = -22.349\,\text{dB}$$

$$A_r(12.7872\,\text{MHz}) = \frac{1}{\sqrt{1 + (Q \times y_2)^2}} = \frac{1}{169.706} = -22.297\,\text{dB}$$

These two equations gave us relative ratio of the maximum power ($+3$ dBm) and each of the tones, and thus we write

$$A_r(12.8128\,\text{MHz}) = 3.01\,\text{dBm} - 22.349\,\text{dB} = -19.338\,\text{dBm}$$

$$A_r(12.7872\,\text{MHz}) = 3.01\,\text{dBm} - 22.297\,\text{dB} = -19.287\,\text{dBm}$$

These results are summarized in Fig. 18.13.

Fig. 18.13
Example 18.2-2

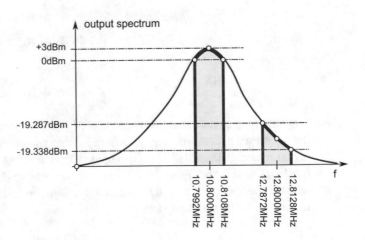

Exercise 18.3, page 355

1. By inspection of the graph in Fig. 18.5, we conclude the following:

1. *Gain*: in the linear part of the transfer characteristics for the input of -50 dBm, the output power is -30 dBm, and hence the gain is 20 dB. The same gain is expected for the input level of -20 dB because the predicted linear output level is 0 dB, i.e. 20 dB above the input level, which is by definition the amplifier gain.
2. 1 dB *compression point*: the linear part of the characteristics extends to approximately -20 dBm of the input power, when the output power becomes -1 dBm instead of the expected 0 dBm. Therefore, the 1 dBcompression point is at -20 dBm of the input power.
3. *The third order intercept point IIP3*: harmonics power of the third harmonic is extrapolated until its intersection with the extrapolated linear gain, and the crossing point is found at the output power of approximately $+9.6$ dBm, which is only *extrapolated* point, not the real measurement point. Keep in mind that the amplifier output *never* reaches that level of output power, as it has already saturated close to the 1 dB compression point level. It should be apparent that 1 dB compression IIP3 points are closely related and can be used interchangeably to quantify the amplifier non-linearity.

2. We define a *memoryless system* as one whose output signal $y(t)$ does not depend on the past values of its input $b(t)$, thus we can define its transfer function by using general polynomial relation:

$$y(t) = a_0 + a_1 b(t) + a_2 b^2(t) + a_3 b^3(t) + \cdots , \qquad (18.13)$$

where a_i are constant in time. Assuming that a single-tone input signal

$$b(t) = B \cos \omega t$$

is placed into (18.13) and after DC term removed (i.e., $a_0 = 0$) from the non-linear transfer function, the output signal waveform is

$$y(t) = a_1 B \cos \omega t + a_2 B^2 \cos^2 \omega t + a_3 B^3 \cos^3 \omega t + \cdots \qquad (18.14)$$

$$= a_1 B \cos \omega t + \frac{a_2 B^2}{2} (1 + \cos 2\omega t) + \frac{a_3 B^3}{4} (3 \cos \omega t + \cos 3\omega t) + \cdots$$

$$= \frac{a_2 B^2}{2} + \left(a_1 B + \frac{3 a_3 B^3}{4} \right) \cos \omega t + \frac{a_2 B^2}{2} \cos 2\omega t + \frac{a_3 B^3}{4} \cos 3\omega t + \cdots$$

$$= b_0 + b_1 \cos \omega t + b_2 \cos 2\omega t + b_3 \cos 3\omega t + \cdots \qquad (18.15)$$

where b_0 is the output signal's DC term. The three experimental data points correspond to the cosine wave input voltage function (the non-inverting amplifier), and thus we find their respective ωt angles by inspection of graph in Fig. 18.14. Because there are three measurement points, we can calculate only three constants $b_0, b_1,$ and b_2.

First, we find

$$V_{in} = V_{max} \quad \therefore \quad \omega t = 0$$

$$V_{in} = V_b \quad \therefore \quad \omega t - \frac{\pi}{2}$$

$$V_{in} = V_{min} \quad \therefore \quad \omega t = \pi$$

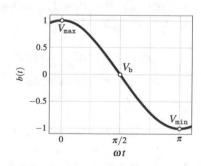

Fig. 18.14 Example 18.3-2

then we substitute the angular arguments into (18.15), where $y(t)$ is $i(t)$, to calculate

$$i(0) = b_0 + b_1 \cos(0) + b_2 \cos(0) \quad \therefore \quad b_0 + b_1 + b_2 = 1 \, \text{mA}$$

$$i(\pi/2) = b_0 + b_1 \cos(\pi/2) + b_2 \cos(\pi) \quad \therefore \quad b_0 - b_2 = 0.01 \, \text{mA}$$

$$i(\pi) = b_0 + b_1 \cos(\pi) + b_2 \cos(2\pi) \quad \therefore \quad b_0 - b_1 + b_2 = -0.95 \, \text{mA}$$

where the solution of the above system of equations is

$$b_0 = 17.5 \, \mu\text{A}, \quad b_1 = 975 \, \mu\text{A}, \quad b_2 = 7.5 \, \mu\text{A}$$

By definition, we write

$$D_2 = \frac{b_2}{b_1} \times 100\% = 0.77$$

$$THD = \sqrt{D_2^2 + D_3^2 + D_4^2 \cdots} = \sqrt{D_2^2} = D_2 = 0.77\%$$

because with only three measurements we can solve for up to the second order term in (18.15). In the case of more measurement, for instance, five measured points would add amplitudes at $V_{\text{in}} (\pm 1/2)$, then we would have $\omega t = \pi/3$ and $\omega t = 2\pi/3$ arguments as well, which would enable us to calculate $b_0, b_1, b_2, b_3,$ and b_4 constants.

3. After textbook definitions of noise power P_n and sensitivity S

$$P_n = -174\,\text{dBm} + 10 \log \Delta f + \text{NF} \quad [\text{dBm}]$$

and

$$S = P_n + \text{SNR}_{\text{desired}} \quad [\text{dBm}] \tag{18.16}$$

we calculate

$$S = -174\,\text{dBm} + 10 \log(1\,\text{MHz}) + 5\,\text{dB} + 10\,\text{dB} = -99\,\text{dBm}$$

which is a relatively typical number for state-of-the-art receivers.

4. By inspection of the equivalent block diagram of a three-stage RF amplifier, Fig. 18.6, we write:

(a) Power gain of the second stage A_2 is found as difference between power levels at node ② and node ① (i.e. $P_2 - P_1$) and by following the signal power from the input node, and thus we simply write

$$P_{\text{in}} = -24\,\text{dBm} \ \text{ and, } \ A_1 = 10\,\text{dB} \quad \therefore \quad P_1 = -24\,\text{dBm} + 10\,\text{dB} = -14\,\text{dBm}$$

$$P_2 = 10\,\text{mW} = 10\,\text{dBm}$$

$$\therefore$$

$$A_2 = P_2 - P_1 = 10\,\text{dBm} - (-14\,\text{dBm}) = 24\,\text{dB}$$

(b) Gain of the last stage in amplifier is found by looking at difference between the output power and power level at node② as

$$P_{\text{out}} = i_{out}^2 R_L = 1\text{W} = 30\,\text{dBm}$$

$$\therefore$$

$$A_3 = P_{\text{out}} - P_2 = 30\,\text{dBm} - 10\,\text{dBm} = 20\,\text{dB}$$

which gives the total gain A_{tot} of the three-stage amplifier as

$$A_{\text{tot}} = A_1 + A_2 + A_3 = 10\,\text{dB} + 24\,\text{dB} + 20\,\text{dB} = 54\,\text{dB} = 251,188$$

The input side bandwidth is limited by LC resonator, and thus the noise power is found as

$$P_n(in) = k\,T\,B_{\text{eff}} = k\,T\,\frac{\pi}{2}\,B = 1.951 \times 10^{-15}\,\text{W} = -117.0976\,\text{dBm}$$

By knowing the input signal power P_{in} and the input noise power $P_n(in)$ by definition, we find the input side SNR_{in} as

$$SNR_{\text{in}} = \frac{P_{\text{in}}}{P_n(in)} = -24\,\text{dBm} - (-117.0976\,\text{dBm}) = 93.0976\,\text{dB}$$

Signal to noise ratio at the output of the first stage is given as $\text{SNR}_o = 90.09655\,\text{dB}$. From these two numbers, we find noise figure of the first stage by definition as

$$\text{NF}_1 = \text{SNR}_o - SNR_{\text{in}} = 3.010\,\text{dB}$$

The noise figure of three-stage amplifier is calculated by apply Friis's formula as

$$\text{NF}_1 = 3.010\,\text{dB} \quad \therefore \quad F_1 = 2$$

$$\text{NF}_2 = 6.020\,\text{dB} \quad \therefore \quad F_1 = 4$$

$$\text{NF}_3 = 9.030\,\text{dB} \quad \therefore \quad F_1 = 8$$

$$\therefore$$

$$F = F_1 + \frac{F_2 - 1}{A_1} + \frac{F_3 - 1}{A_1\,A_2} = 2 + \frac{3}{10} + \frac{7}{2511} = 2.303$$

which means that the total noise figure is $\text{NF} = 10\,\log 2.303 = 3.622\,\text{dB}$.

(c) Therefore, the total output noise power is found as the sum of amplified input noise power plus the internally generated noise, i.e.

$$P_n(out) = P_n(in) + A(tot) + \text{NF}(tot) = -59.4756\,\text{dBm}$$

$$\therefore$$

$$P_n(out) = 1.128\,\text{nW}$$

$$\therefore$$

$$v_n(out) = \sqrt{P_n(out)\,R_L} \approx 336\,\mu\text{V}$$

(d) The total signal to noise ratio at the output node is

$$\text{SNR}_{tot} = \frac{P_s(out)}{P_n(out)} = 30\,\text{dBm} - (-54.4756\,\text{dBm}) = 89.4756\,\text{dBm}$$

Dynamic range of this amplifier is summarized in Fig. 18.15. We conclude that this amplifier must be improved because as it stands it cannot process any useful signal: as its sensitivity is at $-24\,\text{dBm}$, which is as same as the 1 dB compression point. Consequently, $DR = 0\,\text{dB}$.

Fig. 18.15
Example 18.3-4

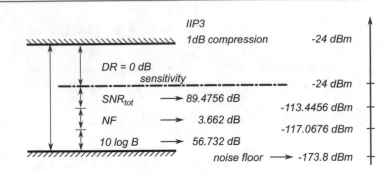

5. Reception of wanted but weak signal from a faraway transmitter in the presence of unwanted but strong signal from nearby transmitter, for example in a crowded bus when two cell phones are next to each other, may result in unexpected side effects.

(a) Quantitative measure of an amplifier's non-linearity is its 1 dB compression point, which is calculated from its non-linear transfer function as

$$A_{1\,dB} = \sqrt{0.145\,\frac{|a_1|}{|a_3|}} = \sqrt{0.145\,\frac{2}{0.267}} = 1.042$$

$$\therefore$$

$$A_{1\,dB} = 10\,\log 1.042 \cong 0.179\,\text{dB}$$

(b) Similarly, the third order intercept point IIP3 is also calculated from the non-linear transfer function as

$$\text{IIP3} = \sqrt{\frac{4}{3}\,\frac{|a_1|}{|a_3|}} = \sqrt{\frac{4}{3}\,\frac{2}{0.267}} = 3.16$$

$$\therefore$$

$$\text{IIP3[dB]} = 10\,\log 3.16 \cong 5\,\text{dB}$$

(c) Sensitivity of a receiver is calculated relative to the thermal noise floor. At room temperature (290 K), given the data, the receiver sensitivity is

$$S = -174\,\text{dBm} + \text{NF} + 10\,\log B + \text{SNR}$$
$$= -174\,\text{dBm} + 20\,\text{dB} + 10\,\log 10^6 + 0\,\text{dB}$$
$$= -94\,\text{dBm}$$

(d) Dynamic range DR is calculated as difference between the sensitivity level (i.e. the noise floor) and either IIP3[dB] or $A_{1\,\text{dB}}$[dB], and hence

$$DR = \text{IIP3} - S = 5\,\text{dB} - (-94\,\text{dBm}) = 99\,\text{dB}$$

$$\therefore$$

$$DR_{\text{eff}} = \frac{2}{3}\,99\,\text{dB} = 66\,\text{dB}$$

$$DR = A_{1\,\text{dB}} - S = 0.179\,\text{dB} - (-94\,\text{dBm}) \cong 94\,\text{dB}$$

$$\therefore$$

$$DR_{\text{eff}} = \frac{2}{3}\,94\,\text{dB} \approx 62.7\,\text{dB}$$

(e) The output signal level $y(x)$ drops down to zero, and therefore the intended receiving signal is blocked, when

$$y(x) = 0 \quad \therefore \quad \left(a_1 + \frac{3}{2}\,a_3 A_2^2\right) = 0$$

$$\therefore$$

$$A_2 = \sqrt{\frac{2a_1}{3(-a_3)}} = \sqrt{\frac{2 \times 2}{3 \times 0.267}} = 2.234\,\text{V}$$

Physical Constants and Engineering Prefixes

Table A.1 Basic physical constants

Physical constant	Symbol	Value
Speed of light in vacuum	c	299 792 458 m/s
Magnetic constant (vacuum permeability)	μ_0	$4\pi \times 10^{-7}$ N/A^2
Electric constant (vacuum permittivity)	$\varepsilon_0 = 1/(\mu_0 c^2)$	$8.854\ 187\ 817 \times 10^{-12}$ F/m
Characteristic impedance of vacuum	$Z_0 = \mu_0 c$	$376.730\ 313\ 461\ \Omega$
Coulomb's constant	$k_e = 1/4\pi\varepsilon_0$	$8.987\ 551\ 787 \times 10^9$ Nm^2/c^2
Elementary charge	e	$1.602\ 176\ 565 \times 10^{-19}$ C
Boltzmann constant	k	$1.380\ 6488 \times 10^{-23}$ J/K

Table A.2 Basic engineering prefix system

tera	giga	mega	kilo	hecto	deca	deci	centi	milli	micro	nano	pico	femto	atto
T	G	M	k	h	da	d	c	m	μ	n	p	f	a
10^{12}	10^9	10^6	10^3	10^2	10^1	10^{-1}	10^{-2}	10^{-3}	10^{-6}	10^{-9}	10^{-12}	10^{-15}	10^{-18}

Table A.3 SI system of fundamental units

Name	Symbol	Quantity	Symbol
Metre	m	Length	l
Kilogram	kg	Mass	m
Second	s	Time	t
Ampere	A	Electric current	I
Kelvin	K	Thermodynamic temperature ($-273.16\,°$C)	T
Candela	cd	Luminous intensity	Iv
Mole	mol	Amount of substance	n

© Springer Nature Switzerland AG 2021
R. Sobot, *Wireless Communication Electronics by Example*,
https://doi.org/10.1007/978-3-030-59498-5

Second-Order Differential Equation

The three basic elements have voltages at their respective terminals as

$$v_R = iR \quad ; \quad v_L = L\frac{di}{dt} \quad ; \quad v_C = \frac{q}{C} \tag{B.1}$$

If they are put together in a series circuit that includes a voltage source $v(t)$, after applying KVL, the circuit equation is

$$v(t) = v_L + v_R + v_C$$

$$\therefore$$

$$v(t) = L\frac{di}{dt} + i\,R + \frac{q}{C} \tag{B.2}$$

However, we know that a current is a derivative of charge with respect to time, hence we have the second-order differential equation

$$v(t) = L\frac{d^2q}{dt^2} + R\frac{dq}{dt} + \frac{1}{C}\,q$$

$$\therefore$$

$$v(t) = \frac{d^2q}{dt^2} + \frac{R}{L}\frac{dq}{dt} + \frac{1}{LC}\,q \tag{B.3}$$

This is solved, starting with its auxiliary quadratic equation

$$0 = x^2 + \frac{R}{L}x + \frac{1}{LC} \tag{B.4}$$

and its general solution with complex roots is

$$r_{1\,2} = \frac{1}{2}\left(-\frac{R}{L} \pm \sqrt{\left(\frac{R}{L}\right)^2 - \frac{4}{LC}}\right) \tag{B.5}$$

© Springer Nature Switzerland AG 2021
R. Sobot, *Wireless Communication Electronics by Example*,
https://doi.org/10.1007/978-3-030-59498-5

Complex Numbers

A complex number is a neat way of presenting a point in (mathematical) *space* with two coordinates or, equivalently, it is a neat way to write two equations in the form of one. A general complex number is $Z = a + jb$, where a and b are real numbers referred to as real and imaginary parts, i.e. $\Re(Z) = a$ and $\Im(Z) = b$. Here is a reminder of the basic operations with complex numbers. Keep in mind that $j^2 = -1$.

$$(a + jb) + (c + jd) = (a + c) + j(b + d) \tag{C.1}$$

$$(a + jb) - (c + jd) = (a - c) + j(b - d) \tag{C.2}$$

$$(a + jb)(c + jd) = (ac - bd) + j(bc + ad) \tag{C.3}$$

$$\frac{(a + jb)}{(c + jd)} = \frac{(a + jb)}{(c + jd)} \frac{(c - jd)}{(c - jd)} = \frac{ac + bd}{c^2 + d^2} + j\frac{bc - ad}{c^2 + d^2} \tag{C.4}$$

$$(a + jb)^* = (a - jb) \tag{C.5}$$

$$|(a + jb)| = \sqrt{(a + jb)(a - jb)} = \sqrt{(a^2 + b^2)} \tag{C.6}$$

It is much easier to visualize complex numbers and operations if we use vectors and the trigonometry of a right triangle, i.e. Pythagoras theorem. The imaginary part always takes its value from the y axis and the real part is always on the x axis (see Fig. C.1).

Fig. C.1 Complex numbers in $[\Re(Z) \ \Im(Z)]$ space, their equivalence to Pythagoras theorem and vector arithmetic

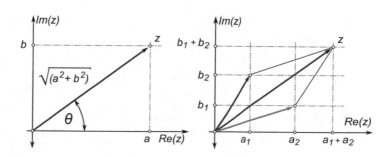

Therefore, an alternative view of complex numbers is based on geometry, i.e.

$$(a + jb) \equiv (|Z| \ \theta) \tag{C.7}$$

© Springer Nature Switzerland AG 2021
R. Sobot, *Wireless Communication Electronics by Example*,
https://doi.org/10.1007/978-3-030-59498-5

where, of course, the absolute value of Z is the length of the hypotenuse and the real and imaginary parts are the two legs of the right-angled triangle, i.e.

$$|Z| = \sqrt{Z\,Z^*} = \sqrt{(a^2 + b^2)} \qquad \theta = \arctan\left(\frac{b}{a}\right) \tag{C.8}$$

where θ is the phase angle. After using Euler's formula, this becomes

$$e^{jx} \equiv \cos x + j \sin x \tag{C.9}$$

and enables us to write a really compact form of complex numbers

$$Z = a + jb = |Z|\,e^{j\theta} \tag{C.10}$$

which leads into another simple way of doing complex arithmetic, by using the absolute values and the arguments in combination with the algebraic rules of exponential numbers, for example

$$\left(A\,e^{j\theta_A}\right)\left(B\,e^{j\theta_B}\right) = AB\,e^{j(\theta_A + \theta_B)} \tag{C.11}$$

and we have the final link,

$$A\,e^{j\theta} \equiv A\,(\cos\theta + j \sin\theta) \tag{C.12}$$

where

$$\Re\left(A\,e^{j\theta}\right) = A\,\cos\theta \qquad \Im\left(A\,e^{j\theta}\right) = A\,\sin\theta \tag{C.13}$$

Basic Trigonometric Identities

D

$$\sin(\alpha + \pi/2) = +\cos\alpha \tag{D.1}$$

$$\cos(\alpha + \pi/2) = -\sin\alpha \tag{D.2}$$

$$\sin(\alpha + \pi) = -\sin\alpha \tag{D.3}$$

$$\cos(\alpha + \pi) = -\cos\alpha \tag{D.4}$$

$$\sin(\alpha \pm \beta) = \sin\alpha\cos\beta \pm \cos\alpha\sin\beta \tag{D.5}$$

$$\cos(\alpha \pm \beta) = \cos\alpha\cos\beta \mp \sin\alpha\sin\beta \tag{D.6}$$

$$\cos\alpha\cos\beta = 1/2\,(\cos(\alpha - \beta) + \cos(\alpha + \beta)) \tag{D.7}$$

$$\sin\alpha\sin\beta = 1/2\,(\cos(\alpha - \beta) - \cos(\alpha + \beta)) \tag{D.8}$$

$$\sin\alpha\cos\beta = 1/2\,(\sin(\alpha + \beta) + \sin(\alpha - \beta)) \tag{D.9}$$

$$\cos\alpha\sin\beta = 1/2\,(\sin(\alpha + \beta) - \sin(\alpha - \beta)) \tag{D.10}$$

$$\sin^2\alpha = 1/2\,(1 - \cos 2\alpha) \tag{D.11}$$

$$\cos^2\alpha = 1/2\,(1 + \cos 2\alpha) \tag{D.12}$$

$$\sin^3\alpha = 1/4\,(3\sin\alpha - \sin 3\alpha) \tag{D.13}$$

$$\cos^3\alpha = 1/4\,(3\cos\alpha + \cos 3\alpha) \tag{D.14}$$

$$\sin^2\alpha\cos^2\alpha = 1/8\,(1 - \cos 4\alpha) \tag{D.15}$$

$$\sin^3\alpha\cos^3\alpha = 1/32\,(3\sin 2\alpha - \sin 6\alpha) \tag{D.16}$$

$$\sin\alpha \pm \sin\beta = 2\sin\left(\frac{\alpha \pm \beta}{2}\right)\cos\left(\frac{\alpha \mp \beta}{2}\right) \tag{D.17}$$

$$\cos\alpha + \cos\beta = 2\cos\left(\frac{\alpha + \beta}{2}\right)\cos\left(\frac{\alpha - \beta}{2}\right) \tag{D.18}$$

$$\cos\alpha - \cos\beta = -2\sin\left(\frac{\alpha + \beta}{2}\right)\sin\left(\frac{\alpha - \beta}{2}\right) \tag{D.19}$$

© Springer Nature Switzerland AG 2021
R. Sobot, *Wireless Communication Electronics by Example*,
https://doi.org/10.1007/978-3-030-59498-5

Useful Algebraic Equations

1. *Binomial formula*

$$(x \pm y)^2 = x^2 \pm 2xy + y^2 \tag{E.1}$$

$$(x \pm y)^3 = x^3 \pm 3x^2y + 3xy^2 \pm y^3 \tag{E.2}$$

$$(x \pm y)^4 = x^4 \pm 4x^3y + 6x^2y^2 \pm 4xy^3 + y^4 \tag{E.3}$$

$$(x \pm y)^n = x^n + nx^{n-1} + \frac{n(n-1)}{2!}x^{n-2}y^2 + \frac{n(n-1)(n-2)}{3!}x^{n-3}y^3 \cdots + y^n \tag{E.4}$$

where $n! = 1, 2, 3, \cdots n$ and $0! \overset{\text{def}}{=} 1$.

2. *Special cases*

$$x^2 - y^2 = (x - y)(x + y) \tag{E.5}$$

$$x^3 - y^3 = (x - y)(x^2 + xy + y^2) \tag{E.6}$$

$$x^3 + y^3 = (x + y)(x^2 - xy + y^2) \tag{E.7}$$

$$x^4 - y^4 = (x^2 - y^2)(x^2 + y^2) = (x - y)(x + y)(x^2 + y^2) \tag{E.8}$$

3. *Useful Taylor series*

$$e^x = \sum_{n=0}^{\infty} \frac{x^n}{n!} = 1 + x + \frac{x^2}{2!} + \frac{x^3}{3!} + \cdots \tag{E.9}$$

$$\sin x = \sum_{n=0}^{\infty} \frac{(-1)^n}{(2n+1)!}x^{2n+1} = x - \frac{x^3}{3!} + \frac{x^5}{5!} - \cdots \quad \text{odd, for all } x \tag{E.10}$$

$$\cos x = \sum_{n=0}^{\infty} \frac{(-1)^n}{(2n)!}x^{2n} = 1 - \frac{x^2}{2!} + \frac{x^4}{4!} - \cdots \quad \text{even, for all } x \tag{E.11}$$

$$\tan x = \sum_{n=1}^{\infty} \frac{B_{2n}(-4)^n(1-4^n)}{(2n)!}x^{2n-1} = x + \frac{x^3}{3} + \frac{2x^5}{15} + \cdots \quad \text{for } |x| < \frac{\pi}{2} \tag{E.12}$$

© Springer Nature Switzerland AG 2021
R. Sobot, *Wireless Communication Electronics by Example*,
https://doi.org/10.1007/978-3-030-59498-5

Bessel Polynomials

1. *Bessel differential equation*

$$x^2 \frac{d^2 y}{dx^2} + x \frac{dy}{dx} + (x^2 - \alpha^2) y = 0 \tag{F.1}$$

2. *Relation with trigonometric functions*

$$\cos(x \sin \alpha) = J_0(x) + 2 \ [J_2(x) \cos 2\alpha + J_4(x) \cos 4\alpha + \cdots] \tag{F.2}$$

$$\sin(x \sin \alpha) = 2 \ [J_1(x) \sin \alpha + J_3(x) \sin 3\alpha + J_5(x) \sin 5\alpha + \cdots] \tag{F.3}$$

$$\cos(x \cos \alpha) = J_0(x) - 2 \ [J_2(x) \cos 2\alpha - J_4(x) \cos 4\alpha + J_6(x) \cos 6 - J_8(x) \cos 8\alpha \cdots] \tag{F.4}$$

$$\sin(x \cos \alpha) = 2 \ [J_1(x) \cos \alpha - J_3(x) \sin 3\alpha + J_5(x) \sin 5\alpha + \cdots] \tag{F.5}$$

3. *Bessel series*

$$J_0(x) = 1 - \frac{x^2}{2^2} + \frac{x^4}{2^2 \cdots 4^2} - \frac{x^6}{2^2 \cdots 4^2 \cdots 6^2} + \cdots \tag{F.6}$$

$$J_1(x) = \frac{x}{2} \left[1 - \frac{x^2}{2^2 \cdots 2} + \frac{x^4}{2 \cdots 2^4 \cdots 2 \cdots 3} + \cdots \right] \tag{F.7}$$

$$J_n(x) = \frac{x^n}{2^n n!} \left[1 - \frac{x^2}{2^2 \cdots (n+1)} + \frac{x^4}{2 \cdots 2^4 \cdots (n+1) \cdots (n+2)} + \right.$$

$$\left. \frac{(-1)^p x^{2p}}{p! \ 2^{2p} \ (n+1)(n+2) \cdots (n+p)} + \cdots \right] \tag{F.8}$$

4. *Bessel approximations*
   For very large x Bessel function reduces to

$$J_n(x) = \sqrt{\frac{2}{\pi x}} \cos \left(x - \frac{n \pi}{2} - \frac{\pi}{4} \right) \tag{F.9}$$

© Springer Nature Switzerland AG 2021
R. Sobot, *Wireless Communication Electronics by Example*,
https://doi.org/10.1007/978-3-030-59498-5

Bibliography

[Amo90] S.W. Amos, *Principles of Transistor Circuits*. Number 0–408–04851–4 (Butterworths, Oxford, 1990)

[BG03a] L. Besser, R. Gilmore, *Practical RF Circuit Design for Modern Wireless Systems I*. Number 1-58053-521-6 (Artech House, Norwood, 2003)

[BG03b] L. Besser, R. Gilmore, *Practical RF Circuit Design for Modern Wireless Systems II*. Number 1-58053-522-4 (Artech House, , Norwood, 2003)

[BMV05] J.S. Beasley, G.M. Miller, J.K. Vasek, *Modern Electronic Communication* . Number 0–13–113037–4 (Pearson, Prentice Hall, Upper Saddle River, 2005)

[Bro90] J.J. Brophy, *Basic Electronics for Scientist*. Number 0–07–008147–6 (McGraw–Hill Inc., New York City, 1990)

[Bub84] P. Bubb, *Understanding Radio Waves*. Number 0–7188–2581–0 (Lutterworth Press, Cambridge, 1984)

[CC03a] D. Comer, D. Comer, *Advanced Electronic Circuit Design*. Number 0–471–22828–1 (John Wiley & Sons Inc., Hoboken, 2003)

[CC03b] D. Comer, D. Comer, *Fundamentals of Electronic Circuit Design*. Number 0–471–41016–0 (John Wiley & Sons Inc., Hoboken, 2003)

[CL62] D.R. Corson, P. Lorrain, *Introduction to Electromagnetic Fields and Waves*. Number 62–14193 (Freeman Co., New York, 1962)

[DA01] W.A. Davis, K.K. Agarwal, *Radio Frequency Circuit Design*. Number 0–471–35052–4 (Wiley Interscience, Hoboken, 2001)

[DA07] W.A. Davis, K.K. Agarwal, *Analysis of Bipolar and CMOS Amplifiers*. Number 1–4200–4644–6 (CRC Press, Boca Raton, 2007)

[Ell66] R.S. Elliott, *Electromagnetics*. Number 66–14804 (McGraw Hill, New York, 1966)

[Fle08] D. Fleisch, *A Student's Guide to Maxwell's Equations*. Number 978–0–521–87761–9 (Cambridge University Press, Cambridge, 2008)

[FLS05] R.P. Feynman, R.B. Leighton, M. Sands, *The Feynman Lectures on Physics*. Number 0–8053–9047–2 (Pearson Addison Wesley, Boston, 2005)

[GM93] P.G. Gray, R.G. Meyer, *Analysis and Design of Analog Integrated Circuits*. Number 0–471–57495–3 (John Wiley & Sons Inc., Hoboken, 1993)

[Gol48] S. Goldman, *Frequency Analysis, Modulation and Noise*. Number TK6553.G58 1948 (McGraw-Hill, New York, 1948)

[Gre04] B. Green, *The Fabric of Cosmos*. Number 0–375–72720–5 (Vintage Books, New York, 2004)

[Gri84] J. Gribbin, *In Search of Schrödinger's Cat, Quantum Physics and Reality*. Number 0–553–34253–3 (Bantam Books, New York, 1984)

[HH89a] T.C. Hayes, P. Horowitz, *Student Manual for The Art of Electronics*. Number 0–521–37709–9 (Cambridge University Press, Cambridge, 1989)

[HH89b] P. Horowitz, W. Hill, *The Art of Electronics*. Number 0–521–37095–7 (Cambridge University Press, Cambridge, 1989)

[Hur10] P.G. Huray, *Maxwell's Equations*. Number 978–0–470–54276–7 (Wiley, Hoboken, 2010)

[II99] U.S. Inan, A.S. Inan, *Electromagnetic Waves*. Number 0–201–36179–5 (Prentice Hall, Upper Saddle River, 1999)

[JK93] W.H. Hayt Jr., J.E. Kemmerly, *Engineering Circuit Analysis*. Number 0–07–027410–X (McGraw Hill, New York, 1993)

[JN71] R.H.Good Jr., T.H. Nelson, *Classical Theory of Electric and Magnetic Fields*. Number 78–137–628 (Academic Press, Cambridge, 1971)

[Jr.89] W.H Hayt Jr., *Engineering Electromagnetics*. Number 0-07-024706-1 (McGraw Hill, New York, 1989)

© Springer Nature Switzerland AG 2021
R. Sobot, *Wireless Communication Electronics by Example*,
https://doi.org/10.1007/978-3-030-59498-5

[KB80] H.L. Krauss, C.W. Bostian, *Solid State Radio Engineering*. Number 0–471–03018–X (Wiley, Hoboken, 1980)

[Kin09] G.C. King, *Vibrations and Waves*. Number 978–0–470–01189–8 (Wiley, Hoboken, 2009)

[Kon75] J.A. Kong, *Theory of Electromagnetic Waves*. Number 0–471–50190–5 (Wiley, Hoboken, 1975)

[LB00] R. Ludwig, P. Bretchko, *RF Circuit Design, Theory and Applications*. Number 0–13–095323–7 (Prentice Hall, Upper Saddle River, 2000)

[Lee05] T.H. Lee, *The Design of CMOS Radio–Frequency Integrated Circuits*. Number 0–521–63922–0 (Cambridge University Press, Cambridge, 2005)

[Lov66] W.F. Lovering, *Radio Communication*. Number TK6550.L546 1966 (Longmans, Harlow, 1966)

[PP99] Z. Popovic, D. Popovic, *Electromagnetic Waves*. Number 0–201–36179–5 (Prentice Hall, Upper Saddle River, 1999)

[Pur85] E.M. Purcell, *Electricity and Magnetism*. Number 0–07–004908–4 (McGraw Hill, New York, 1985)

[Rad01] M.M. Radmanesh, *Radio Frequency and Microwave Electronics*. Number 0–13–027958–7 (Prentice Hall, Upper Saddle River, 2001)

[Raz98] B. Razavi, *RF Microelectronics*. Number 0–13–887571–5 (Prentice Hall, Upper Saddle River, 1998)

[RC84] D. Roddy, J. Coolen, *Electronic Communications*. Number 0-8359-1598-0 (Reston Publishing Company, Prentice-Hall, 1984)

[RR67] J.H. Reyner, P.J. Reyner, *Radio Communication* (Sir Isaac Pitman & Son Ltd, London, 1967)

[Rut99] D.B. Rutledge, *The Electronics of Radio*. Number 0-521-64136-5 (Cambridge University Press, Cambridge, 1999)

[SB00] B. Streetman, S. Banerjee, *Solid State Electronic Devices*. Number 0-13-025538-6 (Prentice Hall, Upper Saddle River, 2000)

[Sch92] R.J. Schoenbeck, *Electronic Communications Modulation and Transmission*. Number 0–675–21311–8 (Prentice Hill, Upper Saddle River, 1992)

[Scr84] M.G. Scroggie, *Foundations of Wireless and Electronics*, 10th edn. Number 0–408–01202–1 (Newnes Technical Books, Boston, 1984)

[See56] S. Seely, *Radio Electronics*. Number 55–5696 (McGraw Hill, New York, 1956)

[Sim87] R.E. Simpson, *Introductory Electronics for Scientist and Engineers*. Number 0–205–08377–3 (Allyn and Bacon Inc., Boston, 1987)

[Sze81] S.M. Sze, *Physics of Semiconductor Devices*. Number 0–471–05661–8 (John Wiley and Sons, Hoboken, 1981)

[Ter03] D. Terrell, *Electronics for Computer Technology*. Number 0–7668–3872–2 (Thompson Delmar Learning, Florence, 2003)

[Tho06] M.T. Thompson, *Intuitive Analog Circuit Design*. Number 0–7506–7786–4 (Newnes, London, 2006)

[Wik10a] Wikipedia.org. *Electromagnetic Wave Equation*. http://en.wikipedia.org/wiki/Electromagnetic_wave_equation. On-line (2010)

[Wik10b] Wikipedia.org. *Waves, wavelength*. http://en.wikipedia.org/wiki/Wave. On-line (2010)

[Wol91] D.H. Wolaver, *Phase–Locked Loop Circuit Design*. Number 0–13–662743–9 (Prentice Hall, Upper Saddle River, 1991)

[You04] P.H. Young, *Electronic Communication Techniques*. Number 0–13–048285–4 (Pearson, Prentice Hill, Upper Saddle River, 2004)

Glossary

This glossary of technical terms is provided for reference only. The reader is advised to further study the terms in appropriate books, for example, a technical dictionary.

1 dB gain compression point   The point at which the power gain at the output of a non-linear device or circuits is reduced by 1dB relative to its small signal linear model predicted value.

Absolute zero   The theoretical temperature at which entropy would reach its minimum value. By international agreement, absolute zero is defined as 0K on the Kelvin scale and as -273.15°C on the Celsius scale.

Active device   An electronic component that has signal gain larger than one, for example a transistor. Compare to *passive device*.

Active mode   Condition for a BJT transistor where emitter-base junction is forward biased, while the collector-base junction is reverse biased.

Admittance   The measure of how easily AC current flows in a circuit (in Siemens [S]). The reciprocal of *impedance*.

Ampere [A]   The unit of electric current defined as the flow of one coulomb of charge per second.

Ampère's Law   A current flowing into a wire generates a magnetic flux that encircles the wire following the "right hand rule" (the thumb points the direction of the current flow and the other curled fingers show direction of the magnetic field). Study *Maxwell's equations* for more details.

Amplifier   A linear device that implements mathematical equation $y = Ax$, where y is the amplified output signal, A is the gain coefficient, and x is the input signal.

Analogue   The general class of devices and circuits meant to process a continuous signal. Compare with *digital* and sampled signals.

Attenuation   Gain lower than one.

Attenuator   A device that reduces a gain without introducing phase or frequency distortion.

Automatic Gain Control   A closed-loop feedback system designed to hold the overall gain as constant as possible.

Average power   The power averaged over one time period.

Bandwidth   The difference between upper and lower frequencies at which the amplitude response is 3dB below the maximum level. It is equivalent to *half-power* bandwidth.

Base   The region of BJT transistor between the emitter and the collector.

Bel [B]   is a dimensionless unit used to express the *ratio* of two powers. A more practical unit is the [dB].

Beta β   The current gain of a BJT. It is the ratio of the change in collector current to the change in the base current, $\beta = dI_C/dI_B$.

Bias   A steady current or voltage used to set the operating conditions of a device.

© Springer Nature Switzerland AG 2021
R. Sobot, *Wireless Communication Electronics by Example*,
https://doi.org/10.1007/978-3-030-59498-5

Breakdown voltage   The voltage at which the reverse current of a reverse biased p-n junction suddenly rises. If the current is not limited, the device is destroyed.

Capacitance   The ratio of the electric charge and voltage between two conductors.

Capacitor   A device made of two conductors separated by an insulating material for the purpose of storing electric charge, i.e. energy.

Celsius [°C]   A unit increment of temperature unit defined as $1/100$ between the freezing point $(0\,°C)$ and boiling point $(100\,°C)$ of water. Compare with Kelvin and Fahrenheit.

Characteristic Curve   A family of I–V plots shown for several parameter values.

Characteristic impedance   The entry point impedance of an infinitely long transmission line.

Charge   A basic property of elementary particles of matter (electrons, protons, etc.) responsible for creating a force field.

Circuit   The interconnection of devices, both passive and active, for the purpose of synthesizing a mathematical function.

Common–base   A single BJT amplifier configuration in which the base potential is fixed, the emitter serves as the input, and the collector as the output terminal. Also known as "current buffer". Equivalent to a "common–gate" configuration for MOS amplifiers.

Common–collector   A single BJT transistor configuration in which the collector potential is fixed, the base serves as the input, and the emitter as the output terminal. Also known as "voltage buffer" or voltage follower. Equivalent to a "common–drain" configuration for MOS amplifiers.

Common–emitter   A single BJT amplifier configuration in which the emitter potential is fixed, while the base serves as the input, and the collector as the output terminal. Also known as the g_m stage. Equivalent to a "common–source" configuration for MOS amplifiers.

Common mode   The average value of a sinusoidal waveform.

Conductivity   The ability of a matter to conduct electricity.

Conductor   A material that easily conducts electricity.

Coulomb [C]   The unit of electric charge defined as the charge transported through a unity area in one second by an electric current of one ampere. An electron has a charge of 1.602×10^{-19}C.

Coulomb's Law   A definition of the force between two electric charges in space.

Current   A transfer of electrical charge through a unit size area per unit of time.

Current gain   The ratio of current at the output terminals to the current at the input terminals of a device or circuit.

Current source   A device capable of providing constant current value regardless of the voltage at its terminals.

DC   See *Direct current*.

DC analysis   A mathematical procedure to calculate the stable operating point.

DC biasing   The process of setting the stable operating point of a device.

DC load line   A straight line across a family of I–V curves that shows movement of the operating point as the output voltage changes for a given load.

Decibel [dB]   A dimensionless unit used to express the ratio of two powers. A decibel is ten times smaller than a *bel* [B].

Device   A single discrete device, for instance a resistor, a transistor, or a capacitor.

Dielectric   A material that is not good in conducting electricity, i.e. the opposite of a conductor. Characterized by the dielectric constant.

Differential amplifier   An amplifier that operates on differential signals.

Differential signal   A difference between two sinusoidal signals of the same frequency, same amplitude, same common mode, and with phase difference of 180°.

Digital   The general class of devices and circuits meant to process a sampled signal. Compare with *analogue* and continuous signals.

Diode   A non-linear, two-terminal device that obeys the exponential transfer function. Used as unidirectional switch.

Direct current (DC)   Current that flows in one direction only.

Discrete device   An individual electrical component that exhibits behaviour associated with a resistor, a transistor, a capacitor, an inductor, etc. Compare with *distributed* components.

Dynamic range   The difference between the maximum acceptable signal level and the minimum acceptable signal level.

Electric field   An energy field generated by an electric charge, detected by the existence of the electric force within a space surrounding the charge.

Electrical noise   Any unwanted electrical signal.

Electromagnetic (EM) wave   A phenomenon exhibited by a flow of electromagnetic energy through space. In the special case of a *standing* wave this definition may need more explanation.

Electron   A fundamental particle that carries negative charge.

Electronics   The branch of science and technology that makes use of the controlled motion of electrons through different media and vacuum.

Electrostatics   The branch of science that deals with the phenomena arising from stationary or slow-moving electric charges.

Emitter   A region of a BJT from which charges are injected into the base. One of the three terminals of a BJT device.

Energy   A concept that can be loosely defined as the ability of a body to perform work.

Equivalent circuit   A simplified version of a circuit that performs the same function as the original.

Equivalent noise temperature   The absolute temperature at which a perfect resistor would generate the same noise as its equivalent real component at room temperature.

Fall time   The time during which a pulse decreases from 90% to 10% of its maximum value (sometimes defined between the 80% and 20% points).

Farad [F]   The unit of capacitance of a capacitor. One farad is very large; the capacitance of the Earth's ionosphere with respect to the ground is around 50mF.

Faraday cage   An enclosure that blocks out external static electric fields.

Faraday's Law   The law of electromagnetic induction. See also *Faraday's cage*.

Feedback   The process of coupling output and input terminals through an external path. Negative feedback increases the stability of an amplifier for the cost of reduced gain, positive feedback boosts gain and is needed to create oscillating circuits.

Field   A concept that describes a flow of energy through space.

Field-Effect Transistor (FET)   A transistor controlled by two perpendicular electrical fields used to change resistivity of the semiconductor material underneath the gate terminal and to force current between the source and drain terminals.

Flicker noise   A random noise in semiconductors whose power spectral density is, to the first approximation, inverse to frequency (1/f noise.).

Frequency   The number of complete cycles per second.

Frequency response   A curve showing the gain and phase change of a device as a function of frequency.

Gain   The ratio of signal values measured at output and input terminals.

Gauss's Law   A law relating the distribution of electric charge to the resulting electric field.

Ground   An arbitrary potential reference point that all other potentials in a circuit are compared against. The difference between the ground potential and the node potential is expressed as voltage at that node. The ground node may or may not be the lowest potential in the circuit.

Henry [H]   The unit of measurement for self and mutual inductance.

Hertz [Hz]   The unit of measurement for frequency, equal to one cycle per second.

Impedance   Resistance of a two-terminal device at any frequency.

Inductance   A property whereby a change in the electric current through a circuit induces an electromotive force (EMF) that opposes the change in current.

Inductor   A passive electrical component that can store energy in a magnetic field created by an electric current passing through it.

Input   Current, voltage, power, or other driving force applied to a circuit or device.

Insertion loss   The attenuation resulting from inserting a circuit between source and load.

Insulator   A material with very low conductivity.

Intermediate Frequency (IF)   A frequency to which a carrier frequency is shifted as an intermediate step in transmission or reception.

Intermodulation Products   Additional harmonics created by a non-linear device processing two or more single-tone signals.

Junction   A joining of two semiconductor materials.

Junction capacitance   Capacitance associated with p-n junction region.

Kelvin [K]   The unit increment of temperature on the absolute temperature scale.

Kirchhoff's Current Law (KCL)   The law of conservation of charge: at any instant, the total current entering any point in a network is equal to the total current leaving the same point.

Kirchhoff's Voltage Law (KVL)   The law of conservation of energy given or taken by a potential field (not including energy taken by dissipation): at any instant, the algebraic sum of all electromotive forces and potential differences around a closed loop is zero.

Large signal   A signal with an amplitude large enough to move the operating point of a device far away from its original biasing point. Hence, non-linear model of the device must be used.

Large signal analysis   A method used to describe the behaviour of devices stimulated by large signals. It describes non-linear devices in terms of the underlying non-linear equations.

Law of conservation of energy   The fundamental law of nature. It states that energy can be neither created nor destroyed, it can only be transformed from one state to another.

Linear network   A network in which the parameters of resistance, inductance, and capacitance are constant with respect to current or voltage, and in which the voltage or current of sources is independent of or directly proportional to other voltages and currents, or their derivatives, in the network.

Load   A device that absorbs energy and converts it into another form.

Local Oscillator (LO)   An oscillator used to generate a single-tone signal that is needed for up-conversion and down-conversion operations.

Lossless   A theoretical device that does not dissipate energy.

Low Noise Amplifier (LNA)   An electronic amplifier used to amplify very weak signals captured by an antenna.

Lumped element   A self-contained and localized element that offers one particular property, for example, resistance over a range of frequencies.

Magnetic field   A field generated by magnetic energy, detected by the existence of a magnetic force within space surrounding a magnet.

Matching   A concept of connecting two networks for the purpose to enable maximum energy transfer between them.

Matching circuit   A passive circuit designed to interface two networks to enable maximum energy transfer between the two networks.

Maxwell's equations   A set of four partial differential equations that relate electric and magnetic fields to their sources, charge density and current density. These equations can be combined to show that light is an electromagnetic wave. Individually, the equations are known as Gauss's law, Gauss's law for magnetism, Faraday's law of induction, and Ampère's law with Maxwell's correction. These four equations and the Lorentz force law make up the complete set of laws of classical electromagnetism.

Metal-oxide-semiconductor field-effect transistor (MOSFET)   Originally, a sandwich of aluminium-silicone dioxide-silicon was used to manufacture FET transistors. Although, metal is no longer used to create gates for FET transistors, the name has stuck.

Microwaves   Waves in the frequency range of 1–300GHz, i.e. with a wavelength of 300–1mm.

Mixer   A non-linear three-port device used for frequency shifting operation.

Negative resistance   The resistance of a device or circuit where an increase in the current entering a port results in a decrease in voltage across the same port.

Noise   Any unwanted signal that interferes with a wanted signal.

Noise Figure (NF)   A measure of degradation of the signal to noise ratio (SNR), caused by the internal noise generated by components in a radio frequency (RF) signal chain.

Non-linear circuit   A system that does not satisfy the superposition principle or whose output is not directly proportional to its input.

Norton's Theorem   Any collection of voltage sources, current sources, and resistors with two terminals is electrically equivalent to an ideal current source in parallel with a single resistor. This is the twin of *Th'evenin's theorem*.

NPN transistor   A transistor with p-type base and n-type collector and emitter.

Octave   The interval between any two frequencies having a ratio of 2:1.

Ohm [Ω]   Unit of resistance, as defined by Ohm's law.

Ohm's Law   The change of current through a conductor between two points is directly proportional to the change of voltage across the two points and inversely proportional to the resistance between them.

Open-loop gain   The ratio of the output signal and the input signals of an amplifier with no feedback path present.

Oscillator   An electronic device that generates a single-tone (or some other regular shape) signal at predetermined frequency.

Output   Current, voltage, power, or driving force delivered at the output terminals.

Passive device   A component that does not have gain larger than one. Compare to *active device*.

Phase   The angular property of a wave.

Phase shifter   A two-port network that provides a controllable phase shift of the RF signals.

Phasor   A mathematical representation of a sine wave by a rotating vector.

Power   The rate at which work is performed.

Quality factor (Q factor)   A dimensionless parameter that characterizes a resonator's bandwidth relative to its centre frequency.

Radio frequency (RF)   Any frequency at which coherent radiation of energy is possible.

Reactance   The opposition of a circuit element to a change of current, caused by the build-up of electric or magnetic fields in the element.

Reactive element   An inductor and capacitor.

Reflected waves   The waves reflected from a discontinuity in the medium in which they are traveling.

Resistance   A measure of an object's opposition to the passage of a steady electric current.

Resistor   A lumped element designed to have a certain resistance.

Resonant frequency   The frequency at which a given system or circuit responds with maximum amplitude when driven by an external single tone.

Root Mean Square (RMS)   The square root of the arithmetic mean (average) of the squares of the original values.

Saturation   A condition in which an increase of the input signal to a circuit does not produce an expected linearly proportional change at the output.

Self-resonant frequency   The frequency at which all real devices or circuits start to oscillate due to the internal parasitic inductances and capacitances.

Signal   An electrical quantity containing information that is carried by a voltage or current.

Single-ended circuit   A circuit operating on single-ended (as opposed to differential) signals.

Skin effect   The tendency of an alternating current (AC) to distribute itself within a conductor so that the current density near the surface of the conductor is greater than at its core. That is, the electric current tends to flow at the "skin" of the conductor, at an average depth called the "skin depth".

Small signal   A low-amplitude signal that occupies a very narrow region that is centred at the biasing point. Hence, linear model always applies.

Small signal amplifier   An amplifier that operates only in the linear region.

Space   The boundless, three-dimensional extent in which objects and events occur and have relative position and direction.

Stability   The ability of a circuit to stay away from the self-resonating frequency.

Standing wave   A wave whose maximum and minimum points remain at a constant position. It can arise in a stationary medium as a result of interference between two waves traveling in opposite directions. For waves of equal amplitude traveling in opposite directions, there is on average no net propagation of energy.

Standing Wave Ratio (SWR)   The ratio of the maximum to the minimum value of current or voltage in a standing wave.

Thévenin's theorem   Any combination of voltage sources, current sources, and resistors with two terminals is electrically equivalent to a single voltage source and a single series resistor. This is a twin of *Norton's theorem*.

Third order intercept point (IP3)   A measure of weakly non-linear systems and devices, for example, receivers, linear amplifiers, and mixers.

Time   A concept used to order a sequence of events.

Transmission line   Any system of conductors capable of efficiently conducting electromagnetic energy.

Tuned circuit   A circuit consisting of inductance and capacitance that can be adjusted for resonance at the desired frequency.

Tuning   The process of adjusting resonant frequency of a *tuned circuit*.

Varactor   A two-terminal p-n junction used as a voltage controlled capacitor.

Volt [V]   A unit of measurement for potential difference.

Voltage controlled oscillator (VCO) An oscillator whose output frequency is controlled by a voltage.

Voltage divider A simple linear circuit that produces an output voltage that is a fraction of its input voltage.

Voltage follower amplifier An amplifier that provides electrical impedance transformation from one circuit to another. Also known as a "voltage buffer amplifier".

Voltage source A device capable of providing a constant voltage value regardless of the current at its terminals.

Wave A disturbance that progresses from one point in space to another.

Wavefront A cross-sectional surface having constant phase.

Wavelength A distance in space between two consecutive points having the same phase.

Wave propagation The journey of a wave through space.

White noise A random signal that consists of all possible frequencies from zero to infinity.

Work The advancement in space of a point under application of a force.

Index

© Springer Nature Switzerland AG 2021
R. Sobot, *Wireless Communication Electronics by Example*,
https://doi.org/10.1007/978-3-030-59498-5

Printed in the United States
by Baker & Taylor Publisher Services